Bridge Loads

To the glory of God
and
to Emma and Michael

Bridge Loads

An international perspective

Colin O'Connor and Peter A. Shaw
Emeritus Professor of Civil Engineering
The University of Queensland, Australia
Senior Engineer
Brisbane City Council, Australia

CRC Press is an imprint of the
Taylor & Francis Group, an **informa** business
A SPON PRESS BOOK

CRC Press
Taylor & Francis Group
6000 Broken Sound Parkway NW, Suite 300
Boca Raton, FL 33487-2742

First issued in paperback 2019

Typeset in Sabon by Wearset, Boldon, Tyne and Wear

ISBN-13: 978-0-419-24600-8 (hbk)
ISBN-13: 978-0-367-44732-8 (pbk)

British Library Cataloguing in Publication Data
A catalogue record for this book is available from the British Library

Library of Congress Cataloging in Publication Data
O'Connor, Colin.
Bridge loads/Colin O'Connor and Peter A. Shaw.
p. cm.
Includes bibliographical references and index.
ISBN 0-419-24600-2 (alk. paper)
1. Bridges – Design and construction. 2. Bridges – Loads. 3. Load and resistance factor design. I. Shaw, Peter A. (Peter Arthur) II. Title.

TG300 .O28 2000
624'.252–dc21

00–023894

Publisher's Note
The publisher has gone to great lengths to ensure the quality of this reprint but points out that some imperfections in the original may be apparent.

Contents

Acknowledgements

This book has relied heavily on the work of others, as a glance at the Bibliography will show. Use of this material is acknowledged in the text. There are some who gave assistance at a personal level and this has been most appreciated; they include Norma Howley, who typed the manuscript and assisted in many ways; Roger Dorton and Ray Wedgwood; Robert Heywood, Andrzej Nowak and John Fisher; Peter Swannell, Colin Apelt, Lew Isaacs, Derek Brady, Michael Gourlay, Peter Dux and John Muller; Gerald Brameld, Robert Blinco, Paul Grundy and Peter Selby-Smith; and Lizl Rigg of the South African Institution of Civil Engineering. It has also been a great pleasure to use material from post-graduate theses by R. Heywood, T. Chan, R. Duczmal, B. Gabel, K. Kunjamboo, R. Pritchard and L.F. Yoe, supervised or part-supervised by one of the authors.

It is appropriate to mention here the sources of some of the bridge photographs: the Qantas hangar, Burke and the Fairfax Photo Library; the Quebec bridge, the Canadian National Railway and the Canadian National Museum of Science and Technology; Kings bridge, the Government Printer, State of Victoria; Westgate bridge, The Melbourne Age; and Ngalimbiu bridge, W. Boyce.

The authors are grateful for all of this, but above all, for the patient and loving care of their wives and families.

Chapter 1

Introduction

A bridge may be defined as a structure used to carry loads over an opening, which may take the form of a valley or stream, a road or railway. One way to cross a valley is to build an embankment which effectively closes the opening. Such a structure is not a bridge, and it is important to recognise that the opening crossed by the bridge generally performs a function in itself, which must be maintained.

The loads mentioned in this definition include the weights of trucks and pedestrians for a road bridge, and of locomotives and rolling stock for a railway bridge. These will be called primary loads, for they express the purpose for which the bridge was required. There are other senses in which the term 'load' may be qualified, such as in the terms 'service', 'design' and 'legal' loads. The second of these is of paramount importance in this book, for before a bridge can be built it is necessary for the designer to choose design loads that are used for the selection of member sizes. However, the choice of design loads cannot be separated from a study of the other two: the service loads, applied to the bridge during its service life, and the legal limits, intended to govern these loads.

There are many loads that the bridge must support in addition to its primary loads, and it is appropriate here to make a preliminary list. Design loads include:

1. the self-weight of the structure;
2. vehicle weights;
3. horizontal vehicle loads, such as those due to braking, centrifugal force, or horizontal surging effects in some vehicles;
4. dynamic vertical loads, caused by dynamic interaction between primary service vehicles and the bridge, and influenced by such factors as road roughness and vehicle suspension characteristics;
5. the weight of pedestrians;
6. loads applied by vehicles or pedestrians to railings and kerbs;

7. stream loads on the sub-structure, or possibly applied also to the superstructure by a stream in flood;
8. wind loads, where in some cases it may be necessary to consider the aeroelastic interaction between wind and bridge, leading possibly to aerodynamic instability;
9. wind buffeting of a bridge that is placed closely down-wind from a neighbouring structure;
10. collision forces, caused either by a service vehicle striking the structure, or by some moving object in the channel beneath the bridge, for the foundation of a bridge may be struck by logs, ice, by a ship, or by a derailed locomotive;
11. earthquake loads; and
12. thermal effects.

These loads will be considered in the chapters that follow.

It has already been mentioned that design loads must be specified for the design of a new bridge. They must bear a proper relationship to vehicles crossing the bridge, and to the legal limits that control the weights of these vehicles. There are economic benefits to the transport industry if these legal limits are increased, and history has shown a gradual increase in these limits. If properly controlled, this can be handled in a satisfactory way in the design of new bridges, but it must be remembered that each road authority possesses a large stock of existing bridges. If legal limits are increased, these bridges may be overloaded. A study of bridge loads must have a regard for the performance of old bridges and this, in turn, means that the structural form of these bridges is still relevant, even if some would be considered uneconomic today. There are, for example, many stone arch bridges still in use around the world. Some are still being designed and built in China, with spans in excess of 100 m, but this is exceptional. In the same way, there are metal truss bridges in service today with geometries that would not now be used.

Design loads must be such that they test the behaviour of all parts of a bridge of every kind. For this reason it is appropriate to review the range of bridge types shown in Fig. 1.1 (see also O'Connor 1971a).

1. The *beam* or *girder* bridge relies primarily on bending actions. It may be (a) simply-supported; (b) continuous; or (c) articulated to form what may be called a cantilever structure.
2. The *truss* may be defined as a triangulated assembly of straight members. Although essentially different from the beam, it bears some points of similarity to it, for the chord members may be regarded as equivalent to the upper and lower flanges of a beam, with the web members forming an open system that replaces the beam's solid web. The strength of a truss relies essentially on the axial load-carrying capacity of its members.
3. A beam may be supported by additional diagonal members to form a *propped girder*.
4. The *arch* is a member shaped and supported in such a way that intermediate transverse loads are transmitted to the supports primarily by axial

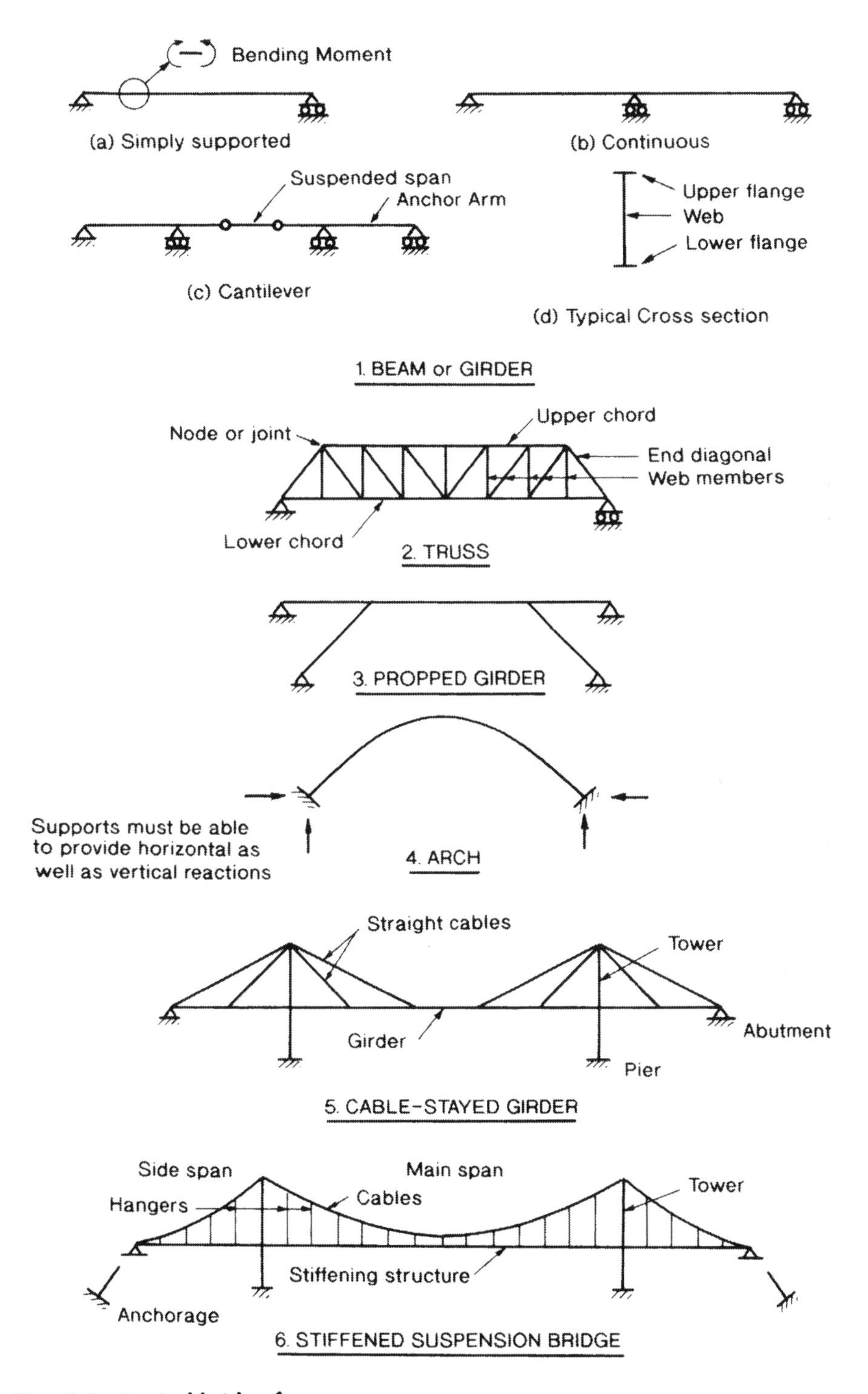

Fig. 1.1 Typical bridge forms

compressive forces in the arch rib. The shape shown in Fig. 1.1 is the simplest form of the arch, but this drawing is incomplete, for it is necessary to supply also a horizontal deck that is supported in some way from the arch.

5. The *cable-stayed girder* consists of a main girder system at deck level, supported on abutments and piers, and also by a system of straight cables passing over one or two towers to the girder.
6. The *stiffened suspension bridge* has some form of stiffening structure – either a truss or girder – hung from continuously curved cables that pass from end anchorages over the full length, and are supported by intermediate towers.

The range of possible bridge geometries is much greater than that shown in Fig. 1.1. The truss, for example, may be used to replace the beam or girder in all its forms, as in the arch or cable-stayed girder; the propped girder has been selected here as a generalised form of what some would call the rigid-frame; and there are many particular forms of the truss, arch and cable-stayed girder. However, these basic examples may be used to illustrate the way in which the selection of design loads may be affected by the form of the bridge.

(a) The maximum load in the cable of a suspension bridge is achieved when all possible vertical loads are placed over the full length of the bridge, including both main and side spans.
(b) The web members of a truss achieve their maximum load when the live load is confined to certain portions of the deck.
(c) In a cantilever bridge, such as that shown in 1(c) of Fig. 1.1, it is necessary to consider possible uplift at the ends. Loads at mid-span will tend to cause the outer ends of the anchor arms to rise, but this is counteracted by the weights of the anchor arms themselves. It may be decided to use the weights of the anchor arms to hold the ends down without the use of vertical, tension anchors. In such a case, the designer must deliberately underestimate the weights of the anchor-arms, while maximising loads on the central span.

In considering these matters, the concept of the *influence line* is most useful. An influence line considers a chosen function, such as the bending moment at a particular point of a bridge girder, and plots the value of this function against the location of a unit load, as this load is moved across the deck. A few typical influence lines are shown in Fig. 1.2.

1. The first graph (a) shows the influence line for bending moment, M_A, at the point A of a simply-supported girder. The + sign is based on the convention that a positive bending moment causes tension in the lower fibres of the girder. It follows from the definition of an influence line that the ordinate shown gives the value of M_A for a unit load located as shown.
2. The second graph (b) shows the influence line for bending moment, M_B, at the point B of a two-span continuous beam. Clearly, to achieve maximum positive bending moment, loads should be placed in the left span, and to

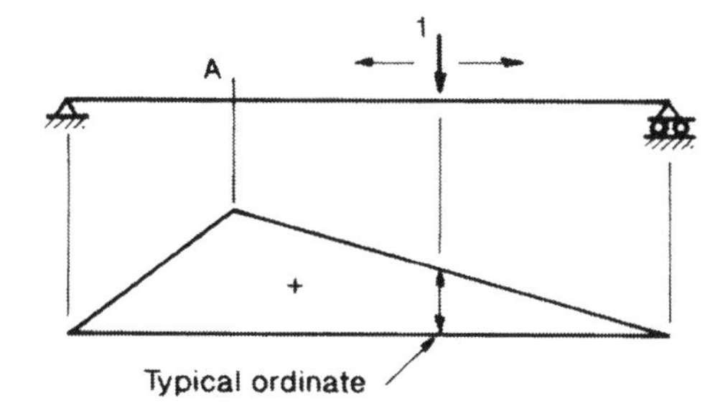

(a) Influence line for M_A

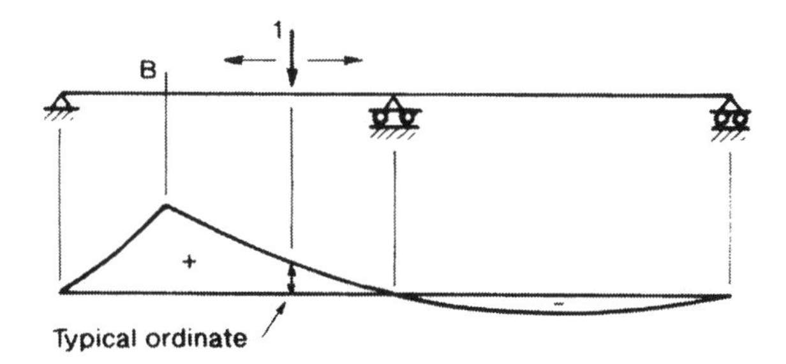

(b) Influence line for M_B

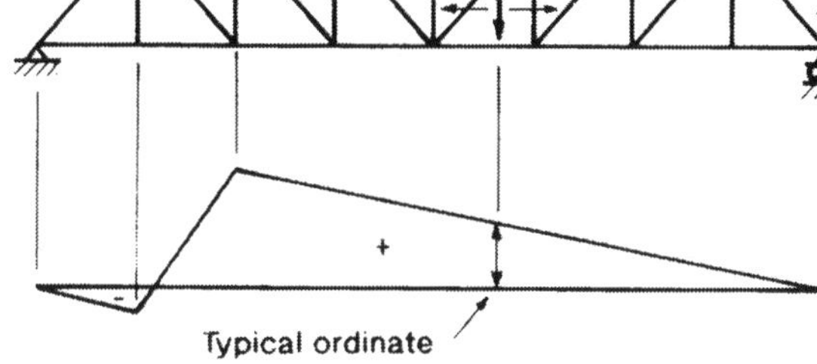

(c) Influence line for axial force in member C

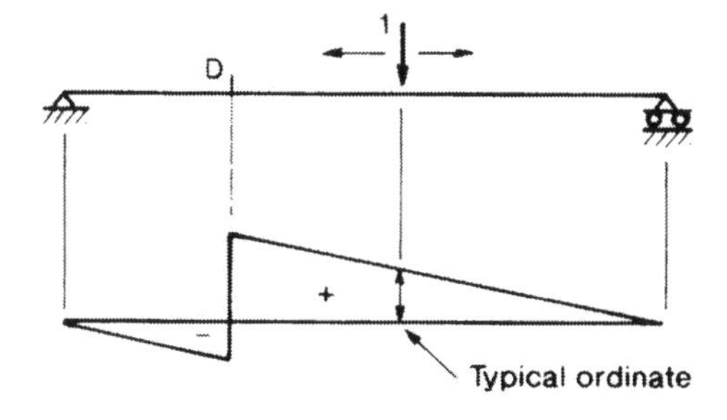

(d) Influence line for shear M_D

Fig. 1.2 Typical influence lines

achieve negative bending moments, they should be placed in the right. Both cases may be of interest.

3. The third example (c) is the influence line for the axial force (P_C, tension positive) in member, C, of a pin-jointed truss. It may happen that the design of this member is based on its maximum compressive force, even though its maximum tension force is larger, for it may buckle in compression. In this case, the design vertical loads should be restricted to the left hand position of the deck. The question that must be faced is: what are reasonable choices for the design loads that may be applied in this restricted area?
4. The last example (d) is the influence line for shear, M_D, at the point D of a simply-supported girder, where a positive shear is assumed to be upwards to the left; that is, if an element of the member at D is considered, the shearing force applied to its left-hand face is upwards. Here there is an abrupt step in the influence line, and it may be argued that for any such influence line this step is unity, i.e. equal in value to the unit load.

These observations must be expanded in a number of ways. The first may be illustrated from the third example (c) of Fig. 1.2; the analysis of this truss is based on two assumptions: (a) that the loads are applied to the truss only at the nodes; and (b) that there is a longitudinal deck, capable of spanning between the lower-chord nodes, and that this deck is simply-supported at these nodes. This assumption of the presence of a deck applies also to other bridge forms. In most cases, the primary structure shown in elevations such as those of Fig. 1.1 consists of a number of parallel systems, and the primary loads must be transferred laterally to these systems by some form of deck. This comment is true both where there is a separate deck system, and where the 'deck' is supplied by some part of the primary structure. To take the simplest example, the case of a separate deck, design loads must be specified with sufficient flexibility to allow the designer to check all components of the deck, and this may require loads that model the effects of a single truck tyre.

This observation leads to a conclusion, namely, that the specification of primary bridge loads must be sufficiently flexible to allow the estimation of the maximum total load on a deck some hundreds of metres in length, and also on an area of 1 or 2 square metres.

A second observation is of less immediate importance, but should still be made. A *member* may be defined as a structural component whose length is great compared with its cross-sectional dimensions. Most components of bridge superstructures satisfy this definition, but not all. The horizontal slab that forms the deck of a road bridge has two major dimensions, its length and width; only its thickness is small. The same may be true for other bridge components. An arch, for example, may consist of a slab curved to form a vault, again with significant width and length, but with a thickness that is relatively small. In a tubular or box girder bridge, the primary structure may have a large width; indeed, its top flange may form the deck. Nevertheless, in all of these cases the vertical dimension of the structure is relatively small, and this simplifies the selection of design loads. In a *member*, the load effects may be suitably expressed, after analysis, by the axial loads, shears, bending moments or twisting moments associated with particular cross-sections of the member, where these functions are sufficient to define all stresses in the member and are readily plotted in the form of influence lines. If the primary structure consisted, say, of a vertical wall whose depth was one half of its span, then its behaviour is more complex and more difficult to summarise. Some parts of a structure do act as deep beams, and this may require consideration, but these cases form the exception rather than the rule.

Further, the shape of an influence line may often be visualised without a detailed analysis: three examples are shown in Fig. 1.3. In the first case, the influence line that is required is for the vertical reaction, V, at the left hand end of a simply-supported beam. Suppose that the member is given a unit vertical displacement at the left hand end, in the assumed positive direction of V; then the displaced shape of the deck is shown. In this case the member remains straight, and the upward displacement, u, at a particular point load may readily be calculated by geometry. Bernoulli's Principle of Virtual Displacements (Marshall and Nelson 1990; Norris and Wilbur 1960: 322; Timoshenko and Young 1965; Todd 1981) states that, for a system of forces acting on a rigid body, in equilibrium, the virtual work done during any virtual displacement is zero. That is,

$$(V \times 1) - Pu = 0$$
$$V = Pu$$

Or, for

$$P = 1$$
$$V = u$$

It follows that the deflected shape of the deck is the same as the influence line for the vertical reaction, V.

The second example shown in Fig. 1.3 concerns the influence line for the bending moment, M_C, in a simply supported beam. Diagram (b) shows an element at C, with a bending moment, M_C, shown in the assumed positive sense, caused by the unit load, P. These moments may be regarded, alternatively, as those applied to the beam on either side of C, as in (c). Now the section of the member at C is capable of transferring three loads: axial loads, shear forces and bending moments. Suppose its capacity to transfer the third only is removed; then this is equivalent to the insertion of a pin at C, as shown in (d).

Suppose that, to some suitable scale, a unit rotation is applied across the pin at C. Then the slope of one end of the member at C with respect to the other is unity. This displacement does not cause stresses in the beam. It may be regarded as a virtual displacement and the virtual work,

$$(M_C \times 1) - Pu = 0$$

that is,

$$M_C = Pu$$

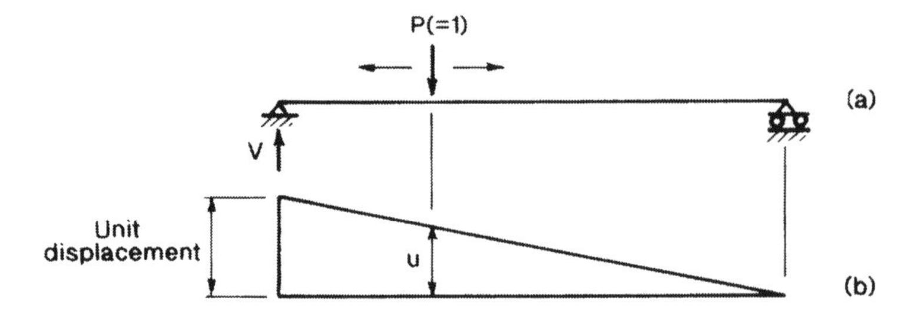

1. Influence line for reaction

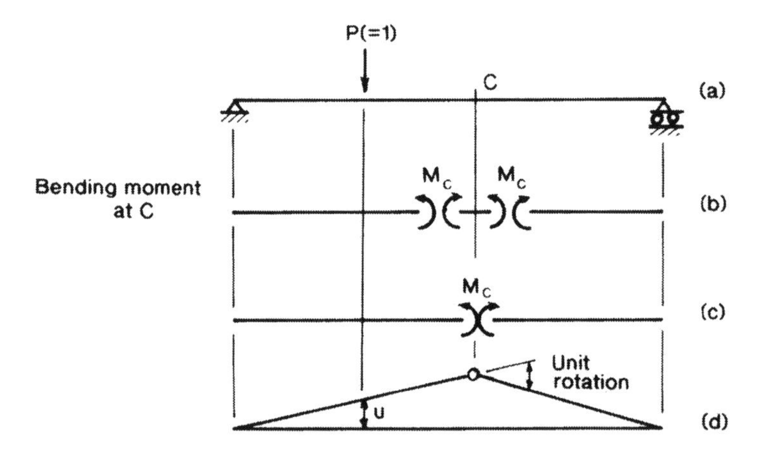

2. Influence line for Bending moment

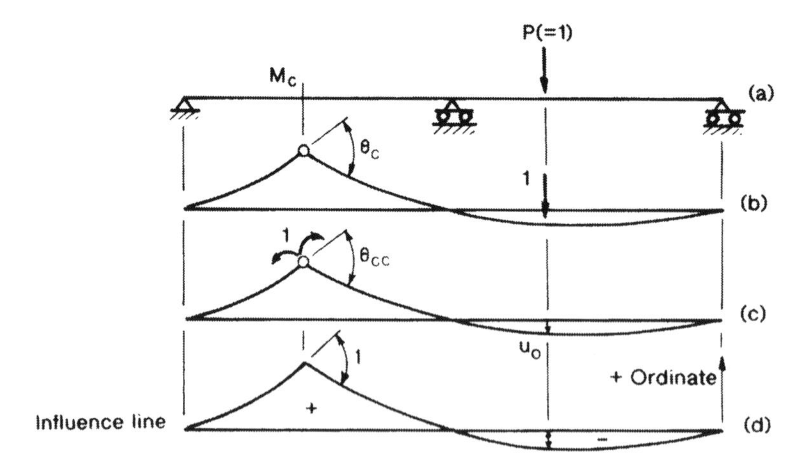

3. Influence line for bending moment in continuous beam

Fig. 1.3 The nature of influence lines

or, for

$$P = 1$$
$$M_C = u$$

Again, the influence line for M_C is identical with the deflected shape (of the deck) produced by this unit rotation. This principle clearly can be extended to other influence lines for determinate structures. For example, the influence line for shear shown in Fig. 1.2(d) corresponds to a unit vertical relative displacement, but with no relative rotation, between the two sides of the member at D.

Consider, further, the case of the two-span continuous beam shown as the third case in Fig. 1.3, and again the influence for the bending moment at C. Let unit value of the load, P, located as in (a), produce the bending moment, M_C. This structure is indeterminate to the first degree. A determinate base structure may be formed by the insertion of a pin at C. Let the load, $P = 1$, produce a rotation across the pin of θ_C. Apply unit moment across the pin as in (c), producing a rotation across the pin of θ_{CC}. Then to restore the structure to its original condition it is necessary to apply a moment, M_C, at C such that

$$\theta_C + M_C\theta_{CC} = 0$$

that is,

$$M_C = -\theta_C/\theta_{CC}$$

By Maxwell's Reciprocal Theorem (Norris and Wilbur 1960: 391), the displacement in the direction of P, caused by a unit pair of moments at C, equals the relative rotation at C caused by a unit load, P. That is, referring to Fig. 1.3(b) and (c),

$$u_o = \theta_{CC}$$

or

$$M_C = -u_o/\theta_{CC}$$

But a pair of moments applied at C of magnitude $1/\theta_{CC}$ will produce a relative rotation there of unity. They will produce also a displacement at P of value u_o/θ_{CC}.

Let

$$u = -u_o/\theta_{CC}$$

then

$$M_C = u$$

That is, the deflected shape produced by a unit relative rotation at C is the influence line for M_C, provided that a positive ordinate to the influence line is taken as upward – as shown in (d).

This conclusion is of general validity: an influence line for the axial load in a member can be produced by imposing a unit axial displacement; for shear, one can impose a unit shear displacement; and for bending moment, one can impose a unit relative rotation. In all cases, this displacement should be applied in the same sense as the force applied to the adjoining structure by a positive elemental force or moment. A further example is shown in Fig. 1.4, which shows the influence line for axial force in one of the cables of a cable-stayed girder, one of a number first published in O'Connor (1971a: 478–81). The various curves correspond to various relative stiffness of the cable ($E_C A_C$) and girder ($E_G I_G$); L is the central span. Any one of the curves shown in Fig. 1.4 can be found by imposing a unit shortening of the cable. The value of the

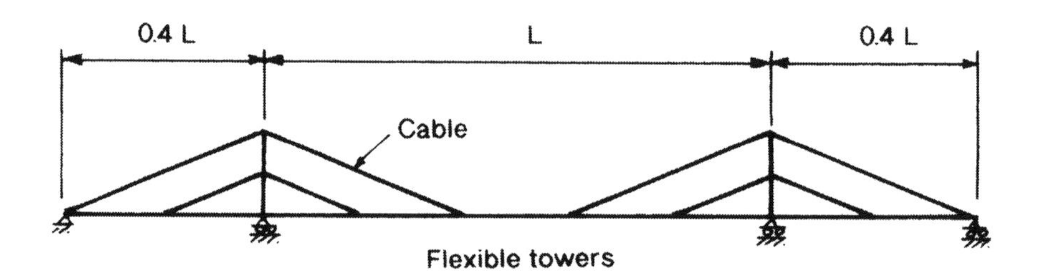

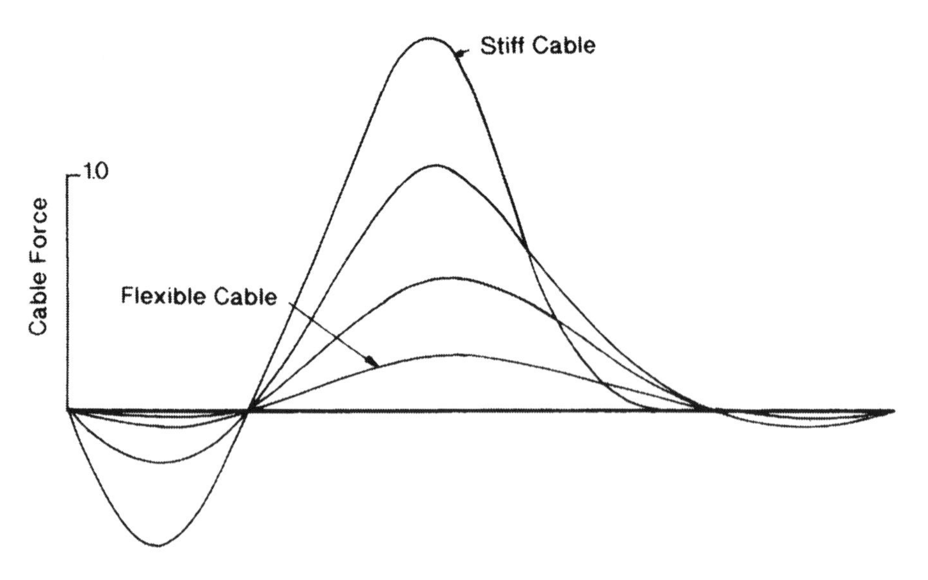

Fig. 1.4 Influence lines for cable force in a cable-stayed girder for various cable stiffnesses

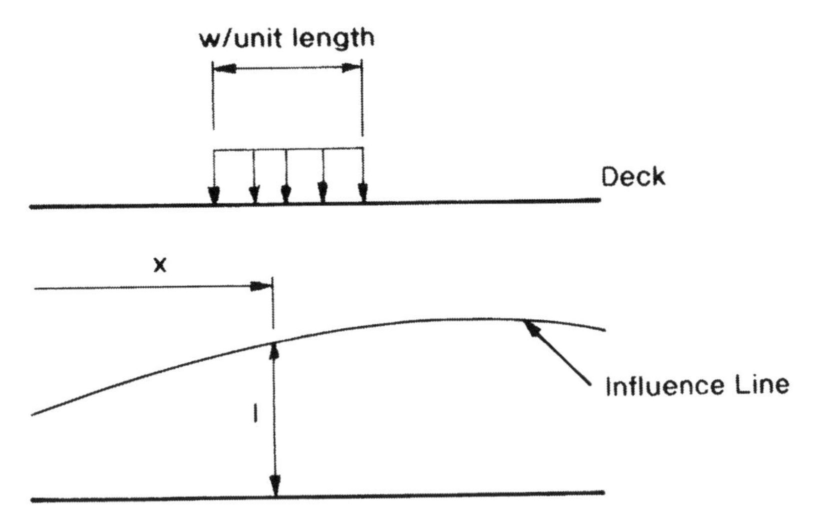

Fig. 1.5 Effect of a uniformly distributed load

method is not so much in the numerical calculation of influence line ordinates, but in the prediction of its shape, for this shape enables one to determine the worst location of a set of moving loads. In Fig. 1.4, for example, the maximum tensile force in this cable under the action of a unit point load moving along the deck occurs when this load lies in the central span at the appropriate maximum ordinate. On the other hand, a load placed in a side span will tend to cause this cable to go slack.

There are a couple of cases that are important. In Fig. 1.5, the deck is loaded with a uniformly distributed load, of intensity w, length ℓ. Consider the effect on a function, F, whose influence line has ordinates I. Then the elemental load, wdx, causes

$$dF = I\,wdx$$

that is,

$$F = \int^{\ell} I\,wdx = w\int^{\ell} I\,dx$$
$$= w \text{ by the area of the influence line under the load}$$

The total load, W, equals $w\ell$. If $\bar{I}$ is the mean influence line ordinate beneath the load, then

$$F = W\bar{I}$$

Consider, second, the maximum bending moment in a simply supported beam caused by the system of moving point loads shown in Fig. 1.6. The shape of the influence line for bending moment at a specific point was shown in

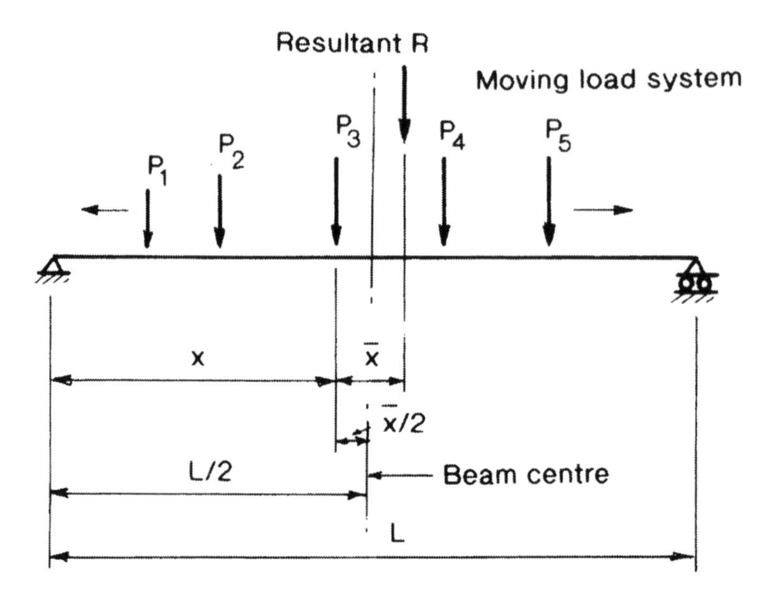

Fig. 1.6 Absolute maximum bending moment in a simply-supported beam

Fig. 1.2(a). The maximum bending moment will occur when one of the point loads is at the maximum ordinate of the influence line. It is not known, however, which point on the beam will suffer the *absolute maximum bending moment*.

The bending moment under load P_3, for example, is given by

$$BM_3 = \frac{R(L - x - \bar{x})x}{L} = \frac{R}{L}(xL - x^2 - x\bar{x})$$

where

R is the resultant force,
x defines the location of P_3 (and hence the load system), and
$\bar{x}$ is the distance from P_3 to R

As x is changed, BM_3 is a maximum when

$$\frac{d}{dx}(BM_3) = 0$$

or when

$$L - 2x - \bar{x} = 0$$
$$x = L/2 - \bar{x}/2$$

This situation is shown in Fig. 1.6. The result may be expressed in words: the maximum bending in a simply supported beam beneath one of a moving system of point loads occurs when the centre of the beam bisects the distance between this load and the resultant of the loads. This is called Muller-Breslau's Principle (Norris and Wilbur 1960: 493).

The theorem should consider only those loads that are actually on the beam. By applying the theorem successively to a number of loads, using for example P_2, then P_3, then P_4 as the chosen load, the absolute maximum bending moment may be determined. In some cases it may happen that the maximum bending moment under one load equals the maximum that may occur beneath another, but this does not alter the principle. For a simply supported span with a single point load, the distance $\bar{x}$ is zero, and the absolute maximum bending moment occurs at mid-span. The theorem does not apply to other beams, such as a two-span, continuous beam (Fig. 1.2(b)). Indeed, in that case, the maximum bending moment under a moving, single point load is not at mid-span.

Finally, the case of the influence line for a deflection, δ_A, should be mentioned. The ordinate to this influence line at point, B, represents the value of δ_A caused by a unit load at B. But, by Maxwell's Reciprocal Theorem, the deflection at A due to a unit load at B equals the deflection at B due to a unit load at A. All values of the influence line ordinates, δ_A, can be found, therefore, by placing the unit load at A and observing the deflections at a variety of points B. The influence line for δ_A is, therefore, identical with the deflected shape of the structure under the action of a fixed unit load at point A.

Chapter 2

Failure

2.1 The nature of failure

In the broadest sense, failure of a bridge occurs whenever it is unable properly to fulfil its function, i.e. to carry the primary loads across an opening. Some examples may be given.

It may be found that a bridge is unable to carry its full design loads and, for this reason, a load restriction is placed on the bridge. This is, in a sense, a failure. Again, the form of the bridge may be such that, in periods of high wind, traffic is unable to cross. An example may be found in the large suspension bridge built in 1966 to cross the Severn estuary in England (O'Connor 1971a: 380ff). Here, the design incorporates a number of features that were advanced at the time, intended to protect the structure against aerodynamic instability. One of these was the adoption of a streamlined cross-section, with light (but strong) roadway railings made up of tensioned, horizontal wire ropes. Unfortunately, this streamlined form causes traffic on the bridge to be exposed to strong winds and, for this reason, the bridge may be closed to traffic (see Section 12.2). It is not sufficient to protect only the bridge from wind; rather, it is required that the traffic should also receive suitable protection.

A bridge may also fail if it interferes unduly with the opening it is designed to cross. This opening may be a stream; O'Connor (1993: 165) lists the ratios of the sum of the clear spans to the total length between abutments of four long Roman bridges in Spain. The distance between abutments ranges from 179 m to 721 m, with ratios varying from 0.80 to 0.62. Three of the bridges are still in service, at Salamanca (179 m; 0.80), Mérida (721 m; 0.64) and Córdoba (274 m; 0.62); the second will be used here as an example. When originally built, about 25BC, the Puente Romano of Mérida had 36 spans and an overall length of about 260 m. Its presence restricted the flow and raised river levels, which then cut through the northern approach. The bridge was extended by adding ten spans to the north, with a length of 129 m. This was still insufficient, for a later flood cut through the southern approach and it was necessary to add another 24 spans, three of which are now buried, with a length of 216 m. The bridge was later modified to give a total of 62 spans. Even so, it is still liable to flood, with the river reaching the deck level as recently as 1942. In some bridges, such a raising of the stream level is tantamount to failure, for the design of the bridge has brought consequences that should not have been permitted.

However, the word *failure* in bridge design is usually taken in a more restricted sense, which may be defined as the onset of unacceptable deflections. These divide into (a) the occurrence of excessive or uncomfortable deflections under service loads, and (b) the collapse of the structure, where the bridge deck falls to the ground. The second type of failure means that the bridge, when viewed after the event, is no longer intact; but it begins with the development of excessive deflections that accelerate with time. In modern terminology the first is grouped with other *serviceability limit states*; the second is regarded as a *collapse limit state*.

Failure is generally a consequence, not only of the geometry of the structure, but also of the nature of the material of which it is composed. Materials include stone, timber, steel and concrete, but only the third and fourth will be considered here.

Steel may be used for structural members, cables, reinforcing rods and prestressing strands or bars. Relevant modes of failure include the following:

(a) tension failure, which may take the form either of the onset of large deformations, or separation, where the material breaks into two parts;
(b) compression failure, either in constrained compression, or by buckling, as evidenced by column buckling, beam and plate buckling;
(c) shear failure, under the action of shear stresses;
(d) fatigue failure, started by the development of cracks under the action of repetitive loads that have a tension component; and
(e) brittle fracture where, in some steels, under the action of tensile stress, a crack will initiate suddenly at some point and proceed rapidly across the structure (see Section 2.4, Case studies).

Concrete may fail

(a) in tension, or
(b) in compression.

The tensile strength of concrete is low and not usually a major factor in the assessment of structural strength. However, tension failure of concrete leads to cracking and this may affect the durability of the structure. Furthermore, the nature of the concrete may change with time, limiting its ability to protect embedded steel reinforcement from corrosion. These and related matters affect the durability of concrete structures, an important subject which will be referred to again briefly in Section 2.4.

2.2 The collapse of a building frame

Neal (1977: 11–3) has given a brief account of the development of the plastic methods of structural analysis now widely used in the design of steel building frames. Kazinczy first observed in 1914 that, for a fixed-ended beam to fail, yielding had to occur at three sections which acted effectively as 'plastic hinges'. In 1940, van den Broek wrote on the 'Theory of Limit Design' in a paper with a substantial discussion; he followed this with a book having the same title (1948). The classical British work is that by J.F. Baker, and his associates J.W. Roderick, M.R. Horne and J. Heyman, commencing in the 1930s (see Baker and Roderick 1938) and culminating in *The Steel Skeleton* (1956) and *The Plastic Design of Frames* (1969–71), both of two volumes. L.S. Beedle (1958) from Lehigh University was another of the earlier writers.

Consider a rectangular steel beam, bent about its major axis, XX, as shown in Fig. 2.1. A typical stress–strain curve is as in (b). For stresses less than σ_y, behaviour is elastic, as expressed by the relationship

$$\sigma = E\epsilon$$

where

E is the elastic constant, or modulus of elasticity, and
ϵ is the strain

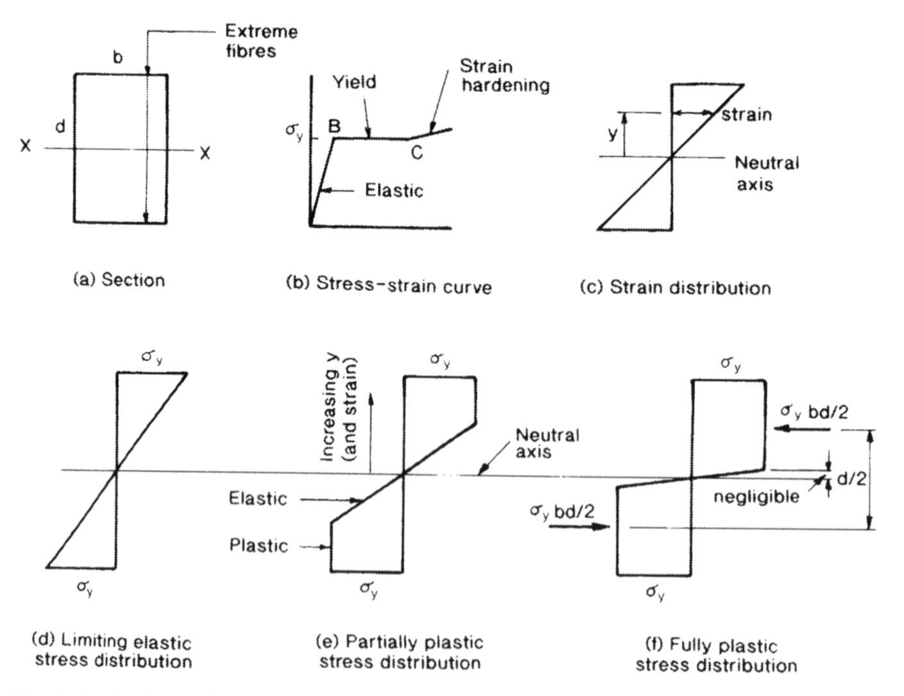

Fig. 2.1 Inelastic bending behaviour

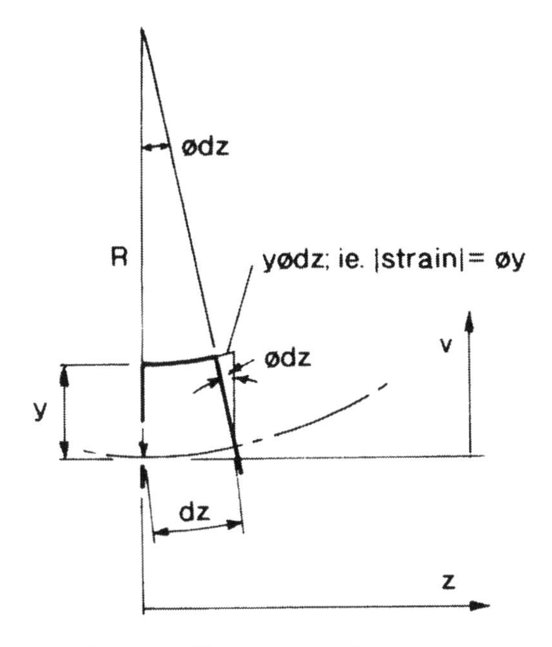

Fig. 2.2 Relationship between fibre strain and curvature

It is assumed that in bending, plane cross-sections remain plane and perpendicular to the longitudinal axis. That is,

$$\epsilon \propto y$$

Let ϕ be the curvature, where

$$\phi = \frac{d^2v}{dz^2}$$

v is a perpendicular displacement of the member axis, and
z is the distance along the member.

Figure 2.2 shows the relationship between fibre strain and curvature; that is,

$$\epsilon = \phi y$$

Assuming that stresses in the plane of the cross-section are small, the stress in a fibre is given by

$$\sigma = E\epsilon = E\phi y$$

For small curvatures, the behaviour is elastic, until the extreme fibre stress equals σ_y. If the corresponding curvature is ϕ_y, then

$$\phi_y = 2\sigma_y/Ed$$

Beyond ϕ_y, the member will yield. The assumed stress–strain curve suggests that at σ_y the material will yield at constant stress. Figure 2.1(e) plots the stress distribution across the member; that is, it plots σ against y. But ϵ equals ϕy. It follows that each half of the stress distribution is similar to the stress–strain curve; in this case, with a constant stress σ_y extending in from each extreme fibre.

As ϕ increases, so the extreme fibre strains increase. The stress–strain curve (b) suggests that, after a sufficient strain, the stress will begin to rise, as the material enters the strain-hardening range. However, the limiting strain (at C) is usually large. It follows that, with increasing ϕ, the stress distribution approaches that shown in (f), where the elastic position near the neutral axis is of negligible extent. Assuming a rectangular stress block, the limiting moment, M_P, is given by

$$\begin{aligned} M_P &= \sigma_y bd/2d/2 \\ &= \sigma_y\, bd^2/4 \end{aligned}$$

It will be remembered that, at first yield,

$$M = M_y = \sigma_y\, bd^2/6$$

where the term $bd^2/6$ is called the elastic section modulus, Z.

M_P is called the plastic hinge moment, and the term $bd^2/4$ is a particular example of the plastic section modulus Z_P. It is evident that, in this case,

$$\begin{aligned} Z_P &= 1.5Z \\ M_P &= 1.5M_y \end{aligned}$$

Within the central, elastic portion of the stress distribution,

$$\epsilon = \phi y$$

Also, here

$$\sigma = E\epsilon = E\phi y$$

It follows that the slope of this portion of the stress–strain curve is proportional to the curvature ϕ. It is possible to plot ϕ against the bending moment M. This will not be done here quantitatively, but it is evident that it has the form shown in Fig. 2.3. As M approaches M_P, ϕ approaches infinity. This situation is, in this context, called collapse.

As M approaches M_P, the curvature rapidly increases, to the extent that a reasonable approximation is that M stays constant at the value M_P as the curvature approaches infinity. This behaviour is analogous to the behaviour of a pin with a friction moment, M_P. This pin is called a plastic hinge. The analogy is not

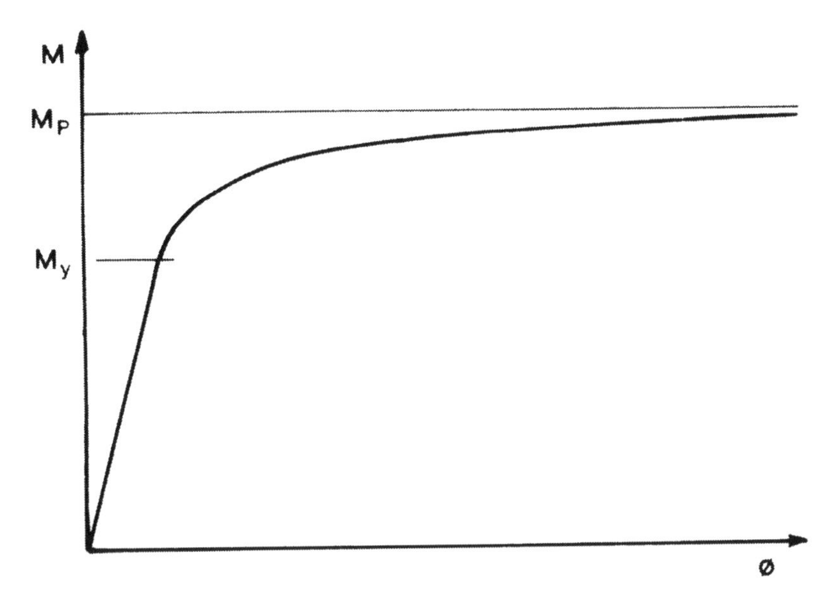

Fig. 2.3 Moment–curvature relationship for steel beam

strictly correct, for the curvature is in excess of the elastic limit on either side of 'the pin'. Nevertheless, the collapse state for a simply-supported beam with a central point load can be shown with reasonable accuracy as in Fig. 2.4(a). Then,

$$M_P = PL/4$$

or the limiting value of P is

$$P = 4M_P/L$$

The insertion of the plastic hinge at point B is sufficient to turn the beam into a mechanism that can collapse as shown. If the beam is indeterminate, more plastic hinges must be formed before collapse is possible. Figure 2.4(b) shows a fixed-ended member under a central point load. The central moment equals

$$PL/4 - M_P$$

For collapse, this also must equal M_P. That is, at collapse,

$$M_P = PL/4 - M_P$$
$$PL/4 = 2M_P$$
$$P = 8M_P/L$$

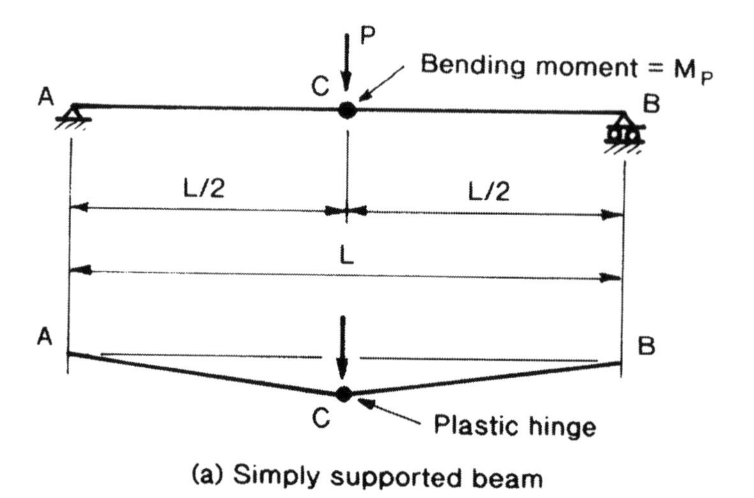

(a) Simply supported beam

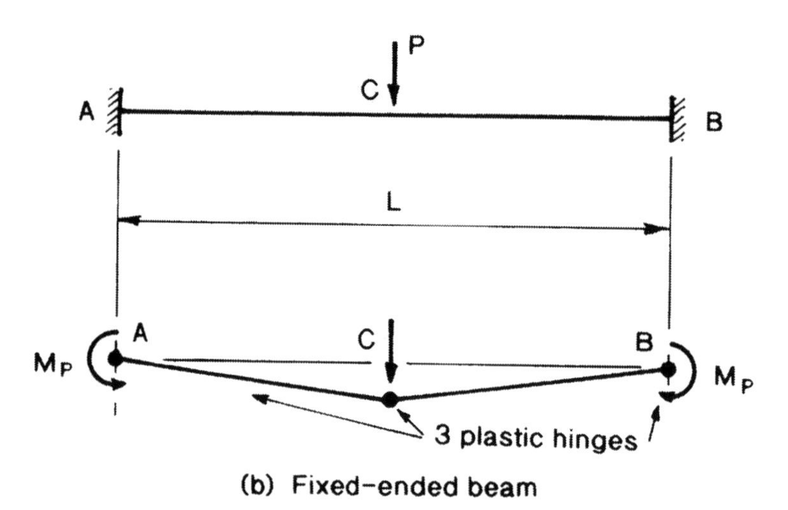

(b) Fixed-ended beam

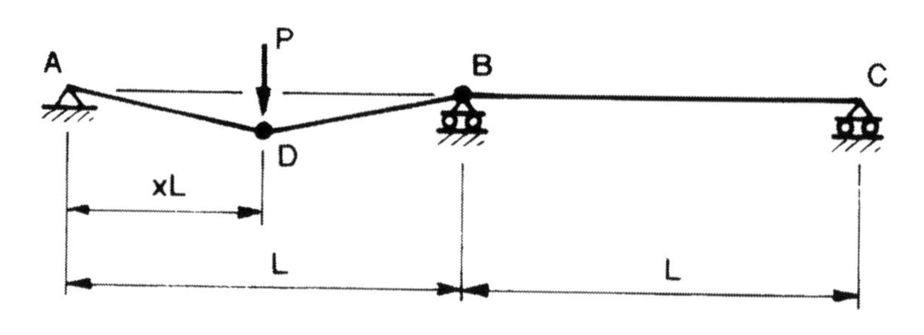

(c) Two-span continuous beam

Fig. 2.4 Collapse mechanisms

A third example is shown in Fig. 2.4(c) – the case of a beam continuous over two equal spans. Assuming plastic hinges as shown, the bending moment at D is given by

$$M_D = Px(1 - x)L - xM_P$$

For this to equal M_P,

$$Px(1 - x)L = M_P(1 + x)$$

$$P = M_P/L \frac{1 + x}{x(1 - x)}$$

Clearly P is a function of x. For example, if x equals 0.5, then P equals $6M_P/L$. However, the value of P that is important is the smallest value that will cause collapse, and it may be shown by differentiation that this least value of P occurs when

$$x = 0.41421$$

Then,

$$P = 5.8284\ M_P/L$$

This analysis was based on the stress–strain curve shown in Fig. 2.1(b), and repeated in Fig. 2.5(a). Suppose that the material is strained to point A at a stress σ_y, and then unloaded. The stress–strain curve on unloading is as shown; and this may be approximated by that in (b), which shows elastic recovery at an effective modulus, E, the same as in the loading curve. Even if the material is loaded to point B in the strain-hardening range, then recovery tends to be elastic at an approximate slope, E.

It follows, therefore, that if a member in bending is loaded to a stage close to the plastic hinge moment, M_P, then the stresses caused by unloading are

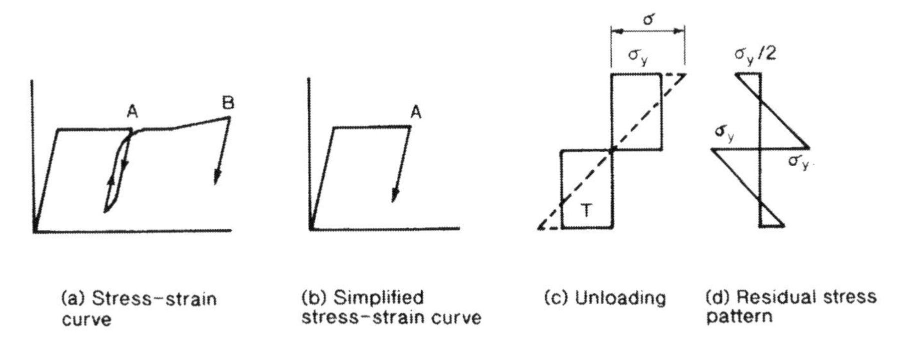

Fig. 2.5 Unloading from the plastic hinge moment

elastic. Apply, therefore, an equal and opposite moment. This will cause extreme fibre stresses in a rectangular section of magnitude

$$\begin{aligned} \sigma &= M_P \div bd^2/6 \\ &= (\sigma_y bd^2/4)(6/bd^2) \\ &= 1.5\sigma_y \end{aligned}$$

The final extreme fibre stress

$$= \sigma_y - \sigma = -0.5\sigma_y$$

The complete pattern of stresses remaining in the section after unloading is as shown in Fig. 2.5(d). These are called residual stresses and are self-equilibrating; that is, they are equivalent to zero resultant moment and axial force.

Suppose now that the section is reloaded with a bending moment that is of the opposite sign to that which was originally applied. Then the extreme fibre stress will be

$$-\sigma - 0.5\sigma_y$$

The simplest assumption is that yield will occur when the absolute value of this combined stress equals σ_y. That is,

$$\begin{aligned} |-\sigma - 0.5\sigma_y| &= \sigma_y \\ -\sigma - 0.5\sigma_y &= -\sigma_y \\ \sigma &= -0.5\sigma_y \end{aligned}$$

The absolute value of the bending moment is

$$|M| = \sigma bd^2/6 = \sigma_y bd^2/12 = \tfrac{1}{3}M_P$$

Whereas on first loading, the moment for first yield was $M_y = \frac{2}{3}M_P$, now on unloading and reverse-loading, the first yield moment is $\frac{1}{3}M_P$.

Let this reverse moment be increased further in magnitude. Yield will again occur, and a typical distribution of the total stresses in the cross-section will be as shown in Fig. 2.6(c). As yield extends into the section, the effects of the stresses near the neutral axis are small and the total stress distribution again approaches that shown in (d). That is,

$$M_P = \sigma_y bd^2/4$$

as on first loading.

Figure 2.7 shows the effect of this on the moment–curvature (M/ϕ) relationship. Collapse moments are unaltered, but deflections prior to collapse will be increased. Point C represents the situation after the initial loading and unloading to a zero bending moment: it is evident that residual curvatures are left in the member. In the case of a simply-supported member, these are accommodated

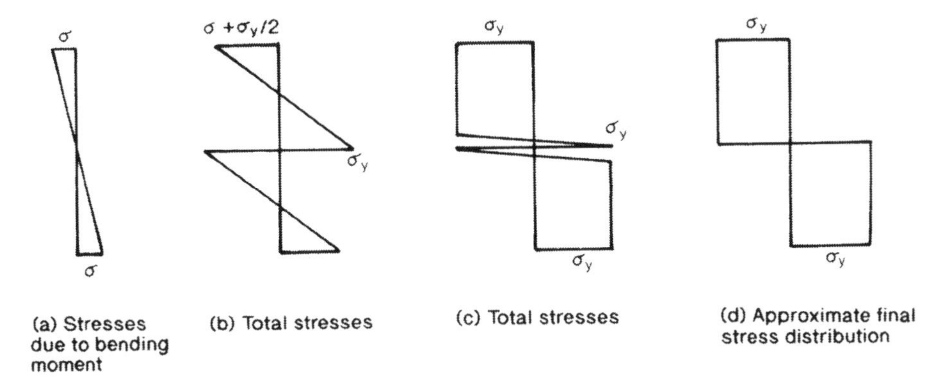

Fig. 2.6 Reversed loading

simply by a change in the member shape. In the case of an indeterminate member, these curvatures may be incompatible with the conditions at the supports and will cause a locked-in system of reactions and bending moments. Figure 2.4(c) showed a two-span continuous beam. Here, this locked-in system must consist of a vertical reaction at *B*, with opposing reactions at *A* and *C*, and a triangular, bending moment diagram. If *P* is a moving load, then the first passage of this load may leave the girder with such curvatures and bending moments; both will modify the response of the member to a second passage of the load.

The assumptions made in this analysis may need to be modified if exact deflections are required for a case of unloading and reloading, particularly if the reloading is in the reverse direction. The slope of the stress–strain curve for

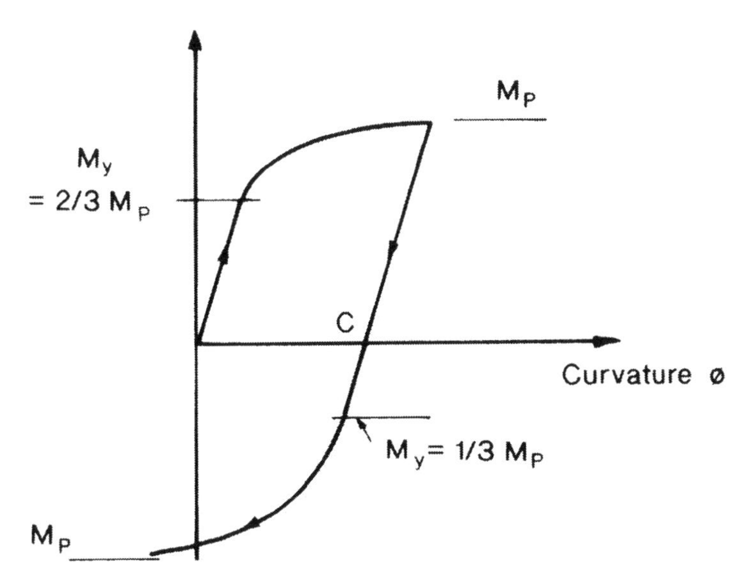

Fig. 2.7 Moment–curvature relationship for reversed bending

unloading may differ from E, and the stress necessary to cause reversed yield may change due to causes such as the Bauschinger effect (Behan and O'Connor 1982; Massonnet et al. 1979: 7; Nadai 1950: 20 etc; Neal 1977: 5; Smith and Sidebottom 1965: 12f, etc).

This discussion is based on the behaviour of a rectangular steel beam, with the stress–strain curve shown in Fig. 2.1(b) and the moment–curvature relationship of Fig. 2.7. For the typical mild steel used in structural members, the ductility is large, with strains to strain-hardening of the order of 10–20 times the elastic strain at first yield. The final elongations at failure are typically much larger again, of the order of 25–30%, with an ultimate stress higher than σ_y. The European Standard ENV 10020: 1988 specifies the following minimum values for 'base steels':

yield stress	360 MPa
ultimate stress	600 MPa
elongation to failure	26%.

There can be considerable variations from this idealised behaviour. The material may be elastic to an upper yield stress that exceeds σ_y, but with the stress dropping to σ_y immediately yield has occurred. On the other hand, first yield in practical members may occur at a lower stress, due either to a change in the material properties, the presence of residual stresses, or some other factor. The geometry may, of course, also vary from the simple rectangle. Neal (1977: 14), for example, suggests that for typical rolled I-beams, the shape is such that M_P is about $1.14M_y$ for bending about the strong axis, or $1.6M_y$ for bending about the weak axis.

Steel is also only one of a number of materials used in bridge design. Darvall and Allen (1987) discuss the moment–curvature relationship for reinforced and prestressed concrete members. Figures 2.8(a) and (b) are based on their Figures 4.2 and 4.3, and show typical curves for under- and over-reinforced concrete beams, and for a prestressed concrete section. These in turn are based on the output of a computer program which includes accurate simulations of material properties. A particular result is presented in Fig. 2.8(c) (based on their p.98), with curvatures calculated only up to the point where the maximum moment is reached.

Referring to Fig. 2.8(b), the moment M_O is defined by Darvall and Allen as the moment when the initial curvature produced by the prestress is reduced to zero. It follows that his curvatures have as their origin the straight member. Alternatively, the curvatures produced by the applied loads (excluding the prestress) have O as their origin. In all cases, there is a bending moment, M_C, when the concrete first cracks, where this moment should be higher for a prestressed member. This is followed by the moment, M_y, corresponding to yield in the conventional reinforcement, followed by M_U, either when the concrete fails in compression, or the reinforcement reaches its ultimate stress. Two observations may be made:

1. the development of curvature beyond the ultimate moment is limited by the ductility of the materials; and

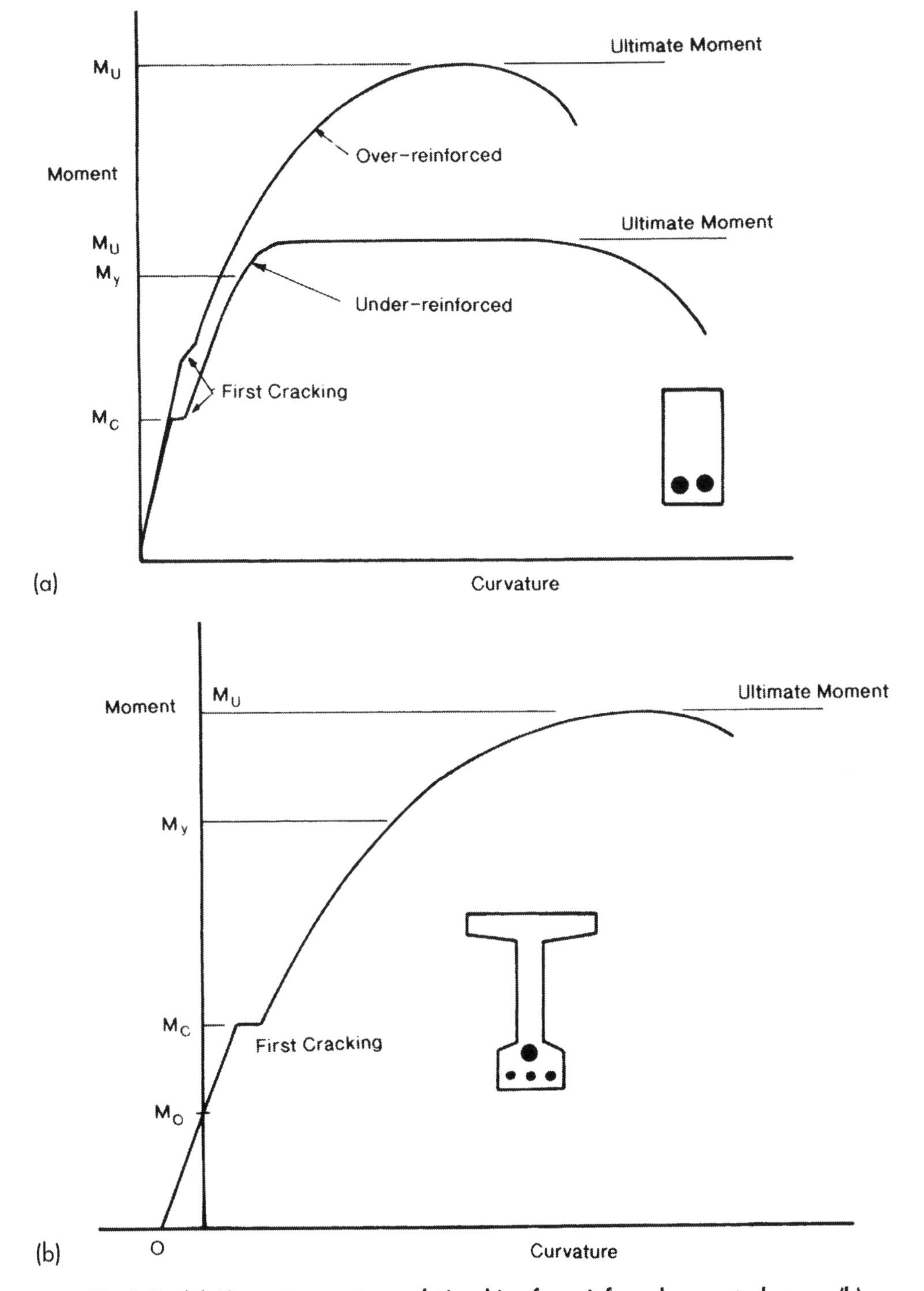

Fig. 2.8 (a) Moment–curvature relationships for reinforced concrete beams; (b) and (c) Moment–curvature relationships for prestressed concrete beams (Redrawn from Darvall and Allen 1987)

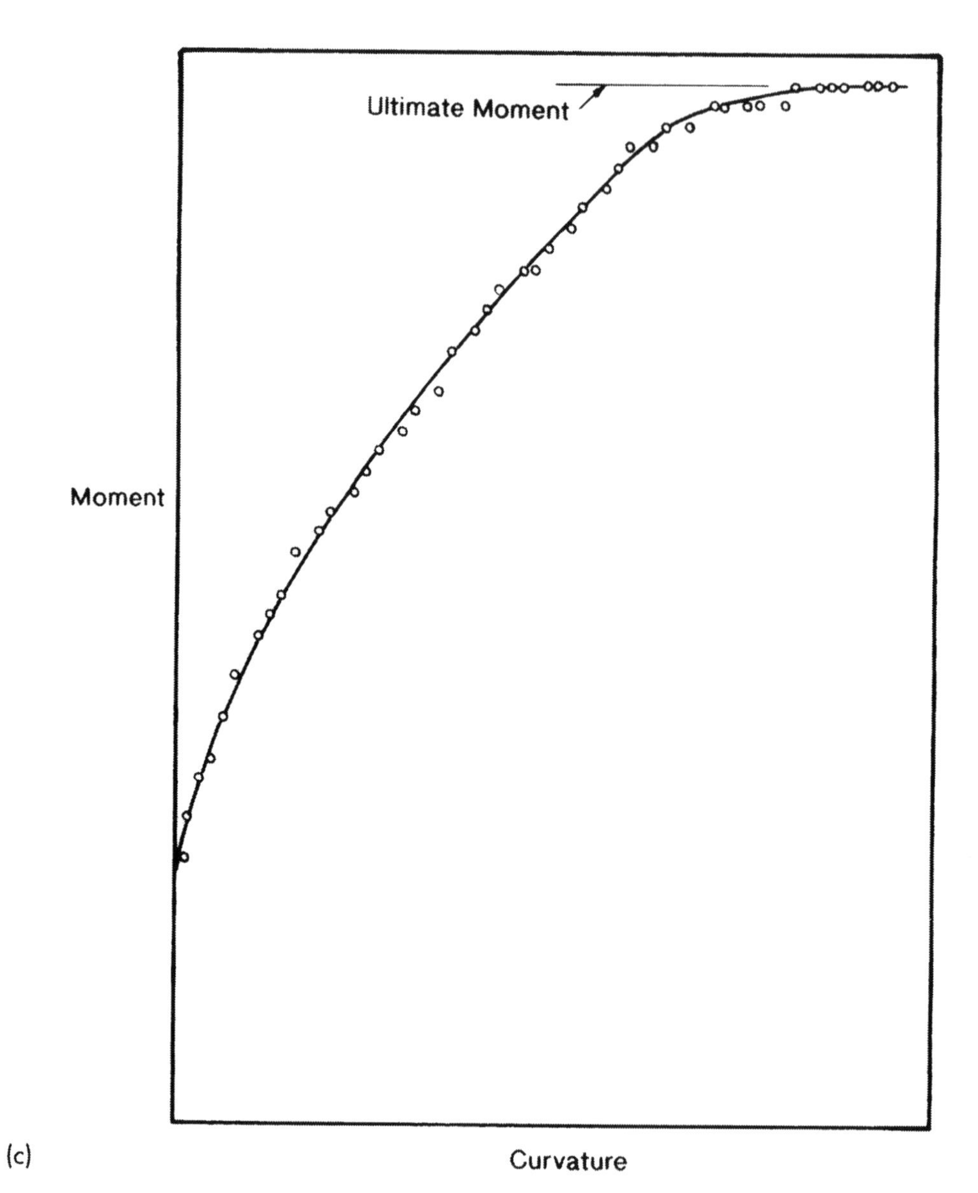

Fig. 2.8 *(continued)*

2. for moments less than the ultimate moment, the shapes of these moment–curvature relationships are broadly similar.

The second of these observations suggests that there is value in proceeding further with the case of the rectangular steel beam to illustrate inelastic behaviour but, in doing so, the first also should be kept in mind. Other computer simulations of the moment–curvature relationship for partially prestressed concrete sections may be found in Cohn and Bartlett (1982).

2.3 Failure under moving loads

It is not sufficiently realised that bridge failure is a dynamic event. This is for two reasons.

1. Primary bridge loads are typically moving loads – they are not stationary loads that increase in magnitude. Indeed, except for the weight of the structure itself, true stationary loads are rare, for even in building design, the use of the structure means that loads will move. Nevertheless, in some cases, such as the floor of a library stack-area, superimposed loads may be modelled as stationary weights that increase with time. This is not the case with bridges.
2. Collapse itself is dynamic, for it implies that the structure will accelerate and move.

Consider a bridge subject to a major vehicle load. Typically the vehicle is moving at high speed. There is an interaction between the suspension system of the vehicle and the vertical shape of the roadway surface. The initial loads supplied to the bridge as the vehicle moves onto the bridge may be well within the strength of the bridge. As the vehicle advances, so its influence increases and parts of the bridge may begin to fail. However, as this happens, the vehicle may move off the bridge and unloading will occur. Excessive loads may therefore leave the bridge intact, but with a permanently deflected shape.

The dynamic interaction between a bridge and a vehicle will be discussed fully in Chapter 7. This section will consider the inelastic behaviour of a bridge girder under a moving load. It is intended only to be introductory, and two simple cases will be considered:

(a) a simply-supported inelastic beam under the action of a moving point load, and
(b) the inelastic behaviour of a two-span continuous beam under the action of a similar load.

In both cases the member will be taken to be of steel, with a rectangular section. For this case, the shape of M/ϕ may be calculated. Figure 2.9(b) shows the stress distribution in a partly yielded section in bending, with y as the distance from the neutral axis to the boundary of the yielded zone. The moment M is given by

$$M = M_P - \sigma_y b y^2/3$$

But

$$M_P = \sigma_y b d^2/4$$

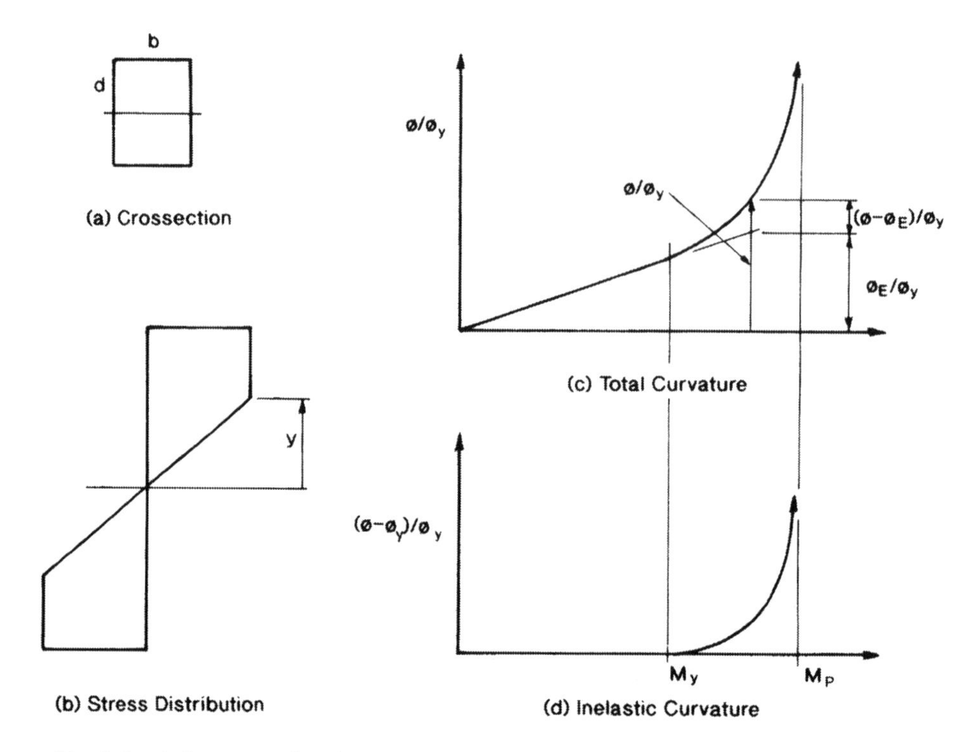

Fig. 2.9 Behaviour of inelastic beam

therefore,

$$M = M_P\left(1 - \frac{4}{3}\frac{y^2}{d^2}\right)$$

or

$$y/d = \sqrt{\tfrac{3}{4}(1 - M/M_P)}$$

Now

$$\sigma_y = E\phi\, y$$

$$\phi = \frac{\sigma_y}{Ed\sqrt{\frac{3}{4}\left(1 - \frac{M}{M_P}\right)}}$$

At first yield,

$$\phi = \phi_y = 2\sigma_y/Ed$$

therefore,

$$\frac{\phi}{\phi_y} = \frac{1}{\sqrt{3\left(1 - \frac{M}{M_P}\right)}}$$

Let ϕ_E be the elastic curvature that would be caused by M if there were no yield. Then

$$\phi_E = M/EI$$

$$\frac{\phi_E}{\phi_y} = \frac{3}{2}\frac{M}{M_P}$$

therefore,

$$\frac{\phi}{\phi_y} - \frac{\phi_E}{\phi_y} = \frac{1}{3\left(1 - \frac{M}{M_P}\right)} - \frac{3}{2}\frac{M}{M_P}$$

This relationship is shown in Fig. 2.9(c) and (d), with the moment plotted horizontally, and the curvature, in this dimensionless form, plotted vertically, as the dependent parameter. It is plotted accurately in Fig. 2.10. This figure also includes other dimensionless parameters that define mean ordinates and the location of the centroid of areas under this curve. In using these it should be remembered that they apply only to the case of a linear variation of M. They can be useful in the calculation of inelastic deflections.

Consider first the case of a point load, W, moving along a simply-supported beam. It was shown previously that a load, $W = W_P = 4M_P/L$, is sufficient to cause collapse, and this will occur when W reaches the location $x = 0.5L$ (see Fig. 2.4(a)).

Let $W = W_P$, with $x < 0.5L$. The first yield moment, M_y, is $\frac{2}{3}M_P$. The bending moment under the load is

$$M = W_P x(1 - x)L$$

Let $M = M_y$. Then these equations reduce to the condition, that

$$x^2 - x - 1/6 = 0$$

with the solution

$$x = 0.2113$$

Successive bending moment diagrams are shown in Fig. 2.11, as x increases from 0.2113 to 0.5; that is, as the load W_P moves from point C to D. The bending moment at a specific point is a maximum when the load is at that

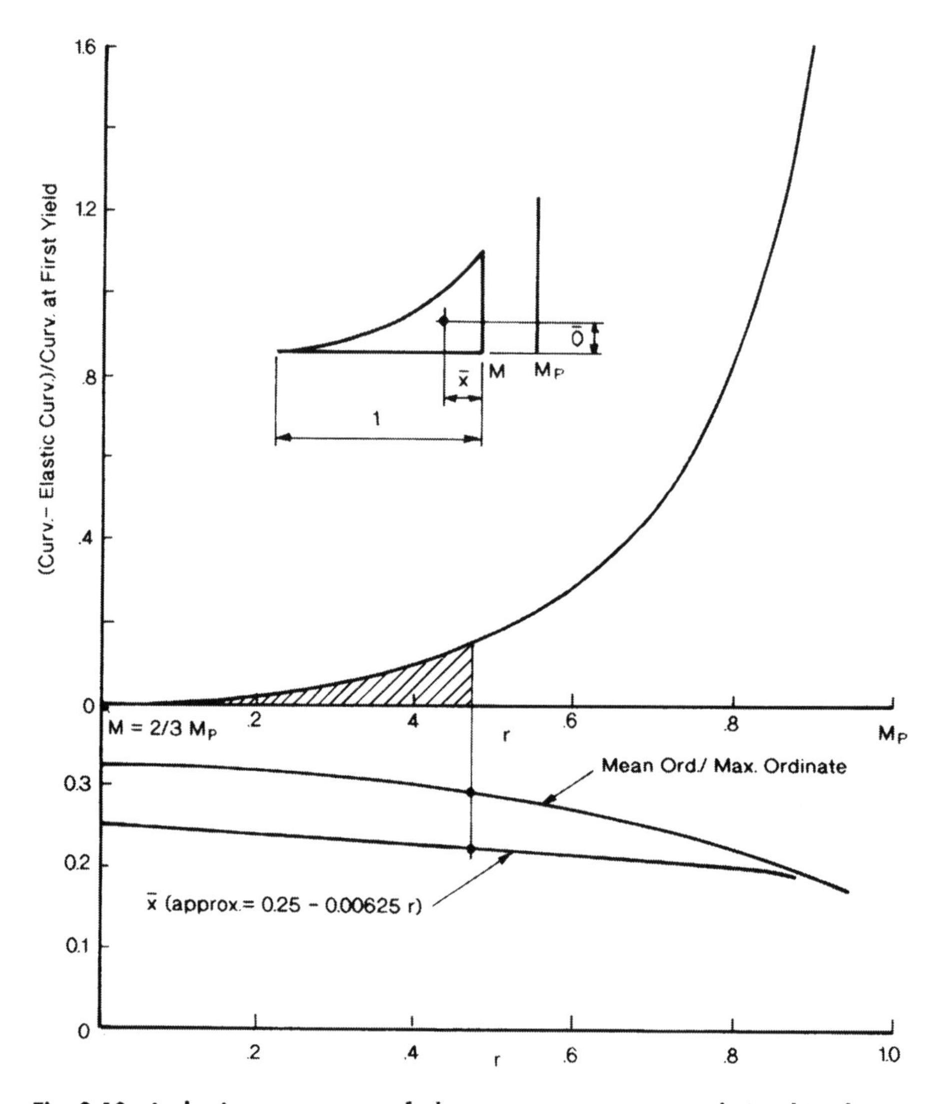

Fig. 2.10 Inelastic component of the moment–curvature relationship for a rectangular steel beam

point. At point *C*, for example, it is a maximum when the load is at *C*, and then reduces as the load moves to the right.

It is important to realise that the inelastic component of the curvature at a point is determined not by the current bending moment, but by the previous maximum bending moment, as shown in Fig. 2.11. For the load at the mid-point, *D*, for example, the curve of previous maximum bending moments is non-linear to the left of *D*, and linear to the right. It follows that, for this case, the plot of the inelastic component of curvature is asymmetrical.

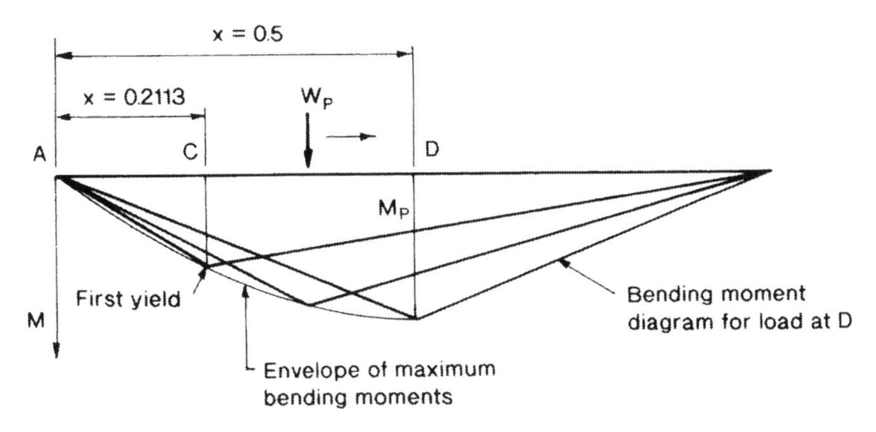

Fig. 2.11 Successive bending moment diagrams as the collapse load, W_P, moves across a simply-supported beam

Figure 2.12 plots the total deflection under the load, W, for three different values of W, expressed as a multiple of L^3/EI.

For W less than or equal to $\frac{2}{3}W_P$ the behaviour is elastic, with a maximum deflection of $WL^3/48EI$ ($= 0.02083L^3/EI$).
For $W = 0.9W_P$, inelastic deflections occur beyond $x = 0.2454$, and rise to a maximum of about $0.024WL^3/EI$. The curve is not symmetrical.
For $W = W_P$, inelastic deflections commence when $x = 0.2113$ and rise to an infinite value as x approaches 0.5. However, deflections larger than the elastic maximum occur only for x in excess of about 0.35, and exceed $0.024WL^3/EI$ only for x greater than about 0.4.

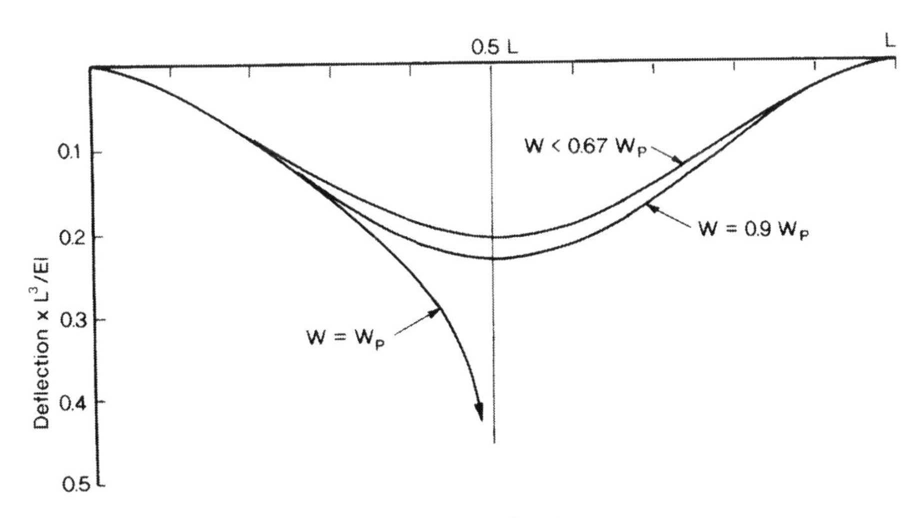

Fig. 2.12 Deflections under a moving point load

Suppose the vehicle velocity is 25 m/sec and the span, 25 m. For the load to travel from $x = 0.4$ to $x = 0.6$ will require 0.2 sec. If the structure were to accelerate downwards at 9.8 m/sec^2, the vertical displacement would be

$$s = \tfrac{1}{2}gt^2 = 0.19 \text{ m} = L/132$$

In fact, the effective acceleration is likely to be far less. It follows that:

(a) a load equal to 0.9 times the collapse value causes only small inelastic deflections;
(b) even if the load were to equal the theoretical collapse value, the dynamic accelerations that occur over the limited period when the load is near mid-span may not cause excessive deflections.

With this analysis, one complete passage of the load, W, over a simply-supported beam is sufficient to cause all sections to reach their maximum bending moment and subsequent passages should not cause yield. This conclusion, however, needs to be qualified. Load W causes deflections of the beam. If it were initially straight, these deflections mean that W passes over a sagging beam. This sag causes W to experience vertical accelerations that slightly increase the load it applies to the beam. For an elastic member, these additional loads are small. Consider, however, the inelastic member. The first passage of a constant load, W, will leave the member with permanent deflections and, during a second passage of the load, these will increase the vertical accelerations, causing slightly larger loads, bending moments, and inelastic deflections. For one or two passages of the load, this effect may be expected to be negligible. However, the problem is that the process is progressive and the analysis raises the possibility of *progressive collapse* (see Chapter 6).

The problem is more complex than that which has been described here. A true analysis would need to take into account other dynamic effects, such as those described in Chapter 7, but the phenomenon needs to be recognised as a possible mode of failure.

Consider second the case of the two-span continuous beam. Collapse occurs with the load $W_P = 5.8284M_p/L$ at location D, with $x = 0.41421$, as shown in Fig. 2.13(a) and (b). This figure also shows the results of some elastic analyses.

Case (c) is the elastic influence line for the bending moment at this point D, with the convention that a positive bending moment will tend to cause tension in the lower fibres. Similarly, (d) is the influence line for bending moment at the central support. They show that the bending moment at D rises to the value, $+0.2071WL$, for the load at D; then reduces with increasing x until M_D is zero with the load at the central support; and continues with negative bending moments, M_D, as the load enters the right-hand span, reaching the extreme value, $-0.0399WL$ as shown. Meanwhile, as x increases from zero, the moment, M_B, at the central support is negative throughout, being $-0.08578WL$ for the load at D, and with an extreme value, $-0.09623WL$, for $x = 0.5774$. This influence line is symmetrical about B. The next diagram, (e), shows the

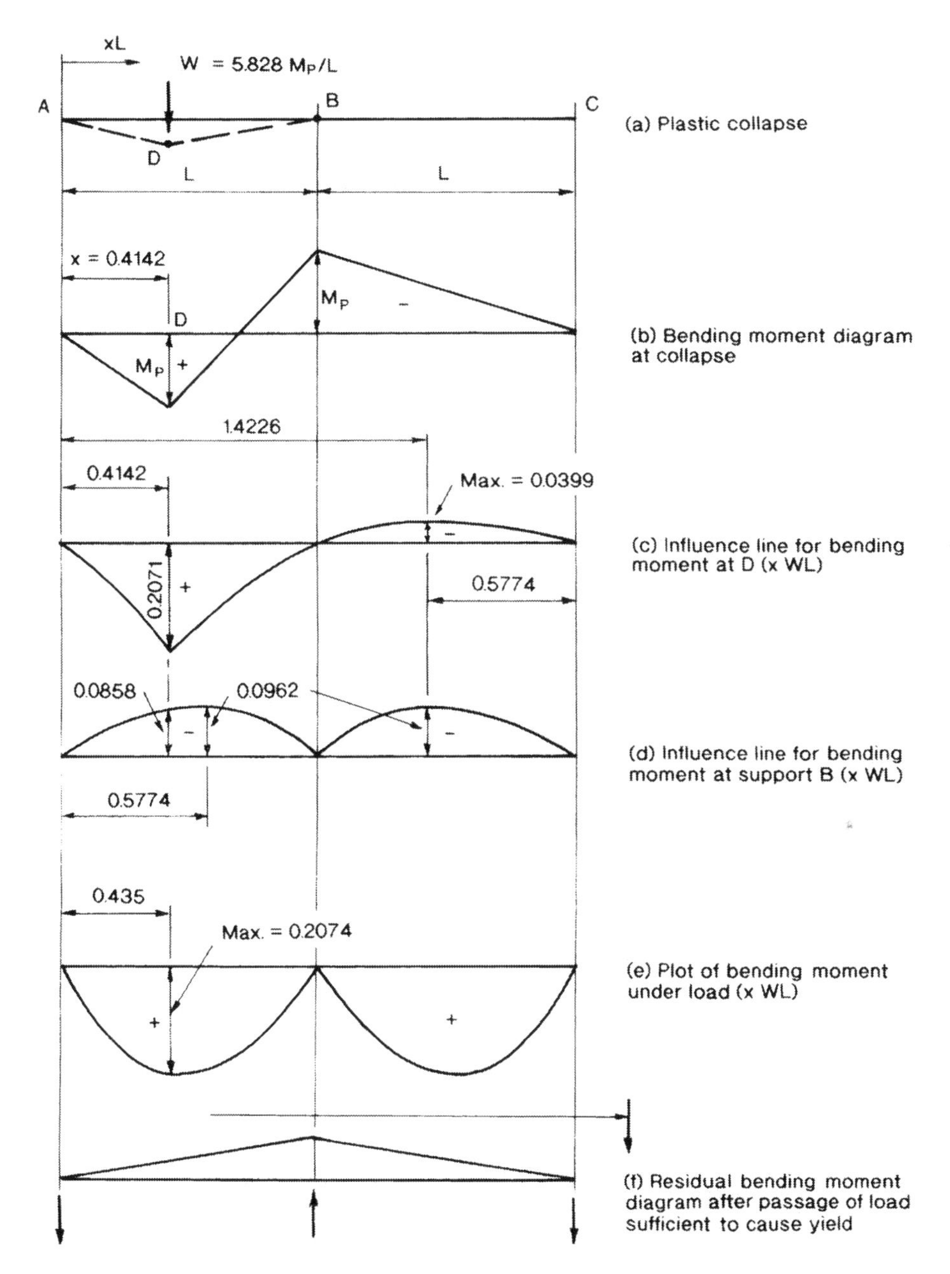

Fig. 2.13 Plastic collapse, elastic and inelastic behaviour of continuous beam

bending moment under the load. Clearly it reaches $+0.2071WL$ for the load at D (as in (c)) but rises slightly to the extreme value of $+0.2074WL$ for the load at $x = 0.435$. For first yield,

$$M_y = 0.2074W_yL$$

But

$$M_y = \tfrac{2}{3}M_P$$

therefore,

$$W_y = 3.2144M_PL, \text{ and}$$
$$W_P = 1.813W_y$$

For a simply-supported beam, this ratio is 1.5. The reserve of strength from first yield to plastic collapse is greater in the case of this continuous beam, which is of constant cross-section. If the cross-section is adjusted at various points along the beam to suit the worst effects at each section, then this ratio will be reduced, and in the extreme case may still approach 1.5.

To return to (c), if $W = W_P = 5.8284M_P/L$, then, for the load at D, the elastic bending moment would be:

$$M_D = 1.207M_P$$

Clearly this will cause yield at D. At the same time, the elastic bending moment at the central support is

$$M_B = -0.08578WL$$

For $W = W_P$ this becomes

$$M_B = 0.5M_P$$

Even with the load W_P at the maximum, E,

$$M_B = 0.09623 \times 5.8284M_P/L$$
$$= 0.561M_P$$

The true inelastic behaviour is that the load, W_P, at D causes yield at D with inelastic curvatures that cause M_B to rise also to the collapse moment, M_P. It may be shown further that, for a rectangular steel beam, as the load, W_P, moves onto the beam, the bending moment under the load reaches the first yield moment, $M_y = \tfrac{2}{3}M_P$, when $x = 0.1381$.

For a load slightly less than W_P, a true inelastic analysis would show that the passage of the load across the double span would leave the member with inelastic deflections that give rise to a locked-in set of reactions and bending moments of the kind shown in Fig. 2.13(f). The same would occur even when the load has reached only the central support, B. Consider this situation:

(a) Sections around D and B have experienced moments well above M_y in the moment curvature diagram shown in Fig. 2.7, with tension in the lower fibres at D and in the upper fibres at B.
(b) Unloading has occurred as W approaches B and for W at B, these locked-in reactions and bending moments will cause some negative bending moments throughout the member.
(c) As the load advances into the second span, it will tend to cause negative bending moments at both D and B – see Fig. 2.13(c) and (d). That is, the moments, M_D, which were originally positive, will become increasingly negative, whereas the support moments, M_B, which have always been negative, will simply become increasingly negative again.

The elastic analysis showed that a load, W, in the right-hand span could cause an extreme value,

$$M_D = -0.0399WL$$

For $W = W_P = 5.8284M_P/L$, this becomes

$$M_D = -0.233M_P$$

Referring again to Fig. 2.7, the question that arises is whether this moment, together with the locked-in bending moment, is sufficient to cause reverse yield; that is, to exceed the value $\frac{1}{3}M_P$. The answer to this question is unclear and will not be resolved here, but this discussion is sufficient to identify two situations which, in principle, may occur.

1. After one or more passages of the load, increases in the yield moment due to prior yield, plus the effect of locked-in moments, may be sufficient to cause the structure to behave elastically.
2. Alternatively, reversed yielding of some parts of the structure may occur under repeated loading. In this case, the stress–strain curve shown in Fig. 2.1(b), which has formed the basis of this discussion, must be modified, for repeated reversed yielding will cause a reduction in the length of the yield plateau, shown as the horizontal line to C. The ductility of the material may be exhausted, stresses will rise into the region shown as the strain-hardening zone in the figure, and fracture may occur.

Clearly this second behaviour is unacceptable. To the first is given the name, *shakedown*. It is discussed in many of the books that deal with the collapse of building frames (see Section 2.2), such as in Neal (1977: 158, 193ff; see also Goh and Grundy 1989).

2.4 Case studies

2.4.1 Qantas Hangar

The first example is the collapse of the Qantas Hangar at Mascot Airport, Sydney, on the evening of 12 January 1957 (Lavery and O'Connor 1957). Although not a bridge, it is included here for it leads to a general conclusion that may often be important. The hangar was 42.7 × 91.5 m in plan, and was designed to house two Boeing 707 aircraft, side by side. Figure 2.14 shows the collapse. The building was incomplete, the roofing was being placed, work ceased about 5 p.m. and the hangar collapsed about 7 p.m. The photograph is taken looking into the front of the building, later to be closed by large horizontal-rolling doors. Parallel to these doors is the ridge, and beneath the ridge is the primary structural element: a steel truss continuous over two equal spans, supported by side columns and a single interior column at mid-length. This supported secondary trusses that cantilevered forward to the plane of the doors. As can be seen, the failure involved only the left-hand span of the main truss.

The central diagonal web member of the main truss sloped down to join the central column, but was supported against buckling in the plane of the truss by a sub-member that came down from the top of the column. It was concluded that the collapse was caused by the failure of this central diagonal, but as the

Fig. 2.14 Collapse of Qantas hangar during construction, on 12.1.1957 (photograph from the Sydney Morning Herald, 14.1.1957, used with permission, Burke/the Fairfax Photo Library)

truss was indeterminate to the first degree, it was necessary for one other member to fail. This was an upper chord, just to the left of mid-span. Both were compression failures.

The central diagonal consisted of two primary rolled steel sections joined by batten plates, provided to achieve a sufficient member strength against out-of-plane buckling. It was concluded that these batten plates were insufficient for their purpose and two factors contributed to this. The number of these batten plates was small. They were placed only at mid-length and at the quarter-points, but at mid-length one of the batten plates was omitted to allow for the junction of the sub-member. Further, the mode of supply was that the individual components were fabricated overseas, galvanised, and then assembled at the site. The batten plates were joined to the main material with bolts in clearance holes, and the galvanising would have minimised any transfer of shear by friction. Under these circumstances, the batten plates would have been ineffective until large relative slip rotations had occurred. Not only so, but on inquiry it was found that this diagonal member had suffered earlier damage, being dropped from a crane and then cold-straightened. This straightening would have left a damaging pattern of residual stresses in the main material. It was not clear who had given permission for the cold-straightened member to be re-used. The judgment reached in this report was that creep buckling of the web member occurred, casting overload onto the upper chord, which also failed.

The general conclusion that is an important consequence of this study is that it took a combination of events to cause collapse, here (a) deficient use of batten plates bolted to the main material; (b) the presence of galvanised interfaces; and (c) the prior bending and cold-straightening of the member. It is rare that a single circumstance leads to collapse.

2.4.2 Tay Bridge

The second case study is the Tay Bridge disaster of December 1879 (Thomas 1972). This long railway bridge crossed the River Tay to Dundee, on the northern bank. It had about 91 spans totalling 2 miles, as it was said or, more precisely, either 3015 or 3122 m. The central spans in the final design were 69, 11 × 74.7 and 69 m, flanked at each side by a large number of 36.6 m spans, and some others (Shepherd and Frost 1995: 35 give somewhat different dimensions; see also Koerte 1992: 30). These central spans were through trusses, at a clear height of the order of 25 m. The bridge was completed by 26 September 1878 and officially opened on 31 May 1879. Seven months later, on 28 December 1879, it collapsed. A passenger train entered the bridge from the south at 7 p.m. A wild storm had hit the area about 6 p.m. and it was dark. Wind forces on the train were so large that witnesses reported seeing a continuous stream of sparks caused by friction between the flanges of the wheels and the rails. The collapse occurred as the train approached the northern end of the fourth high span, and the train with its 75 occupants plunged into the river; there were no survivors.

There was an official inquiry into the collapse. It concluded that 'The bridge was badly designed, badly constructed and badly maintained' (Thomas 1972: 177). This conclusion applied particularly to the high piers which, above water level, consisted of six cast-iron columns, 0.46 m in diameter, located at the corners of a hexagon elongated in the direction of the stream flow, and braced by a system of diagonal wrought-iron tie-bars. The cast-iron columns were not full-height, but each consisted of seven segments joined by flanged joints. Each diagonal was connected by bolts in shear to a vertical fillet plate that was supposed to be cast integrally with the column and its flange. In fact, in at least some cases, these fillet plates or lugs were added later by a process called 'burning on', with moulds placed on the column and molten metal poured in. There was evidence that some of the tie-bars had worked loose. The designer, Sir Thomas Bouch, believed that the wind had caused at least one of the carriages to be derailed, striking a truss member and causing it to fail, but evidence brought to the inquiry did not support his suggestion. He also stated that he had been advised by the Astronomer Royal to design the bridge for a horizontal wind load of 'ten pounds per square foot' (0.5 kPa), a figure that is low compared with modern practice. Two conclusions may be reached here.

(a) Horizontal wind loads and defective bracing in the high columns were the primary reasons for the collapse.
(b) Although the primary load – the weight of the traffic – was not a principal cause of the collapse, yet it was present. A major consequence was that 75 people were killed. The failure of a bridge has the capacity to kill or harm many people.

2.4.3 Quebec Bridge

Not far south of Dundee is the great Firth of Forth Railway Bridge, completed in 1890 with two main spans of 518 m. It is a cantilever truss bridge. The Quebec Bridge (*The Quebec Bridge*, Opening Booklet 1918; Modjeski et al. 1919), for which tenders were called in 1899, was essentially similar in form, but with a single main span of 549 m, side-spans of 153 m, and a suspended span of 206 m. It was decided to erect this suspended span by cantilevering out each half from the main trusses. On the south side, this stage was almost completed when, on 25 August 1907, it collapsed (see Fig. 2.15). The structure was completely redesigned, with spans of 157, 549 and 157 m. This time the suspended span was assembled on falsework in the river near the bank, and then floated out beneath its final position. As it was being lifted, on 11 September 1916, part of a temporary supporting bearing fractured and caused the structure to fall into the river. It was decided that there was no fault with the second design of the bridge itself. A new suspended span was fabricated and successfully lifted into position on 21 September 1917. The first train passed across the bridge 26 days later. This information is taken from the official booklet prepared for the opening of the bridge, as are the photographs shown in the figure. They are included as a spectacular illustration of the consequences of failure.

Fig. 2.15 First collapse of the Quebec Railway Bridge, 25.8.1907 (Opening booklet 1918; courtesy Canadian National Railway and the Canadian National Museum of Science and Technology, Ottawa)

The initial cause of the first collapse was the failure of a steel compression member. The Royal Commission appointed to look into the disaster was chaired by John Galbraith, and carried out full-scale tests of large compression members that had a significant effect on later design procedures for these members.

2.4.4 Tacoma Narrows Bridge

The fourth case study is the failure of the Tacoma Narrows Bridge by aerodynamic instability on 7 November 1940 (Shepherd and Frost 1995: 40ff). The bridge was a suspension bridge with a main span of 854 m, distinguished by its use of stiffening girders (2.2 m deep, 11.3 m centre-to-centre) in place of the normal stiffening trusses (see Fig. 12.4), and was opened to traffic on 1 July 1940. It misbehaved both before and after opening and, for this reason, F.B. Farquharson, Professor of Civil Engineering of the University of Washington, was engaged as early as February 1939; this account is taken from one of his reports (Farquharson 1949). His studies of the bridge included both observations and measurements and a motion-picture record. To quote him, 'These records ... by means of a transit and by motion-picture camera ... cover fairly adequately the complete life of the bridge, including its final catastrophic throes', and his film has been seen by many. He also says that 'With the experience at the Bronx-Whitestone and other bridges with comparatively shallow stiffening girders, the engineers of the Tacoma Narrows Bridge anticipated, before starting erection of the suspended structure, the oscillations which later occurred'. The bridge included some remedial measures installed in June 1940, and tie-downs to the side-spans were added in October 1940.

On numerous occasions before the collapse, vertical oscillations of the deck had been observed, with various modes and frequencies, typically either 0.13 or 0.2 Herz, corresponding to periods of 7.5 or 5.0 sec. A double amplitude of 1.52 m had been recorded and 'travellers reported that cars following each other over the bridge observed the car ahead completely disappearing from view, only to reappear several times while making one crossing'. On the final day, the bridge oscillated for some hours in a relatively unusual vertical mode, with perhaps eight half waves in the central span, at about 0.6 Herz. The wind velocity was measured as 68 kph. Suddenly the motion changed to a torsional mode and became violent. Within 8 or 10 min there was evidence of damage to lamp posts and to the concrete side-walks. During a temporary lull, an attempt was made to remove a car stalled at about the quarter-point of the bridge. 'As far as the eye and ear could detect, the double amplitude of approximately 28 feet (8.5 m) was causing no distress in the girder.' On the other hand, there was some evidence of failure in the lateral bracing system beneath the deck. Perhaps 15 min later, 'a distinct lateral failure was observed ... about the quarter point ...' of the bridge, at one end and then the other. The subsequent collapse was rapid, with progressive failure either of the suspenders, or of their connections to the deck, and finally a large portion of the deck of the main span fell into the stream. One factor in the collapse was longitudinal movement between the deck and main cables at mid-span, of the order of '1 to 2 yards'

(0.9–1.8 m), causing damage to the main cable by the sliding of a cable clamp, and excessive forces in the short suspenders near mid-span. It is possible that the collapse was triggered by some change in the direction of the wind, whose angle-of-attack on the bridge was influenced by adjacent cliffs.

An important element in the collapse was the change to a torsional mode, and one of the strategies used in later designs has been to increase the torsional stiffness. However, Farquharson's account of the collapse includes some points that have not commonly been observed: notably, the failure of the lower lateral system and of the central suspenders. Torsion of the deck itself would not have caused forces in the lateral members. More truly, torsion of the structure as a whole into two half-waves caused equal and opposite longitudinal forces to be applied to the deck at mid-span. For equilibrium, these would have required horizontal reactive forces at the ends, parallel to the stream flow, and it is these that caused the forces in the laterals. The effect of the torsional mode on the short central suspenders is also important.

The failure clearly has had a major effect on suspension bridge design; see, for example, O'Connor (1971a: 424ff). However, the problem of aerodynamic instability was not previously unknown. Farquharson (1949, Part 1: Appendices 1, 2), for example, includes accounts by Provis (1842) and Russell (1841) of wind-induced vibrations in the Menai Straits Bridge and the Brighton Pier. The subject as a whole will be discussed further in Chapter 12.

2.4.5 Kings Bridge

Kings Bridge, Melbourne, is a welded steel girder bridge, opened to traffic on 12 April 1961. It consists not only of spans across the Yarra River, but a long southern approach, which also includes a span across a major road. The whole is divided into separate north- and south-bound carriageways, and there are additional lanes across parallel separate structures across the Yarra River. It has recently been modified. The south approach was contained between concrete curtain walls. On 10 July 1962, a loaded semi-trailer entered the bridge from the south. When it reached span W14 (on the western side, the span being numbered 14 from the northern end, of a total of 16 spans), that span suddenly collapsed, due to brittle fracture in all of its four girders. However, the deflection of the span was held to about 0.3 m by the deck slab bearing on the side walls. A Royal Commission was established; it reported in 1963, and this account is taken from its *Report* (see also Shepherd and Frost 1995: 46). The photographs of Fig. 2.16 correspond to Figures 2 and 3 of the *Report*, and the drawing is from their Figure 11. The girders that fractured each had spans of 30.5 m, simply-supported between cantilevered end girders. The web plate was 1.5 m × 11 mm; there was a single upper flange plate, 356 × 16 mm; and the lower flange consisted of a plate 406 × 19 mm, with a single cover plate that was 305 × 19 mm on the outside girders, and 356 × 13 mm on the inner pair. It is evident that fracture occurred at the ends of the cover plates. The vehicle had a total weight of 47 tonnes, producing calculated stresses that were within permissible limits. The successful tender was said to have 'involved the use of high-tensile steel of Australian origin for the first time in a welded bridge'. The order for the steel was deficient in that it did not require appropriate testing of the high strength steel – Izod tests were not required (see Section 2.5). The steel actually supplied was variable in quality, and it was found that, with respect to the welding, 'the standard of inspection was high but not high enough'. After the failure, a study was made of 168 cover-plate ends. It was found that 86, or 51%, were cracked. The span that failed had cracks in at least seven of the eight cover-plate ends. Very soon after welding, and before painting, some of these cracks extended through the lower flange and into the web plate. Before the final collapse, the crack in Girder W14-2 (see Fig. 2.16) had 'completely severed the bottom flange' and 'continued 44 inches (1.12 m) up the web; that is, three quarters of the web depth. These cracks had not been observed. Although the initiation of the cracks was at a particular design detail (a transverse fillet weld) it has since become apparent that the quality of the material is the major parameter in controlling brittle fracture. At the time of the collapse, two welded steel bridges of comparable size were being fabricated elsewhere in Australia, one of mild steel and the other classified as a notch-ductile steel, NDIB, to the British Standard BS 2762. Some test results on these steels are quoted in Section 2.5 (O'Connor 1964, 1966, 1968, 1969). Kings Bridge was not the first to fail by brittle fracture. Boyd (1970: 17) refers to the failure of a series of Vierendeel girder bridges in Holland about 1938, including the Hasselt Bridge of 74.5 m span, and also the Duplessis Bridge in the province of Quebec, about 1950.

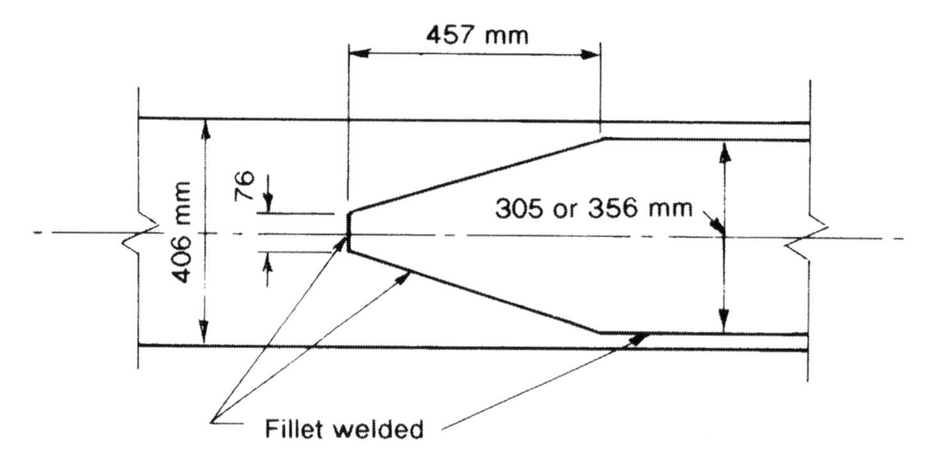

Fig. 2.16 Failure of Kings Bridge, Melbourne, on 10.7.1962 (from Report of the Royal Commission, 1963, Figures 2, 3 and 11, Crown Copyright Material, reproduced by permission of the Government Printer, State of Victoria, 29.4.1999)

Fig. 2.16 *(continued)*

Fig. 2.16 (*continued*)

2.4.6 Steel box girders

In the period 1970–1, there occurred the failure of steel, box-girder bridges at Milford Haven in the United Kingdom on 2 June 1970, Westgate, Melbourne, Australia on 15 October 1970, and Koblenz, Germany, November 1971. Although the structures at the point of failure were similar in form, there were also substantial differences.

The Milford Haven Bridge had a single steel box, and was designed to be continuous over seven spans, from 76.8 to 213.4 m. It was of welded steel construction. The first span had been erected. The second was being cantilevered forward from Pier 1; it had almost reached Pier 2 when the lower flange buckled near Pier 1, causing the cantilevered portion to bend and fall to the ground. There was no doubt as to the cause of the failure: the steel diaphragm above Pier 1 had failed in compression, causing the lower flange to buckle (Merrison et al. 1971, 1973).

In its final form, Westgate Bridge was to be cable-stayed, with main spans of 112, 144, 336, 144 and 112 m. There were, in addition, extensive approaches with the main spans extending from Piers 10 to 15, and the cables from Piers 11 to 14. These main spans were of welded steel box-girder construction. During erection of the first span, from Piers 10 to 11, this span collapsed, bringing down Pier 11 with it. The final cross-section of the bridge was a three-cell box. The initial error was to divide the cross-section down the centreline for erection. The span was erected in two halves, each consisting of an outer box, with the top and bottom flanges cantilevering inwards to a free edge along the centreline. It was found that once the self-weight was transferred to these halves, the outstanding edges buckled. Also, through faulty fabrication, one half had a larger camber than the other. To bring them into line, kentledge, in the form of ten concrete blocks each of about 8 tonnes, was placed on the structure. The edges still did not align and, to aid in their alignment, some of the bolts in a transverse, bolted splice were removed, causing the collapse and killing 35 men (Barber et al. 1971; Hitchings 1979). Figure 2.17 shows the cross-section of the bridge in photographs taken by one of the authors a few weeks before the collapse. It also shows the collapse, with the fallen Pier 11. The open longitudinal joint in the top flange can be seen, together with an open transverse joint.

The failure at Koblenz was superficially similar to that at Milford Haven. During erection, while being cantilevered forwards, the steel box bent by buckling of the lower flange and fell into the river. The difference was that the failure occurred away from a pier.

There were, again, lessons from these failures. The Royal Commission into the failure of Milford Haven carried the title, *Inquiry into the Basis of Design and Method of Erection of Steel-Box Girder Bridges*, and looked also at Westgate and Koblenz. It led to the production of new rules for the design of these structures, issued as an Appendix to the final report (see also Rockey and Evans 1981). These, in turn, led to other studies, such as those contained in the Proceedings of an International Conference on steel box-girder bridges, held by the

Fig. 2.17 Westgate Bridge, Melbourne: photographs before collapse showing units joined down the centreline (O'Connor), and after the collapse (Melbourne Age photograph, used with permission, The Age Photo Sales)

Fig. 2.17 *(continued)*

Institution of Engineers, London, in 1973. Their report also commented on the role of Engineer and Contractor (final report, 1973: 25):

> A clear division of responsibilities between Engineer and Contractor is vitally important ... A certified independent check of the erection method should be performed by engineers other than the design Engineer, who should nevertheless be given full details of the erection method and have the right of veto over it.

This comment tends to reflect the conditions that were common in Britain at the time – that the Engineer designed the bridge and it was the Contractor's job to erect it. More truly, the tasks cannot be fully separated, and it would be better to say that the Engineer should include in the design at least one feasible method of erection, but should also be prepared to consider alternatives proposed by the Contractor.

2.4.7 Quinnipiac River Bridge

Quinnipiac River Bridge has been chosen as an example of fatigue cracking in a steel girder bridge. This three-span structure has a suspended span of length 50.3 m and depth 2.81 m. The welded girders are non-composite and have a web 11 mm thick, flanges 559 × 35 mm and two longitudinal stiffeners, each consisting of a flat plate 114 × 10 mm, welded to one side of the web, and located about 550 mm from a flange. The bridge was opened to traffic in 1964; in October 1973, a large crack was discovered in an outside girder, about 10 m from one end. It initiated within a faulty butt weld in the lower stiffener; when discovered it had extended over about one half the depth of the web, and had proceeded from the web about halfway across the width of the lower flange. Fisher et al. (1980) have reconstructed stages in the development of the crack. The defective butt weld in the stiffener caused a complete crack through the stiffener at an early age. This then propagated by fatigue until it fully penetrated the adjacent web. It then spread suddenly by brittle fracture above and below the stiffener, probably during the winter of 1972–3, with low temperatures contributing to the fracture. This fracture was arrested when it reached a zone of compressive residual stresses at the flange-web junction, but then initiated further fatigue cracking in this area. During the 9 years of service, the girder received about 14,500,000 cycles of random truck loading.

2.4.8 Tasman Bridge

The next case study is of another Australian bridge, the Tasman Bridge, which crosses the Derwent River at Hobart. A long bridge, it has 22 spans, most of 42.7 m, but with a central group of 60, 94.5 and 60 m. On Sunday 5 January 1975, a freight ship struck the bridge at 9.30 p.m., destroying two piers and causing the collapse of 127 m of the deck. The ship sank and is still present, submerged beneath the bridge in 24 m of water. So sudden was the collapse that four cars ran into the gap, killing five of their occupants, with seven seamen lost from the ship. Another two cars were left, tilted precariously over the edge. Although there was this loss of life, it would have been much worse in peak-hour traffic. The work by Whelan et al. (1976) is unusual in that it studies the social effects of the collapse. A temporary Bailey bridge was built a few kilometres upstream and opened in early May. In the interim, residents who lived on the east bank of the river used an emergency ferry system. The site of the Bailey bridge became the site of the later Bowen Bridge, put into service later with the repaired Tasman Bridge, for it was decided that the city should not again be divided by such a disaster. Even today, a system of traffic lights operates to ensure that no traffic uses the Tasman Bridge during the passage of a ship. Figure 2.18(a) shows the Tasman bridge after its repair, with the doubled span over the missing pier. The disaster led to the development of rules for the design of piers for ship impact, essentially so that the piers would stand and, if necessary, the ship would sink after the impact. Figure 2.18(b) also shows the Bowen Bridge, with piers designed according to this rule.

A somewhat similar disaster occurred at Granville, Sydney, on 18 January 1977, when a derailed locomotive hit the supports of a heavy concrete overpass, causing it to collapse onto carriages filled with passengers, with great loss of life.

Fig. 2.18 (a) Tasman Bridge, Hobart, after reconstruction, showing longer span, with missing pier, at left (upper photograph, O'Connor); (b) Bowen Bridge, Hobart, with piers designed to resist ship impact (lower photograph, O'Connor)

2.4.9 The Welsh Ynys-y-Gwas Bridge

The Welsh Ynys-y-Gwas Bridge was a prestressed concrete girder bridge, with a single clear span of 18.3 m, constructed in 1953. On the morning of 4 December 1985, it was found to have collapsed by bending at mid-span; it may represent the first collapse of a prestressed concrete highway bridge in the United Kingdom. It is believed that there was no traffic on the bridge at the time. In cross section, the bridge was made up of edge units beneath the footways, and then nine precast I-section beams, with the flanges touching, joined by transverse prestressing. In the longitudinal direction, these beam units were 2.44 m long, with longitudinal, post-stressed tendons passed through ducts formed in the segments. Typically these ducts were unlined, although it had been intended that metal sleeves, about 100 mm long, should have been placed across the transverse joints. These joints were 25 mm thick and packed with sand and high alumina cement. The ducts were grouted. Woodward and Williams (1988) made a thorough study of the collapse, and concluded that the primary cause was corrosion of the post-stressed tendons where they crossed the joints. In most cases, cardboard tubes had been used instead of the intended metal sleeves at these locations, but there was no evidence that the performance of the metal sleeves was better. The study suggested that the use of de-icing salt was probably the chief cause, although there was some evidence that dune sand with a salt content may have been used in the mortar in the joints. The collapse may have been initiated by the passage of a heavy vehicle at some time during the life of the bridge, causing the transfer of its load to adjacent tendons, leading to progressive failure.

2.4.10 Ngalimbiu River Bridge

The final case study to be considered here concerns a bridge opened in May 1984 across the Ngalimbiu River in the Solomon Islands (Boyce 1987, 1989). It was a composite girder bridge with a concrete deck on flange-plated universal beams and five spans, each of 21 m. On 19 May 1986 a flood occurred in the river, resulting from the tropical cyclone, Namu. The rainfall caused land slips over 50% of the upper catchment. These brought down trees that were then carried down by the stream as 'a solid wall of timber some 200–300 m wide, standing 3–4 m above the water surface' (Boyce 1989: C-15) and approached the bridge at about 20 kph (5.5 m/s). The result is shown in Fig. 2.19. The flood bypassed the bridge on the near-side, leaving the debris wall packed against the upstream side (to the right in the photograph). Severe damage was done to the bridge, with two of the spans carried downstream. The case serves as a reminder of the major forces that may be applied to a bridge by a stream in flood. In some cases, the piers of a bridge, if inclined in plan to the flow, may act as an aerofoil, generating horizontal forces equivalent to both lift and drag. In a rising flood, water may be trapped beneath a bridge, causing buoyancy forces that lift the bridge and cause its destruction. High flows may also cause scour around the piers, causing them to fall. These matters will be discussed further in Chapter 11.

Fig. 2.19 Ngalimbiu River Bridge, showing debris mat, May 1984 (photograph courtesy W. Boyce)

This group of case studies forms a selected set that is not statistically based. Larger samples may be found in OECD (1979: 13) and in the paper by Smith (1976); the latter includes an extensive set of references, as does Shepherd and Frost (1995). It is nevertheless valid to draw the following conclusions.

1. Often, a number of factors combine to cause a collapse.
2. It is rare for a collapse to be caused simply by an excessive set of primary loads, such as by the weight of a group of highway vehicles.
3. Rather, collapse has often been caused by loads that would have been overlooked by the designer, or regarded as secondary, or by a faulty erection procedure, or by faulty materials.
4. Although modern prestressed concrete structures appear generally to perform well, it is desirable to be aware of the possibility of durability problems in both reinforced and prestressed members; see, for example, Mallett (1994), Sommer (1995) and Springenschmid (1994).

2.5 Brittle fracture

It is appropriate to consider first the phenomenon of ductility or yield, which acts as an escape mechanism allowing the material to deform plastically, rather than fracture. Consider, first, the problem in three dimensions. The von

Mises-Hencky yield criterion suggests that, under the action of three mutually perpendicular principal stresses, σ_1, σ_2 and σ_3, yield will occur if

$$(\sigma_1 - \sigma_2)^2 + (\sigma_2 - \sigma_3)^2 + (\sigma_3 - \sigma_1)^2 = 2\sigma_0^2$$

where

σ_0 is a material constant (Nadai 1950, vol.1: 209f)

If the left-hand side is less than $2\sigma_0^2$, then the material will not yield. Suppose first that $\sigma_2 = \sigma_3 = 0$. Then, at yield,

$$\sigma_1 = \sigma_0$$

or, expressed alternatively, σ_0 equals the yield stress in uni-axial tension, σ_y.

Consider instead the case with $\sigma_1 = \sigma_2 = \sigma_3$. Then the left-hand side equals zero and the material will not yield, even for large values of these stresses. However, if these tri-axial stresses are tensile, then the material may fracture, with separation occurring.

For stresses less than the yield condition, the behaviour is elastic, with strains of the form

$$\epsilon_1 = 1/E\{\sigma_1 - \nu(\sigma_2 + \sigma_3)\}$$

where

ν is Poisson's Ratio, and for steel will typically have a value of 0.3

For $\sigma_2 = \sigma_3 = 0$,

$$\epsilon_1 = 1/E\ \sigma_1$$
$$\epsilon_2 = \epsilon_3 = -\nu/E\ \sigma_1$$

ϵ_2 and ϵ_3 are of such sign as to minimise the volume change in the body.

For stresses in excess of yield, the magnitude of the inelastic or plastic strains, which add to the elastic strains, is undefined, but the normal assumption is that these inelastic strains cause zero volume change. In uni-axial tension, if ϵ_{P_1} is the plastic component strain in the direction of σ_1, then, possibly

$$\epsilon_{P_2} = \epsilon_{P_3} = -\epsilon_{P_1}/2$$

If a circular rod is stretched in tension, these plastic strains will occur, but will give rise to an instability of behaviour. As has been said, the magnitude of the plastic component of strain is undefined. If, for some reason, an increased plastic strain occurs at some point in the direction of σ_1, this will cause a reduction in diameter and an increase in stress at that point. Further plastic strains occur at this point leading to the development of a waist and final fracture.

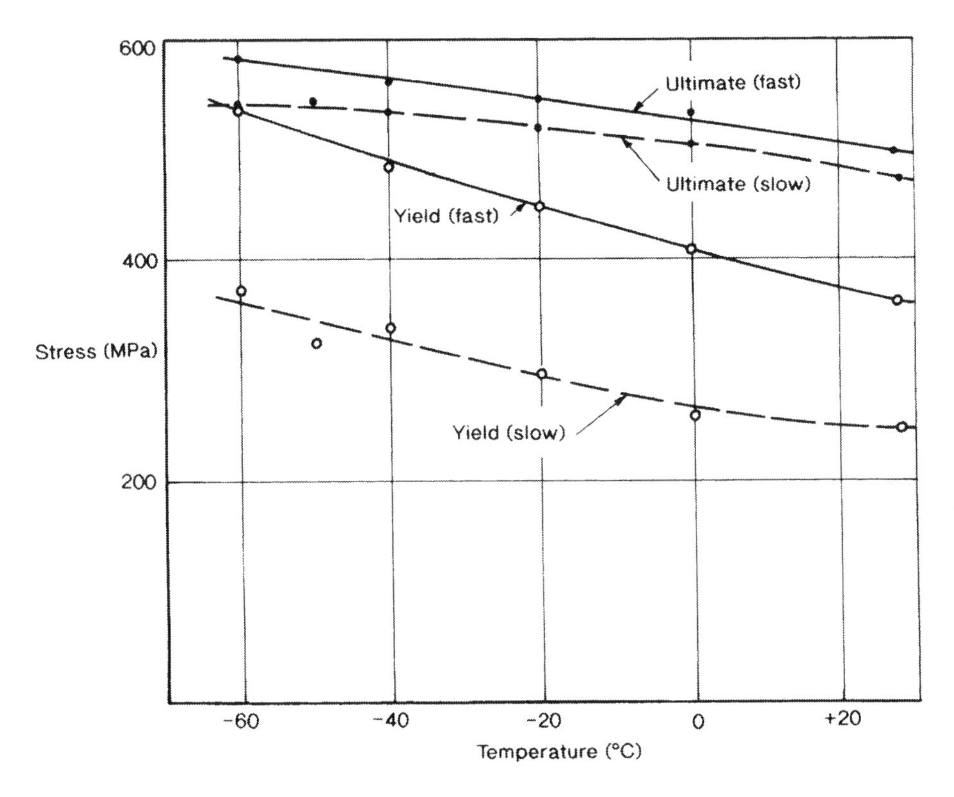

Fig. 2.20 Effect of strain rate on yield stress (O'Connor 1968)

The yield stress in structural steel depends on the strain rate and the temperature. Figure 2.20 presents some results from simple tension tests carried out on specimens with a net section 15 × 5 mm, taken from a single length of Australian girder plate, 305 × 29 mm, at the time of the failure of Kings Bridge (O'Connor 1968). It clearly shows the increase in yield stress as the temperature is reduced, or when the strain rate is increased, from 17.2 to 29,600 MPa/sec in this case.

Consider the notched specimen shown in Fig. 2.21(a). The notch is semicircular; a uniform tensile stress, σ, applied to the ends of the member will cause an elastic stress concentration, of the order of 3σ, at the base of the notch, the behaviour being not unlike that beside a hole in a tension member, as described by Timoshenko and Goodier (1970: 90ff). Allen and Southwell (1949; see also Henrickson et al. 1958) developed an inelastic analysis for this member. At first yield, $\sigma = 0.33\sigma_y$. As σ is increased, yield spreads slowly from the notch. At the same time, the stress combination at a point on the longitudinal axis, some distance away from the net section, also approaches yield. Suddenly the two yielded zones combine and, theoretically, the member fails, with $\sigma = 0.7\sigma_y$. The maximum tensile stress within the member is $1.78\sigma_y$. Similarly, for the notch shown in (b), the maximum stress, theoretically, is $2.3\sigma_y$. Clearly, therefore, a notch has the capacity to increase the maximum tensile stress well above σ_y. The

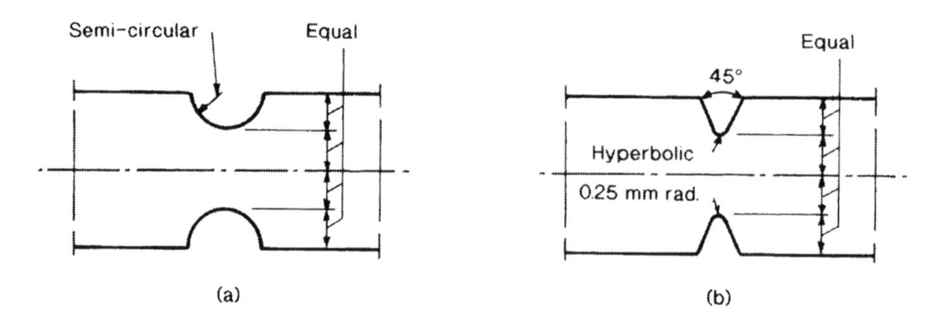

Fig. 2.21 Notch geometries

situation is even worse if the notch is in the form of a crack, transverse to the applied stress. The classic analysis of this case is Griffith's crack theory, which suggests that, for an initial crack of length, c, the average applied stress to cause fracture is proportional to $\sqrt{1/c}$, but this is based on a different failure criterion than that assumed above (Nadai 1950, vol.1: 197f).

The so-called notch ductility of a steel has commonly been measured by either the Charpy or Izod tests. In both, the specimen is 10 mm square, with a 45° notch, having a 0.25 mm root radius, cut across one face, 2 mm deep, leaving an 8 mm net section. In the Izod test (BS 131.1: 1961) the specimen is mounted as a vertical cantilever and struck sharply at a horizontal line 22 mm above the notch, by a tool mounted on a pendulum. The work done in fracturing the specimen is determined from the difference in maximum pendulum height, before it is released to strike the specimen, and after impact. In the Charpy test, the specimen is mounted as a horizontal beam, 55 mm between supports, and is struck at its vertical centre line (BS EN 10045.1: 1990). Tests may be carried out at various temperatures and the energy to failure plotted against temperature. High energy levels represent ductile behaviour and, typically, at a particular temperature there is a marked transition to lower energy levels and brittle fracture. For example, one set of three Charpy tests, reported in O'Connor (1964: 23) gave the values:

− 5°C	7 ft lbf =	9 joules
+ 4°C	25 ft lbf =	34 joules
+24.5°C	30 ft lbf =	41 joules

The corresponding transition temperature lies roughly between ±5°C. The energy level to failure also depends on the quality of the material. The code BS EN 10045.1 (1990) classifies low, medium, high and very high energy levels respectively as <30, 30–110, 110–220 and >220 Joules.

The Izod and Charpy tests are only two of many that can be used to test the susceptibility of a material to brittle fracture. Tipper (1962: 56ff, 107) suggested the use of notched tensile tests, using the full plate thickness, with an emphasis on tests at various temperatures and the observation of the mode of fracture (percent crystallinity). Some of these tests, using slow strain rates, were

Fig. 2.22 (a) Tipper tests of 38 mm steel plate at 20°C; (b) The effect of welded patches on fracture in direct tension (O'Connor test report, 11.9.1962)

described in O'Connor (1964). There is value in observing here some elements of behaviour, as shown in Fig. 2.22. These photographs are taken from a test report (by O'Connor) dated 11 September 1962, on mild steel similar to that used in one of two bridges that were being fabricated at the time of the Kings Bridge collapse (10 July 1962). Figure 2.22(a) shows the results of two Tipper tests, both on 38 mm thick material, tested at 20°C. The one on the right is of the material as supplied, and shows the cross section after fracture; the notches can be seen at both sides of the specimen; the fracture would be described as 'crystalline'. The one on the left is of the material after heat-treatment. The specimen had two notches, but only one can be seen – at the right. The change in thickness shows that there was considerable ductility. The right-hand half of the fracture is ductile. At the left is a brittle fracture, whose point of initiation can clearly be seen by lines radiating out from about the centre of the specimen. After initiating at this point, the fracture proceeded to the left, away from the left hand notch. The photograph in Fig. 2.22(b) is of samples of the parent material that were unnotched, but with six patches of 13 mm material welded on. At the centre were two patches on the main faces, and towards each end the patches were attached to the edges, in all cases using 4.8 mm fillet welds. Fracture occurred near the centre. At the left of the photograph are the two parts of a specimen tested at +5°C; the initiation point is indicated by an arrow and is along one of the side welds. The patch plates are still present on the lower piece, and this gives some idea of the welds that were used. The sample at the right was similar, except that it was tested at +20°C. In this case the upper patch plate has been removed; the upper fracture is partly stained by oil. This behaviour is clearly disturbing. Welding has caused some change in the basic metallographic structure of the main material. Either a crack has also been formed at that time, or the application of loading caused a crack, which led to brittle fracture. It was decided to post-stress the tension flange of the bridge.

Tipper (1962: 5ff) has an excellent summary of failures due to brittle fracture. This includes statistical data on ship fractures during the 1939–45 war, with photographs of welded ships that completely fractured, one in port and another at sea; in both cases the ship broke into two separate parts. There were many more such failures, through which the nature of the problem became widely known.

This discussion has been intended simply to introduce the problem. There has been much later work, such as can be found in Anderson (1995), Boyd (1970), Knott (1973) and Meguid (1989). It would appear that the primary protection against brittle fracture should be in the selection of the main material, but detailing and supervision are also important. Service behaviour is influenced greatly by temperature, but strain rate is also significant, as shown in Fig. 2.20. The stresses required to produce fracture are often high, but this is not always the case.

2.6 Fatigue

Fatigue differs from brittle fracture in that it is caused by repeated applications of stress, or a repeated range of stress. It initiates at either a notch or internal defect, and the behaviour involves changes within the crystal structure. Nevertheless, it can be seen as the imposition of repeated cycles that cause repetitive and reversed plasticity at some point in the material. These eventually cause the initiation of a crack which spreads. The net section of the member is reduced and finally failure occurs, either by brittle fracture or in a ductile way. Figure 2.23 illustrates the difference in appearance between the failure surfaces caused by fatigue and brittle fracture. The specimens, which are of mild steel, form part of the test programme, mentioned earlier, following the collapse of Kings

Fig. 2.23 Fatigue and brittle fracture (O'Connor test report, 22.11.1962)

Bridge. Each specimen was notched, and then loaded in a fatigue machine to give a fatigue crack extending part-way across the section. It was then loaded in direct tension. Some of the specimens failed in a ductile fashion, but others through brittle fracture. The cases included in Fig. 2.23 show, at the left, combined fatigue and brittle fracture in a specimen cut from 38 mm plate, and at the right, similar failures in 25 mm material. The two failure surfaces are quite different. A typical fatigue failure has a smooth surface, generally with concentric lines that are perpendicular to the direction of crack propagation. The brittle fractures, on the other hand, present a rough, crystalline appearance, often with discontinuities that radiate out from the point of initiation.

In this case, initial fatigue cracks developed from notches that were cut either parallel or perpendicular to the rolled surface. They extended over 17–67% of the cross-sectional area beneath the original notch. The specimens were then tested in direct tension at either 0°C, or at temperatures in the range 27–35°C. The stress to cause failure, calculated on the uncracked area, was remarkably constant, ranging from 350 to 430 MPa, for both temperatures. The apparent applied stress, or stress on the gross cross-section, was somewhat more. For comparison, the ultimate stress in a standard tension test was 463 MPa.

Fatigue performance is commonly represented on a graph of stress range (S) against the number of cycles to failure (n) such as those shown in Fig. 2.24. These curves include those for (a) a typical rolled beam, (b) a welded girder, and (c) a rolled beam with cover plates, the stress being calculated at the cover plate end; they are taken from Fisher (1977: 19). A few points may be made:

(a) When plotted on these scales, both parts of a typical S/n curve for steel are linear.
(b) These graphs show an endurance limit; that is, for each case, a stress below which a fatigue failure will not occur.
(c) The number of cycles until the endurance limit is reached varies here from 2 to 9×10^6.
(d) The stress concentration at the cover plate end has caused a marked reduction, both in the endurance limit and in the number of cycles to cause failure.

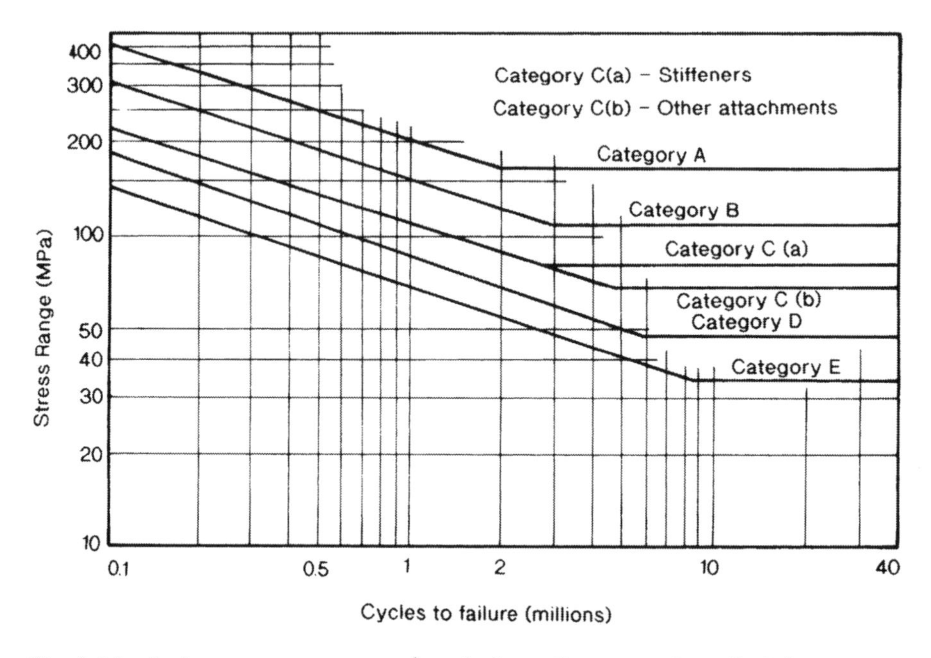

Fig. 2.24 Fatigue stress ranges for design: Category A, rolled beams; B, welded plate girders; C, welded plate girders with (a) stiffeners or (b) 50 mm welded attachments; D, with 100 mm welded attachments; and E, coverplated beams (Redrawn from Fisher 1977; used with permission)

Fisher also discusses behaviour under variable stress cycles and the prediction of fatigue life. Fatigue is considered in most of the references cited at the end of Section 2.5. Some other recent works are those by Klesnil and Lukas (1992) and Monahan (1995). Fisher (1981) has also written on the inspection of bridges in service. The estimation of the number of service loads within the life of a bridge is discussed in Section 6.3, together with some other general issues relating to fatigue.

Chapter 3

Design philosophies

3.1 Design at working loads

The older bridge design codes were based on (a) the selection of reasonable, upper bound estimates of normal working loads; (b) the use of elastic methods of structural analysis; and (c) the provision of some margin in strength by the selection of allowable working stresses separated by a *factor of safety* from some critical stress, such as the yield or ultimate stress of the material. This procedure is also called Allowable Stress Design (ASD).

For example, the 1973 edition of the American *Standard Specifications for Highway Bridges* (the American Association of State Highway Officials, or AASHO Code), in its section on working stress design, specified for structural carbon steel a minimum yield stress of 240 MPa, a minimum tensile strength of 400 MPa and a basic allowable stress of 138 MPa. The implicit factor of safety on yield was 1.8. Earlier, for many years an allowable stress of 124 MPa had been used, corresponding to a factor of safety of 2.0.

However, these factors of safety were not the same for all materials. For example, for a reinforced concrete beam, the allowable stress in compression in the extreme fibre of the concrete was specified as $0.4f'_c$, where f'_c was the ultimate strength of the concrete as determined by tests on concrete cylinders tested at an age of 28 days. The implicit factor of safety was 2.5, providing for a greater variability in the strength of this material, as compared with steel. A typical concrete strength at that time was 20 MPa.

The 1990 American Railway Engineering Association (AREA) *Specifications for Steel Railway Bridges* (AREA *Manual*: Ch. 15) gave the same basic allowable stress for steel (138 MPa). However, it is interesting to observe that older bridge design codes had allowable stresses that varied with the nature of the loading. Skinner (1908: 57ff) gives extracts from 25 different codes then in use by the bridge companies of the United States. A typical 'working stress' for steel (p.64) was

$$55(1 + Min/Max)\,\text{MPa}$$

where

Min represented the minimum stress caused by the applied loads;

and

Max, the maximum

Often, *Min* corresponded to the effects of dead load alone, and *Max* to dead load plus live load. This clause adds: 'For combined live, dead, wind and centrifugal stresses, increase the preceding unit stress 30% above live and dead load unit stresses'.

For the effects of dead load alone (i.e. $Max = Min$), the allowable stress became 110 MPa, or for live load alone ($Min = 0$), 55 MPa. A more typical result would be with $Max = 2 \times Min$, giving an allowable stress of 83 MPa. These values are low compared with later standards, partly because of the use of lower strength steels (Blockley 1992: 57), but the principle is of interest: effects such as impact, fatigue and the greater variability of live load, as compared with dead load, caused the codes to increase the factor of safety for live load. This also allowed for the unlikelihood of the combination of extreme values of wind and live loads.

3.2 Design for collapse

As mentioned in Section 2.2, it was in the 1940s that the so-called plastic methods of structural analysis came to be used in the design of building frames, with the emphasis shifted from behaviour at working loads to collapse. The typical model was that of a frame loaded with stationary loads, whose magnitude increased until collapse occurred. In 1971, in comparing methods, one of the present authors wrote (O'Connor 1971a: 15f): 'If this rigid choice (between ultimate and working load methods) is the only one available, it is possible to argue for the adoption of the conventional, elastic, working-load method in bridge design', but went on:

> It may be unreasonable, however, to be bound to the rigid choice discussed above. The ideal design procedure should include analyses at both working and ultimate loads and should use both for separate and clearly defined purposes.... A further possibility is that statistical methods of design may be adopted in which the emphasis is on probability of failure or distress.

These latter options have now been adopted in most bridge design codes; they will be discussed further in the following section.

3.3 Limit states design

The Limit States Design Method, as currently used in structural design, has two basic characteristics:

1. it tries to consider all possible limit states; and
2. it is based on probabilistic methods.

The simplest limit state is the failure of a component under a particular applied load. This depends on two parameters: the magnitude of the load as it impinges on the structure, here called the load effect, and the resistance or strength of the component. If the load effect exceeds the resistance, then the component will fail. However, both the magnitude of the load effect and the resistance may be subject to statistical variation. Concerning the resistance, suppose a large number of tests is carried out; then these may be used to calculate the mean anticipated resistance, R_M, and some measure of its variation. For a general population of some function x, the standard deviation is defined as

$$\sigma = \sqrt{\Sigma(x - \overline{x})^2/N}$$

where $\overline{x}$ is the mean, and N the number of samples. The variance equals σ^2 and the coefficient of variation is

$$c.v. = \sigma/\overline{x}$$

This population may be divided up. The number of samples whose values lie between x and $x + \delta x$ may be n. Then these values are said to have a probability of occurrence, P, where

$$P = n/N$$

If the interval is small and is δx, the value, $p = P/\delta x$, may be plotted against x as shown in Fig. 3.1. The resulting curve is called a probability density function (or probability distribution function) where, in general, such a function $p(x)$ is defined by the condition that the probability, P, of a value between $x = a$ and b is given by

$$P = \int_a^b p dx$$

or, for $(b - a)$ small and equal to δx,

$$P = p\delta x$$

The shape of this function will vary. The term Normal Distribution (Devore 1995: 155) is given to the function

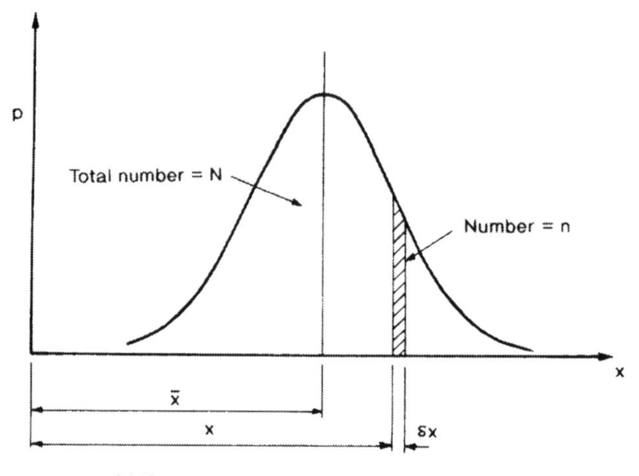

(a) Typical Probability Distribution Function

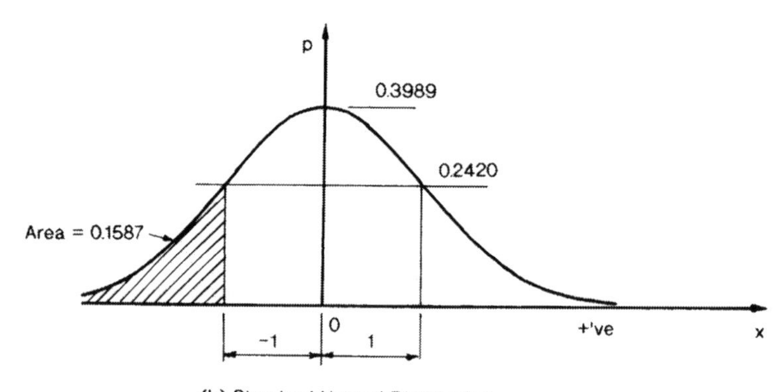

(b) Standard Normal Distribution

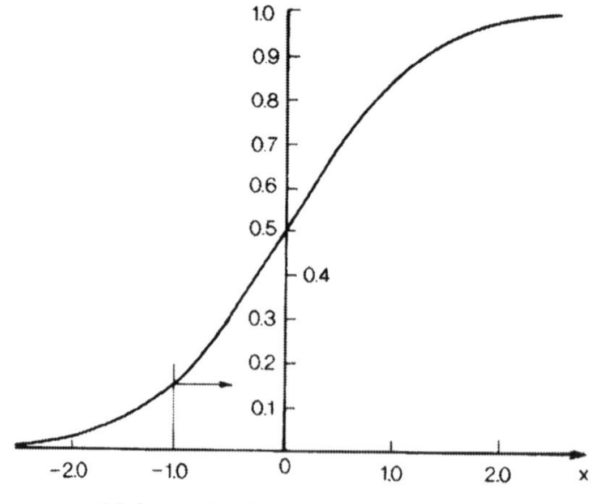

(c) Cumulative Standard Normal Distribution Curve

Fig. 3.1 Probability distributions

$$f = \frac{1}{\sqrt{2\pi\sigma}} e^{-(x-\bar{x})^2/(2\sigma^2)}$$

where f is symmetrical about $\bar{x}$.

If, in this function, $\bar{x}$ is set equal to zero, and σ equal to unity, the result is the Standard Normal Distribution (Fig. 3.1(b)),

$$f = \frac{1}{\sqrt{2\pi}} e^{-\bar{x}^2/2}$$

Areas beneath this curve have been tabulated by Benjamin and Cornell (1970), Devore (1995: 704f), Pearson and Hartley (1966) and others. From these tables it may be found, for example, that the area between $x = -1$ and 1 is 0.6826; that is, if N is the total number of samples, then $0.6826N$ would be expected to have values lying between $\pm$ the standard deviation from the mean. More directly, these tables list values of ϕ, the areas under the curve from $-\infty$ to the current value. For $x = -1$, $\phi = 0.1587$; or of the total of N values, $0.1587N$ would be expected to lie below the mean by an amount in excess of the standard deviation. Such a value is called a cumulative probability. This area is shown in Fig. 3.1(b), and is also marked on Fig. 3.1(c), which shows the cumulative Standard Normal Distribution curve.

To return to the failure of a component under a particular applied load, let it be assumed that both the resistance and load effect vary in a way that is consistent with a normal distribution; then their probability distribution functions p_R and p_L may be as plotted in Fig. 3.2(a). Consider any value of F. Then the probability of the load effect, L, lying between F and $F + \delta F$ is $p_L \delta F$. The probability that the resistance R is less than F is the cumulative probability given by

$$\int_{-\infty}^{F} p_F \, dF$$

The combined probability is

$$p_L \delta F \int_{-\infty}^{F} p_F \, dF$$

All such cases must be examined, for even large values of R will be exceeded by some L. The total probability of failure, p_F, therefore becomes

$$p_F = \int_{-\infty}^{\infty} (p_L \int_{-\infty}^{F} p_R dF) dF$$

This integral, called a convolution integral, may be evaluated as follows (Galambos, T.V. Design Codes, 47–71 in Blockley, 1992: 63; see also below):

1. Calculate $x = -(R_M - L_M)/(\sigma_R^2 + \sigma_L^2)^{\frac{1}{2}}$
   where R_M and L_M are the mean values of R and L, and σ_R and σ_L their standard deviations.

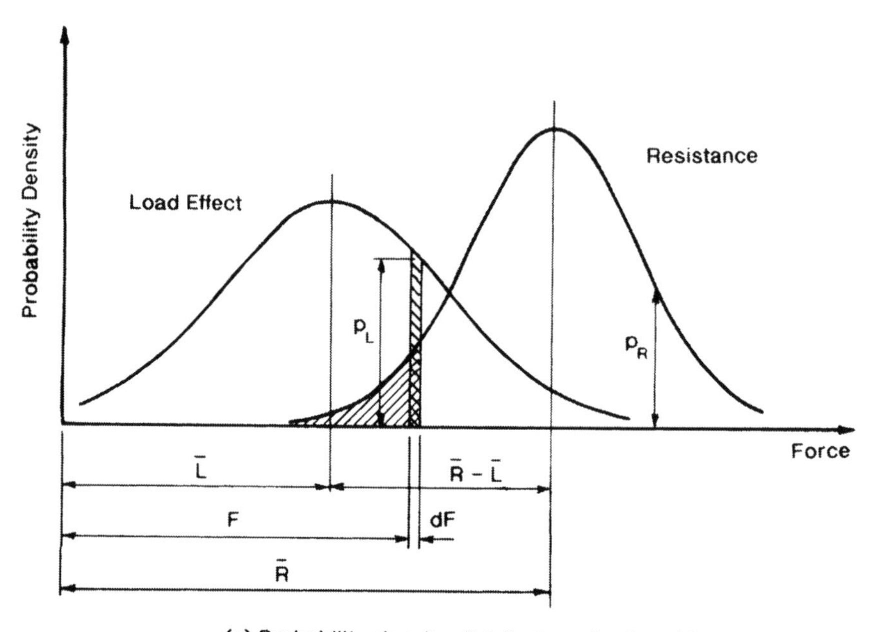

(a) Probability density distributions for R and L

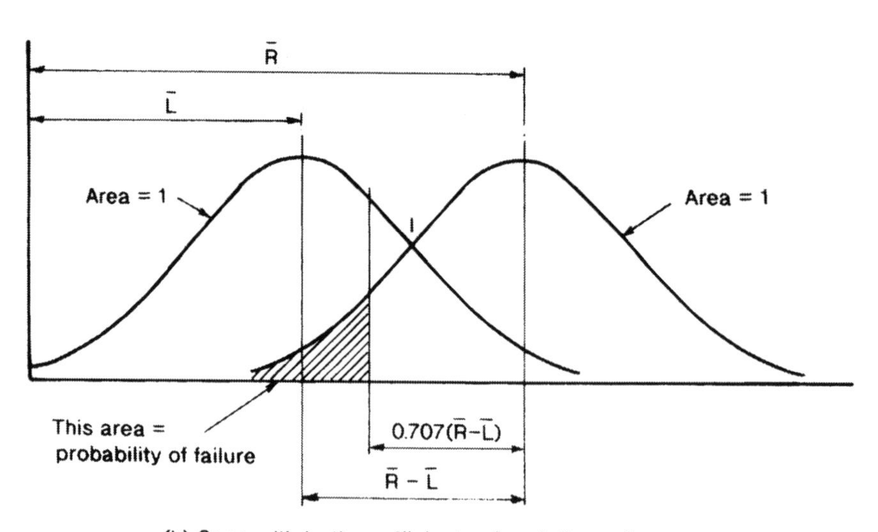

(b) Case with both coefficients of variation = 1

Fig. 3.2 Probability density distributions for load and resistance

2. Then $p_F = \phi(x)$.
   That is, p_F equals the area beneath the Standard Normal Distribution curve, from $-\infty$ to a negative value of x (see Fig. 3.1). If, for example, $\sigma_R = \sigma_L = 1$, then x equals $-0.707(R_M - L_M)$, as shown in Fig. 3.2(b). The corresponding distance x to the intersection point, I, is 0.5. The probab-

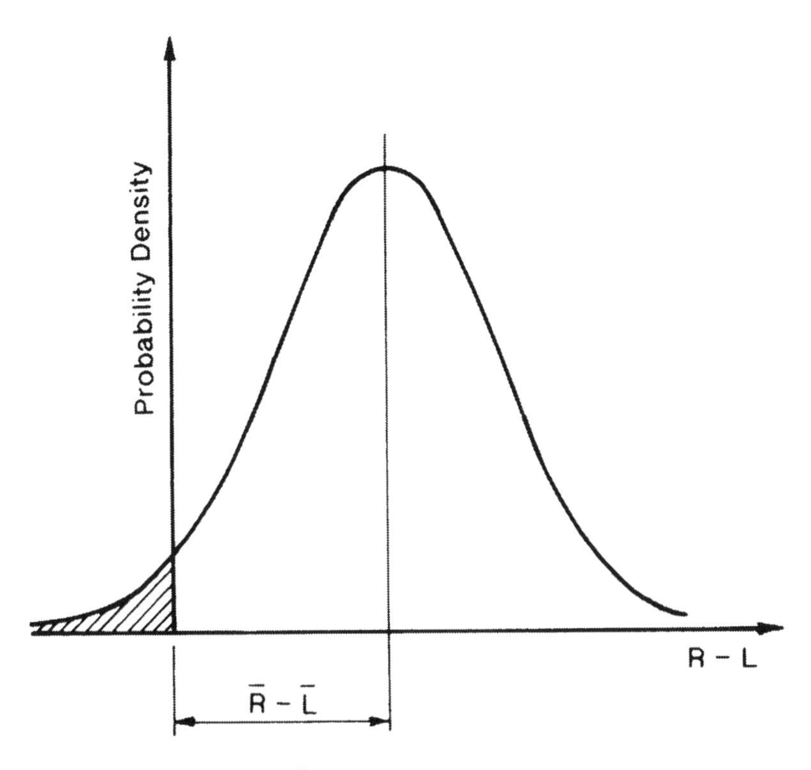

Fig. 3.3 Normal distribution for $(R - L)$

ility of failure, p_F, is represented by the area shown, and is somewhat less than the ratio of the area beneath the intersection of the two curves, divided by their combined area (which is 2). The area of intersection is, at best, a rough visual indicator of the probability of failure. For example, for $R_M - L_M$ equal to 1, 2, 3 and 4 times σ_R or σ_L, the true values of p_F, followed in brackets by those computed from the area of intersection, are 0.240 (0.3085), 0.0787 (0.1587), 0.0170 (0.0668) and 0.0023 (0.0228).

This problem can be expressed differently. Consider the function $(R - L)$; then if the standard deviations, σ_R and σ_L, were zero, $(R - L)$ would equal $R_M - L_M$. If R and L vary independently, each according to a normal distribution, then $(R - L)$ will also be normally distributed, with a mean value $R_M - L_M$, and a standard deviation σ_{R-L}, as shown in Fig. 3.3. This curve is symmetrical. The part of its mean square deviation to the right of the centreline represents cases with $R > R_M$, and/or $L < L_M$. The sum of the squares of these deviations in $(R - L)$ equals the sum of the squares of the deviations for $R > R_M$, plus those for $L < L_M$. It follows that

$$\tfrac{1}{2}\sigma_{R-L}^2 = \tfrac{1}{2}\sigma_R^2 + \tfrac{1}{2}\sigma_L^2$$

or

$$\sigma_{R-L} = \sqrt{\sigma_R^2 + \sigma_L^2}$$

Failure occurs when $(R - L)$ is negative; that is, the limiting situation occurs when, on this curve,

$$(R - L) = -(R_M - L_M)$$

and the probability of failure, p_F, equals the shaded area in Fig. 3.3 divided by the total area beneath the $(R - L)$ curve. Let

$$\beta = \frac{R_M - L_M}{\sigma_{R-L}} = \frac{R_M - L_M}{(\sigma_R^2 + \sigma_L^2)^{\frac{1}{2}}}$$

Then

$$p_F = \phi(-\beta)$$

The term, β, is called the reliability index and is directly related to p_F by the table of cumulative areas beneath the Standard Normal Distribution curve (Devore 1995: 704f; and others). This solution is clearly the same as that given previously. Some values of p_F were quoted earlier, but these were based on the ratio of $R_M - L_M$ to either σ_R or σ_L, with these equal. To give another typical value, for $\beta = 3.0$

$$p_F = 0.00135 = 1/740$$

The anticipated number of failures, n, equals $p_F N$. It follows that n is a function of N, $R_M - L_M$, σ_R and σ_L. The only one of these parameters that is directly under the control of the designer is $R_M - L_M$. Inverting,

$$R_M - L_M \text{ must be } \geq \beta(\sigma_R^2 + \sigma_L^2)^{\frac{1}{2}}$$

that is,

$$R_M \geq L_M + \beta(\sigma_R^2 + \sigma_L^2)^{\frac{1}{2}}$$

This result depends on the assumption that the resistance, R, and the load effect, L, each vary in accordance with the normal distribution, and this may or may not be correct. Devore (1995: 174ff), for example, discusses other continuous distributions, one of which is the Lognormal Distribution. Suppose a positive function x is such that $\ln(x)$ is normally distributed, then the probability density function of x is

$$p = \frac{1}{\sqrt{2n}\sigma x} e^{-[\ln(x)-\mu]^2/(2\sigma^2)}$$

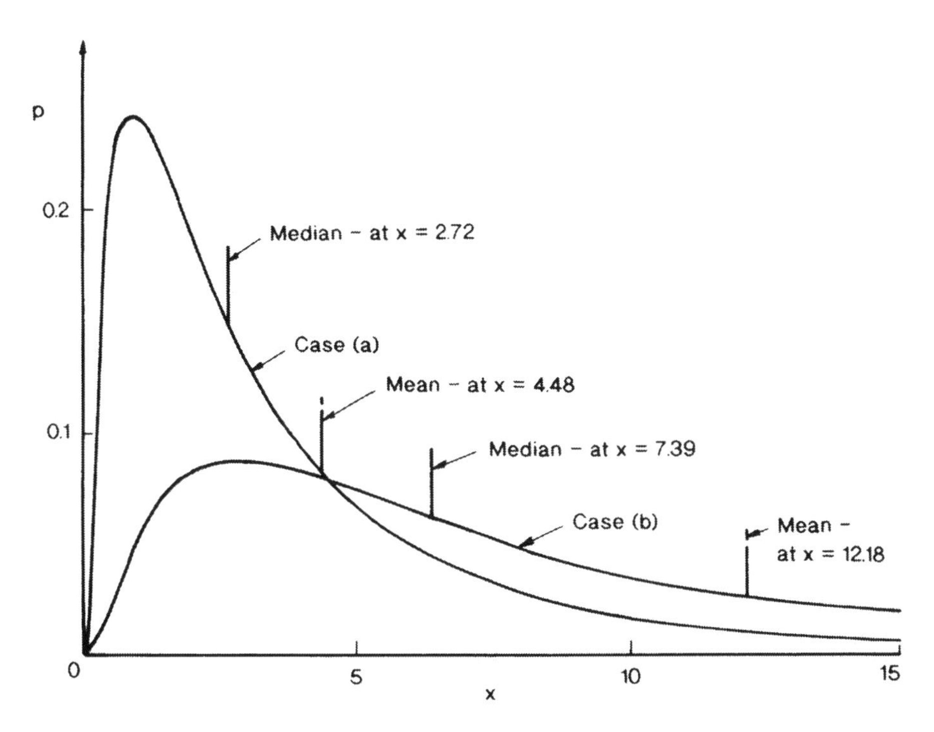

Fig. 3.4 Typical lognormal probability density distributions

where

> μ is the mean of $\ln(x)$, and
> σ is its standard deviation

Two curves of p are plotted in Fig. 3.4, for $\mu = \sigma = 1$; and $\mu = 2$, $\sigma = 1$. They are asymmetrical, with a long upper tail, and are said to have a positive skew. The mean value of x is e^a, where $a = \mu + \sigma^2/2$. The median value occurs when the cumulative probability is 0.5, at $x = e^\mu$; both points are shown. The maximum probability is not at the point $x = \mu$. The variance is:

$$e^{2a}(e^{\sigma^2} - 1)$$

This distribution is still a function of two variables. Other distributions, such as the Weibull Distribution (Devore 1995: 174f) may involve more than two, and are classified as Second Order Reliability Methods (SORM; Galambos, T.V. Design Codes, 47–71, and Ellingwood B.R. *Probabilistic Risk Assessment*, 89–137, in Blockley 1992: 65, 90–5). The Normal Distribution is the basis of First Order Reliability Methods (FORM), or First Order, Second-Moment Reliability Methods (FOSM), where the term, second-moment, reflects the description of the data only by the mean and the standard deviation, the latter being a second order term, using the squares of the deviations. Alternatively, analyses based on FORM are called Level 1, and on SORM, Level 2. Galambos states that current structural design codes are Level 1, although some of the factors used in these codes have been derived using Level 2 methods.

There are other complexities that arise when the principles of the present example come to be used in practice. One of these is the question of combined or joint probability. This may be seen most readily in the case of the load effect, for this may result from the combination of a number of loads of differing kinds, such as the combination of dead load with the primary live load and wind or stream load. Not only this, but even in the simplest case it has been necessary to know the standard deviations for both resistance and load. Freudenthal, one of the earliest exponents of the use of probabilistic methods in the assessment of structural safety (Freudenthal 1960; 1961; 1968a, 1972; Freudenthal et al. 1966; etc) wrote in 1968 (Freudenthal 1968b: 16) of the following problems:

> the existence of non-random phenomena affecting structural safety which cannot be included in a probabilistic approach;
> the impossibility of observing the relevant random phenomena within the ranges that are significant for safety analysis, and the resulting necessity of extrapolation far beyond the range of actual observation . . .;
> the assessment and justification of a numerical value for the 'acceptable risk' of failure; and
> the codification of the results of the rather complex probabilistic safety analyses in a simple enough form to be usable in actual design.

The normal answer to his final problem has been to specify a range of 'partial factors', i.e. load factors that are used to increase the load effects, and corresponding capacity reduction factors, used to reduce the nominal resistances.

These factors may be deduced from a study of the variability of the loads and resistances, although Freudenthal's second problem should always be kept in mind. The study of this relationship is continuing, but it is important to observe Blockley's (1992: 451) honest assessment of the situation as it was in Britain:

> In fact in the implementation of the first generation of limit state codes in the United Kingdom, the recommended values for the nominal, or characteristic, values Q_u (demand), R_u (capacity) and for partial factor values were not chosen by the use of statistical methods – they were chosen subjectively by a committee.

He goes on to say, 'The design of the second generation of codes has tried . . . to make the whole process more rational'.

The limit states considered by modern bridge design codes are divided into ultimate and serviceability limit states, where the first refers broadly to incipient collapse, and the second to undesirable behaviour that does not involve collapse of the primary structure. However, the first of these must immediately be qualified. It was seen in Section 2.2 that the collapse of a redundant or indeterminate building frame required the development of uncontrolled plastic deformation at two or more sections. There is a reserve of strength beyond initial failure, and this involves a redistribution of actions between the various parts or members of the structure. Many current bridge codes do not allow for this type of redistribution, and failure is deemed to occur when the first section reaches the ultimate capacity.

With this in view, possible ultimate limit states correspond to the conditions when:

(a) one or more parts of the structure reach their ultimate capacity, i.e. they are incapable of carrying additional load;
(b) parts of the structure move by sliding, uplift, or overturning; or
(c) parts of the structure are on the point of failure because of deterioration caused by corrosion, cracking or fatigue.

The serviceability limit states typically include:

(a) dynamic movements that cause discomfort or public concern;
(b) dynamic movements that cause damage to ancillary parts of the structure, such as lamp standards or hand railings;
(c) permanent deformations, either of the structure itself or its foundations, that cause public concern or make the structure unfit for use;
(d) damage by scour;
(e) the flooding or scour of adjacent properties; and
(f) damage due to corrosion or fatigue that is sufficient to cause a significant reduction in the strength of the structure or in its service life.

Nowak, in the recent volume, *Safety of Bridges* ('Application of bridge reliability analysis to design and assessment codes' 42–9, in Das 1997), has listed load factors adopted for the 1994 AASHTO Code (American Association of State Highway and Transportation Officials). It carries the title, 'LRFD bridge design specifications', where LRFD stands for 'Load and Resistance Factor Design'. The factors quoted here are for use with the ultimate limit state and are illustrations of modern practice:

dead load of the bridge itself (D)	1.25
dead load of an added wearing surface (D_A)	1.50
live load, including impact ($LL + I$)	1.75

The figures for dead load are for load effects where the dead and live load are additive. In some cases where the live load dominates, and the effects of the dead load are of opposite sign, it may be necessary to use a lower bound for the dead load, with a load factor of the order of 0.85 to 0.9. Also, the figure of 1.25 may be intended for use with concrete structures rather than steel, where a lower factor may be applicable, such as 1.1.

Nowak also quotes values of the capacity reduction factors adopted for this code:

moment or shear in steel girders	1.00
moment or shear in reinforced concrete T-beams	0.90

moment in prestressed concrete AASHTO beams	1.00
shear in prestressed concrete AASHTO beams	0.90

To give a typical example, these rules would require

$$1.25D + 1.5D_A + 1.75(LL + I) \text{ to be } \leq 0.9R$$

where R may be the calculated nominal ultimate bending strength of a reinforced concrete member.

The choice of these factors was based on the desire to achieve a target value of the reliability index, $\beta_T = 3.5$. It is appropriate to compare the form of the above expression with that obtained previously from the analysis of the probability of failure for the simple case, with both R, the resistance, and L, the load effect nominally distributed. Assume only a single loading, L, with a load factor, a, and a resistance, R, with a capacity reduction factor, b. The code expression would require

$$aL \leq bR$$

or

$$(a/b)L \leq R$$

The earlier expression, rearranged, gives

$$L + \beta\sqrt{\sigma_R^2 + \sigma_L^2} \leq R$$

where the design values, L and R, have been used in place of the mean values, $L_M - R_M$.
Suppose, for example,

$$(a/b) = 1.75/0.9 = 1.94$$

Then, for equivalence

$$\beta\frac{\sqrt{\sigma_R^2 + \sigma_L^2}}{L} = 0.94$$

Rearranging, and setting β equal to 3.5

$$3.5\sqrt{1 + \sigma_R^2/\sigma_L^2}\; c.v._L = 0.94$$

where

$c.v._L$ is the coefficient of variation of L, equal to $\sqrt{\sigma_L}/L_M$

The variability of the resistance is likely to be much less than that of the load. Suppose $\sigma_R = 0.5\sigma_L$, then the above expression suggests that, for β equal to 3.5,

$$c.v._L = 0.24$$
$$c.v._R = 0.12$$

In fact, R and L may not be normally distributed, and this calculation is only an indication of behaviour which will be discussed further in Chapter 5. It suggests, however, that the form of the code rule is reasonable.

The chief advantages of the probabilistic-based, Limit States Design Method are:

(a) the recognition of the different variabilities of the various loads, such as the dead load versus the live load, for the old working stress method encompassed both in the same factor of safety;
(b) the recognition of a range of limit states; and
(c) the promise of uniformity by the use of statistical methods to relate all to the probability of failure.

Their chief disadvantage, as forecast by Freudenthal (1968a and b), lies in the necessity to choose an acceptable risk of failure; it is very difficult, for example, to quantify the acceptability of some risk that involves only structural collapse, with a risk that leads to loss of life. Furthermore, the probability of failure must be applied to the number of events that may occur during the life of the structure, where, in the case of a bridge, this life is commonly set at about 100–120 years. There is an essential difficulty in predicting an event that may not occur until close on 100 years from the point of design.

Chapter 4

Codes of practice

4.1 Introduction

This chapter has two objectives: to review current codes of practice, and to outline the history of their development. The treatment will be by country, with reference to design philosophy and primary design loads for both road and rail bridges. Code provisions for pedestrian bridges and other loads will be discussed in later chapters.

The order chosen is somewhat arbitrary, with Britain followed by the United States and then Canada, for the 1979 *Ontario Highway Bridge Design Code* (OHBDC) was of major importance, both in the development of the limit states format and in its methodology for the choice of a design vehicle and its association with legal limits. The treatment goes on to include other European countries and the Eurocode, Australia and South Africa. The selection is incomplete, but is probably a fair sample for the present purpose, and one that illustrates a range of design vehicles or equivalent loads. The relative magnitudes and forms of these loadings will be discussed later in the chapter.

4.2 British codes for highway bridges

Henderson (1954) has given a brief account of early highway bridge design loads in Britain. This uses the old Imperial units of tons, pounds-force and feet, and in these units includes simple figures, such as 5 tons (49.8 kN) or 10 feet (3.05 m). Here they have been converted to SI units.

About 1875, Jenkins suggested a uniformly distributed load of 5.4 kPa, together with an axle load of 199 kN. Early in the twentieth century, Unwin proposed 5.75 kPa, with a four-wheeled vehicle loading of 100 to 249 kN, but

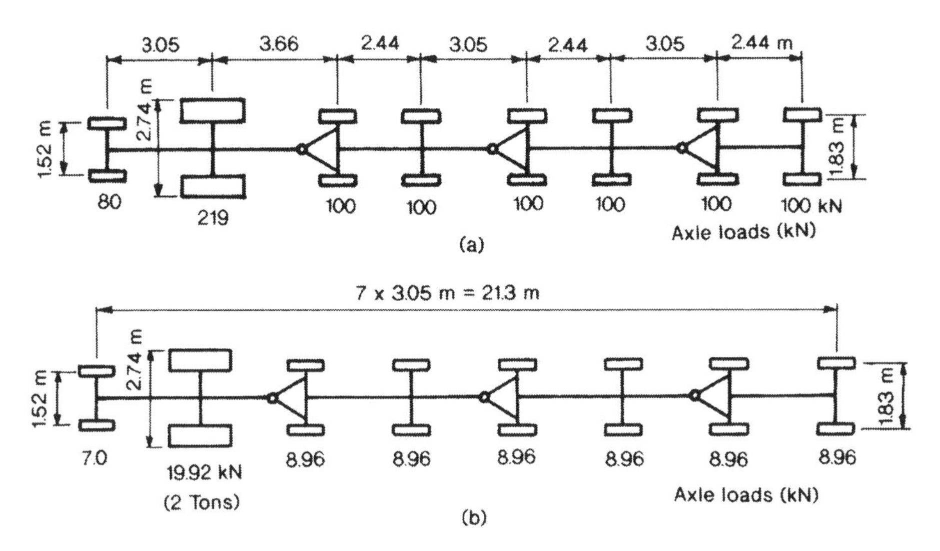

Fig. 4.1 (a) Ministry of Transport Standard Loading Train (1922); (b) BS 153 Unit Loading Train (1923)

later revised this to a distributed load of 3.83 to 5.75 kPa and, in manufacturing districts, a four-wheeled vehicle loading of 299 kN.

The first 'modern' loading was the Ministry of Transport Standard Loading Train, introduced in 1922 and shown in Fig. 4.1(a). It consisted of a tractor and four trailers, with a major axle load of 219 kN, followed by a succession of 100 kN axles, and included an impact value of 50% (i.e. ×1.5). It occupied a lane width of 3.05 m, but if the total road width exceeded a multiple of this figure, the total load was increased proportionately. One Standard Loading Train occupied a length of 22.9 m (75 feet), but for longer spans, additional trains would normally be added.

In 1931, the Ministry of Transport (MOT) Standard Loading Curve was introduced, consisting of a specified 'knife-edge load' (seen as a point load in side view) together with a uniformly distributed load whose intensity varied with the loaded length, as shown in Fig. 4.2. Chettoe (see Chettoe and Adams 1933: 88ff) has described the development of the Loading Curve, which is based on the 1922 Standard Loading Train. Much of its 22.9 m length was taken up by 100 kN axle loads spaced alternately at 2.44 and 3.05 m. Taking the latter figure, Chettoe spread a 100 kN axle over an area of 3.05 × 3.05 m, to give a load per unit area of 10.7 kPa. In Imperial units, this was rounded down to 10.5 kPa which became the standard distributed load for loaded lengths from 3.05 to 22.9 m. The major axle, of 219 kN, was 119 kN heavier than this 100 kN trailer axle. This excess was spread over the 3.05 m width to give a knife-edge load of 39.2 kN/m, rounded up slightly to 39.4 kN/m. The Standard Loading Train included a 50% impact value (×1.5). For spans less than 22.9 m, this was retained, but for longer spans it was reduced to 15% at 122 m, and zero at 762 m, and included in the loading curve. Also, the load intensity was reduced for the longer spans, down to 3.4 kPa for lengths equal to or greater

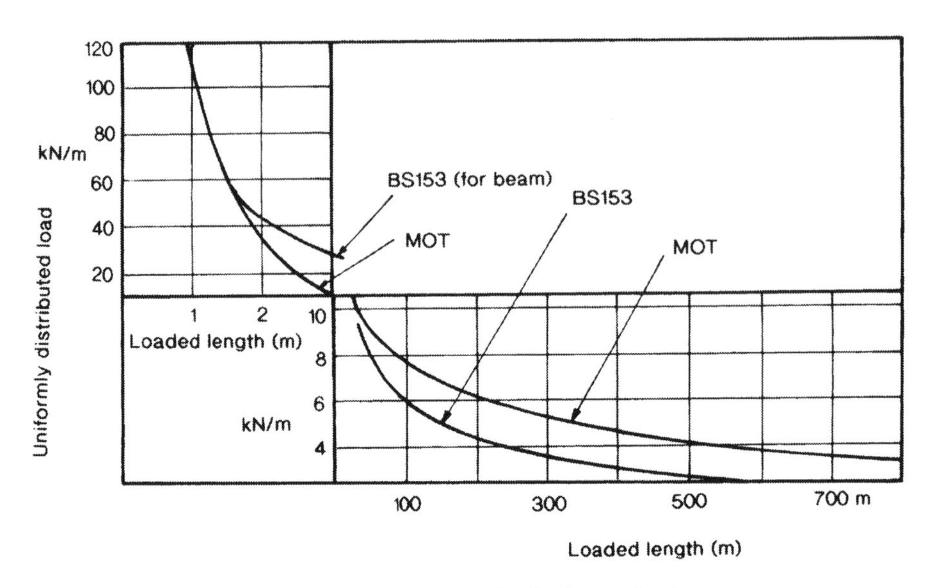

Fig. 4.2 MOT and BS 153 loading curves for highway bridges

than 762 m. The loading also provided for a single wheel load of 199 kN, and this increased the equivalent distributed load for the shorter spans. The Ministry of Transport Standard Loading Curve was widely used until 1958 for the design of highway bridges in Great Britain.

Also current at this time was the British Standard BS 153, first published in 1923, which specified the Unit Loading Train shown in Fig. 4.1(b), to be factored up by a stated number. Eleven of these units give a load that is close to the Ministry of Transport Standard Loading Train, but it should be borne in mind that the MOT loading included impact, whereas the BS 153 loading did not. Impact values were specified, but these reduced with increases in span. Henderson (1954: 326) states that fifteen of these BS 153 units were specified 'for conditions similar to those applying in Great Britain'.

About 1958, a new edition of BS 153 was introduced, including a revised set of loads, divided up into 'normal traffic' and an 'abnormal unit loading train'. The development of these loads was described by Henderson (1954). The normal loading was based on trains of four-axle trucks, with axle spacings of 1.22, 3.05 and 1.22 m, in lanes of width 3.05–3.66 m. The spacing of successive vehicles, between axles, was 2.13 m, i.e. each vehicle occupied 7.62 m. Various combinations of these vehicles were specified for various loaded lengths: for example, three trucks, each applying a total load of 219 kN, in 22.9 m, and in longer spans, up to five such trucks, with additional vehicles of 100 kN or 50 kN. These vehicles were used to obtain equivalent loads, in the manner of the Ministry of Transport loads, but expressed as lane loads, or loads per unit length of lane, rather than loads per unit of area. The equivalent distributed loads are shown with the MOT loads in Fig. 4.2; they include impact. For a loaded length of 22.9 m, the specified distributed load is 32.1 kN/m. For a lane width of 3.05 m, this gives a load per unit area of 10.5 kPa, as in the Ministry of

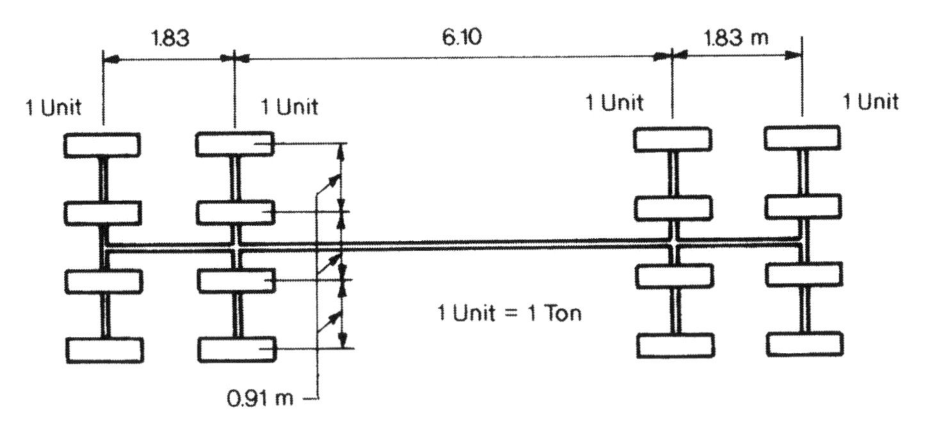

Fig. 4.3 BS 153: 1958 Abnormal Unit Loading Train

Transport load, but for longer lengths the specified distributed load is much smaller. For example, for a loaded length of 152 m, the distributed load is 14.6 kN/m, equivalent to 4.8 kPa for a lane width of 3.05 m. The MOT distributed load was 6.7 kPa for this length. The new knife-edge load was 120 kN, the same as the MOT loading for a 3.05 m lane. Figure 4.3 shows the Abnormal Unit Loading Train specified by BS 153: 1958, applied only to particular bridges. For main roads, 30 units of this loading could be specified, but with a stress overrun for dead load plus this live load of 25%.

These earlier British codes were all based on allowable working stress design. The specified loads were quite heavy, as compared with at least some other countries. In the discussion of Henderson's paper (1954: 355–9), it was pointed out that the standard American AASHO loading, described below, had effects that were equivalent to about two-thirds of the British loading over the whole range of spans. Earlier, in 1933, Chettoe (p.84) had said, 'The average intensity of traffic in Great Britain is heavier than in any other country in the world', where his reference was to loading intensity.

The first British limit states design code was BS 5400, issued in 1978. As in the older BS 153, there is a normal traffic and an abnormal vehicle load, identified as the HA and HB loading. The HA loading consists of a 120 kN knife-edge load, as in the earlier codes, together with a uniformly distributed load, W. For loaded lengths (L) up to 30 m, W is defined as 30 kN/m; for L greater than 30 m,

$$L = 151(L)^{-0.475}\ \text{kN/m}$$

but not less than 9.0 kN/m ($L \geq 380$ m).

The HA loading does not differ greatly from the older BS 153; for example, at $L = 30.5$ m it gives $W = 29.8$ kN/m, compared with the earlier 29.2 kN/m; and at $L = 152.4$ m, it gives 13.9 compared with 14.6 kN/m. It differs in that W is not allowed to go below 9.0 kN/m, even for the longest spans, and there is the cut-off for L less than 30 m. However, this cut-off is apparent rather than real, for it was required that all bridges carrying public highways should be designed also to carry at least 25 units of the HB, or abnormal vehicle, loading. The abnormal vehicle is essentially similar to that shown in Fig. 4.3, but with the central axle spacing (previously 6.1 m) specified as 6, 11, 16, 21 or 26 m, whichever produces the most severe effect, and with 10 kN per axle. The transverse wheel spacing is also 1.0 instead of 0.91 m (3 feet). In addition to these loads, it is required that the bridge be able to carry a single wheel load of 100 kN, distributed over an area of 9×10^4 mm^2 (either circular or square). For carriageway widths in excess of 4.6 m, the number of traffic lanes may be found by dividing by 3.8 m, and rounding up. This implies that a 4.6 m width should be divided into two 2.3 m lanes. Another rule applies for widths less than 4.6 m. For multi-lane bridges, the full specified HA loading is applied to two traffic lanes, with one-third of the standard loading applied concurrently to other lanes.

The HA loading was chosen to represent the effects of closely spaced 24 t (235 kN) vehicles in each of two traffic lanes, for lengths up to 30 m, with additional 10 t (98 kN) and 5 t (49 kN) vehicles in the longer lengths. An impact allowance of 25% 'on one axle or pair of adjacent wheels' was included in the loading. Typical load factors are:

(a) ultimate limit states:
$1.05\,DL + 1.75\,SDL + 1.5\,LL$; or
$1.05\,DL + 1.75\,SDL + 1.25\,LL + 1.1\,WL$;
(b) serviceability limit states:
$1.0\,DL + 1.20\,SDL + 1.20\,LL$; or
$1.0\,DL + 1.20\,SDL + 1.0\,LL + 1.0\,WL$,

where DL indicates the dead load, SDL the superimposed dead load, LL the live load and WL the wind load.

BS 5400 is issued in a number of parts. As each Part is revised and reissued it is given a new date, thus these dates may vary between Parts. For example, the version current in 1997–8 had Parts with dates ranging from 1978 (for 'Part 2. Specification for loads') to 1990. The 1988 General Statement refers to 'a design life of 120 years'; it has a section headed '7. Analysis' which speaks of 'Plastic methods' allowing for the possibility of plastic hinges (or yield lines in slabs). It permits 'plastic methods ... for ... redistribution of moments and shears' subject to certain qualifications, but adds that even with the use of these methods the supports should still be 'capable of withstanding reactions calculated by elastic methods', and 'changes in geometry due to deflections (should) not significantly influence the load effects or are fully taken into account'. However, review of the design loads continued; in 1986 there had been a decision to adopt uniform rules for legal weights and dimensions throughout the European Economic Community. This meant an increase in Britain in the maximum legal vehicle mass, from 38 to 40 tonnes, a review of the design loads and a proposed new HA loading (Chatterjee 1991: 65).

In 1988 the British Department of Transport required the use of this loading for the design of all of their highway bridges and issued the Departmental

Standard BD37/88, 'Loads for Highway Bridges', which included as an appendix a new 'Composite Version of BS 5400: Part 2: 1978'. The revised *HA* loading continues the form of the previous loading with, per lane, a knife-edge load of 120 kN (as before) and a distributed load given by

$$W = 336(L)^{-0.67} \text{ kN/m, for } L \leq 50 \text{ m,}$$

and

$$W = 36(L)^{-0.1}, \text{ for lengths greater than 50 m and up to 1600 m.}$$

The length L is generally 'the full base length of the adverse area'; and in cases where a number of separate portions are loaded simultaneously, L is the sum of their lengths. The specification gives some guidance for the selection of L in cases with 'highly cusped influence lines', such as may arise in the case of a cross-member in the deck. It requires also that bridges on motorways and trunk roads be able to carry 45 units of the 1978 HB loading; on principal roads, 37.5; and on all other public roads, 30.

The commentary refers to 3.65 m as the standard lane width, but permits the use of narrower lanes; for example, carriageways from 5 to 7.5 m wide are designed to carry two notional lanes, as in the previous code. However, the code provides for reduced loads on narrow lanes for bridges with L less than 40 m; for example, for $L < 20$ m, a factor of $0.274 \times$ the lane width is applied to both the first and second lane loading. For longer spans, the factors are typically 1.0 for lanes 1 and 2, and 0.6 for additional lanes. Dynamic effects are included within the loadings as before, but a comment is of interest: 'Recent research has shown that the impact effect of an axle on highway bridges can be as high as 80% of the static axle weight and an allowance of this magnitude was made in deriving the HA loading.' Load factors of 1.5 and 1.2 are applied to the HA loading when considered in conjunction with dead load for ultimate and serviceability limit states; and 1.33 and 1.1 for the HB loading.

Both BS 5400: Part 2: 1978 and the 'Composite Version' of 1988 tabulate values of the distributed load included in the HA loading. For values of L between 25 and 60 m, the new loading represents an increase of the order of 10%, but for greater lengths the increase may be substantial; for example, 56% at $L = 150$ m. For shorter lengths, values of W in the new code are much higher, but in these cases the abnormal or HB loading will tend to govern.

Departmental Standard BD37/88 states that 'the type HA loading is the normal design loading for Great Britain where it adequately covers the effects of all permitted normal vehicles – as defined in The Road Vehicles ... Regulations 1986 (S.I.1986/1078) and subsequent amendments'. It was issued by the Department of Transport rather than by BSI (the British Standards Institution) 'as an interim measure, pending a long term review of BS 5400 as a whole and bearing in mind the current work on Eurocodes'. The draft Eurocode I will be discussed in Section 4.6. In the period before its finalisation, BD37/88 continued in use as the defining document for British highway bridge loads.

4.3 Standard specifications for American highway bridges

The American Association of State Highway Officials (AASHO) *Standard Specifications for Highway Bridges* were first published in 1931, following a period of development that commenced in 1921, and have been widely used for the design of highway bridges in the United States of America and elsewhere. In 1973, the term AASHO was changed to AASHTO, reflecting the change to State Highway and Transportation Officials in the title of the issuing body. They have, for many years, included in their loading the HS20 truck and semi-trailer combination, previously called the H20-S16, shown in Fig. 4.4, taken from the 1973 (eleventh) edition. The designation given there is, more precisely, the HS20-44 loading, indicating that it was introduced in 1944.

The first two axles of this loading (at the left) form the H20 truck, so-called because its total weight is 20 US tons (one US ton is 2000 lbm, corresponding to a load of 8.90 kN). The leading axle is 4 US tons (35.6 kN), followed by one of 16 US tons (142 kN), with an axle spacing of 4.3 m. To this is added the rear, semi-trailer axle, also of 16 US tons (142 kN), with an axle spacing that can be varied by the designer from 4.3 to 9.1 m. This description of the two major 142 kN loads is qualified for deck design by a footnote in the code: if indeed there is only one axle, then the maximum axle load is 107 kN; alternatively, the load of 142 kN may be applied, but divided equally between two axles, spaced 1.22 m apart. The 1976 AASHTO Interim Specifications modified this by requiring that bridges on Interstate Highways should also be able to

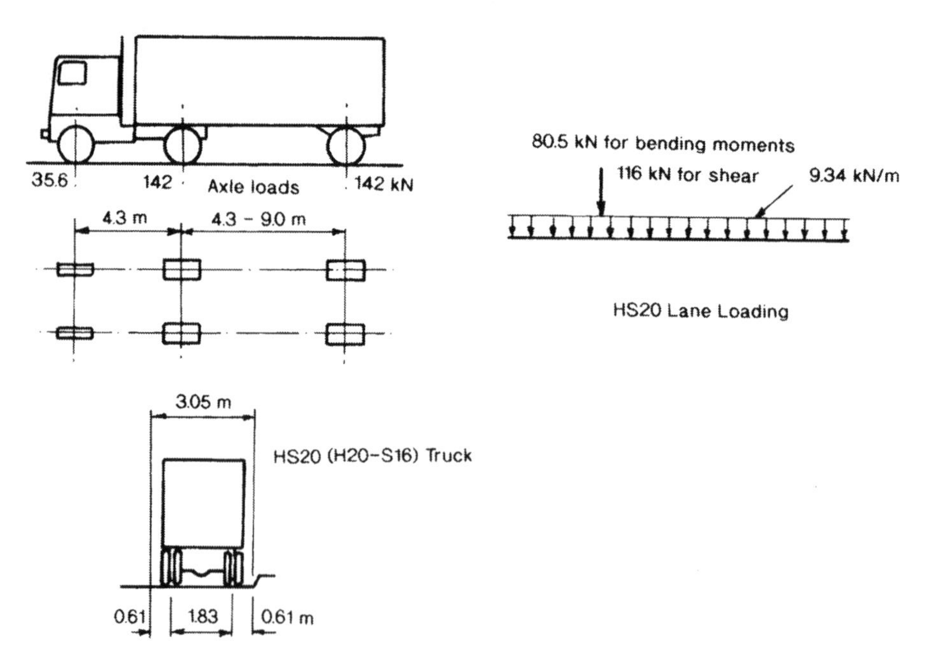

Fig. 4.4 American AASHO HS20–44 loading

carry an 'Alternate Military Loading' consisting of two axles, 1.22 m apart, with each axle applying a load of 106.8 kN. At a later date, this was changed to two 110 kN axle loads at a spacing of 1.20 m.

The HS20 truck semi-trailer combination was not, however, the only 1944 loading. Figure 4.4 shows also the HS20 equivalent Lane Loading, consisting of a concentrated (or knife-edge) loading that is 80 kN for bending moments, and 116 kN for shear, together with a uniformly distributed load of 9.34 kN/m, where these have the same magnitudes for all span lengths. However, the Introduction to the code indicates that it should be used only for spans less than 152 m. The distributed load may be interrupted, with no loading applied to part of the deck if this causes a larger effect; and in a bridge that is structurally continuous over two or more spans, in estimating the continuity moment two concentrated loads may be applied, one in each span.

Each loading corresponds to a lane width of 3.05 m; the total roadway width is used to determine a number, N, of design traffic lanes, based broadly on this dimension, but with some provision for additional width in multi-lane bridges. For example, a 12.8 m roadway is assumed to carry only three lanes; widths over 12.8 m and up to 16.5 m carry four lanes (a table is given). Fractional lane widths are not used, but the design loading, of width 3.05 m, may be moved transversely anywhere within its design lane to achieve maximum effect. Two lanes are assumed to carry their full design loads concurrently; with three lanes, all are loaded, but with a reduction factor of 0.9 applied to all; and with four lanes, this reduction factor reduces to 0.75. Only one truck may be applied within any one lane, over the whole length of the bridge. It is believed that it may be faced in either direction. To allow for dynamic effects, all loads are multiplied by $(1 + I)$, where the impact factor, I, is given by

$$I = \frac{50}{L + 125}$$

where

L is the loaded length in feet.

The expression with L in metres is

$$I = \frac{15.24}{(L + 38.1)}$$

I is not to exceed 0.3.

The code also specified the proportionately smaller HS15 truck semi-trailer and HS20, H15 and H10 trucks. Provision for abnormal vehicles was made by requiring the application of a single HS20 standard truck and semi-trailer, increased by 100%, placed anywhere on the bridge, but with the allowable combined dead and live load stresses increased by a factor of 1.5.

The 1973 edition of the AASHO code was essentially based on allowable stress design, and the magnitude of these stresses may be gauged by observing that the basic allowable stress in steel was 138 MPa for a yield strength of 248 MPa. The nominal factor of safety in this case was 1.8. However, it did allow for the ultimate strength design for reinforced concrete sections, which by this time had been generally adopted for building design; it specified the following load factors for use with the HS20 live loads:

$$1.30[DL + 5/3(LL + I)]$$

A typical capacity reduction factor was 0.90 for a reinforced concrete section in bending.

An important review of design loads for American bridges was carried out in 1981 by a Committee on Loads of the American Society of Civil Engineers (ASCE), chaired by Buckland (Buckland 1981; 1982; et al. 1980; 1981; see also Section 6.1). It recommended no change in the basic traffic loads for short-span bridges (less than 152 m), but for bridges of longer span recommended a uniformly distributed load, U, to be used with a concentrated load, P, where both are loads per lane, as shown in Fig. 4.5. The point load P (equivalent to the knife-edge load in British practice) increases linearly from 0 at 15.2 m to 747 kN for a loaded length of 1950 m. The uniformly distributed load (U) depends on the percentage of heavy vehicles and varies as shown in the figure. The loaded length may be broken into parts of the deck, so as to produce maximum effect, but U is determined from the total length. Only one point load was to be applied, and this within an area covered by the distributed load. Zero impact was to be added. For multiple traffic lanes, the full load was to be applied in one lane, plus 0.7 of this in a second lane, plus 0.4 in succeeding lanes. This loading has later been called the ASCE Loading.

In 1986, AASHTO commissioned a further major review of US bridge design practice, leading to the 1994 first edition of their *LRFD Bridge Design Specifications*, based on load and resistance factor design. Although this study revealed some 'discernible gaps ... (and) ... inconsistencies' in the previous code, and led to a code based on a new design philosophy, it left American designers with 'a choice of two standards to guide their designs', the older, working stress design specifications and the new 1994 LRFD code (quotations are from the Foreword to this document), that is, it was seen as adding to rather than necessarily replacing the old specifications. It involved considerable work, some of which has been described by Nowak in Das (1997: 42ff) and other papers (Nowak 1995; Nowak and Hong 1991; Tabsh and Nowak 1991; see also Section 5.3, and Kulicki and Mertz 1991). It was also based on a major review of Truck Weight Limits, conducted by a group led by Cohen (see Cohen and Hoel 1990).

The new AASHTO loading consists of a design truck, or a tandem, coincident with a design lane load. The design truck is effectively the old HS20 truck semi-trailer combination adjusted to rounded SI units. The old axle loads of 35.6, 142 and 142 kN become 35, 145 and 145 kN. The axle spacings, which were previously 4.3 and 4.3–9.1 m, are now 4.3 and 4.3–9.0 m. The design tandem, which could be used instead of the truck semi-trailer, consists of two 110 kN axle loads, spaced 1.2 m apart. The tyre contact area for either the

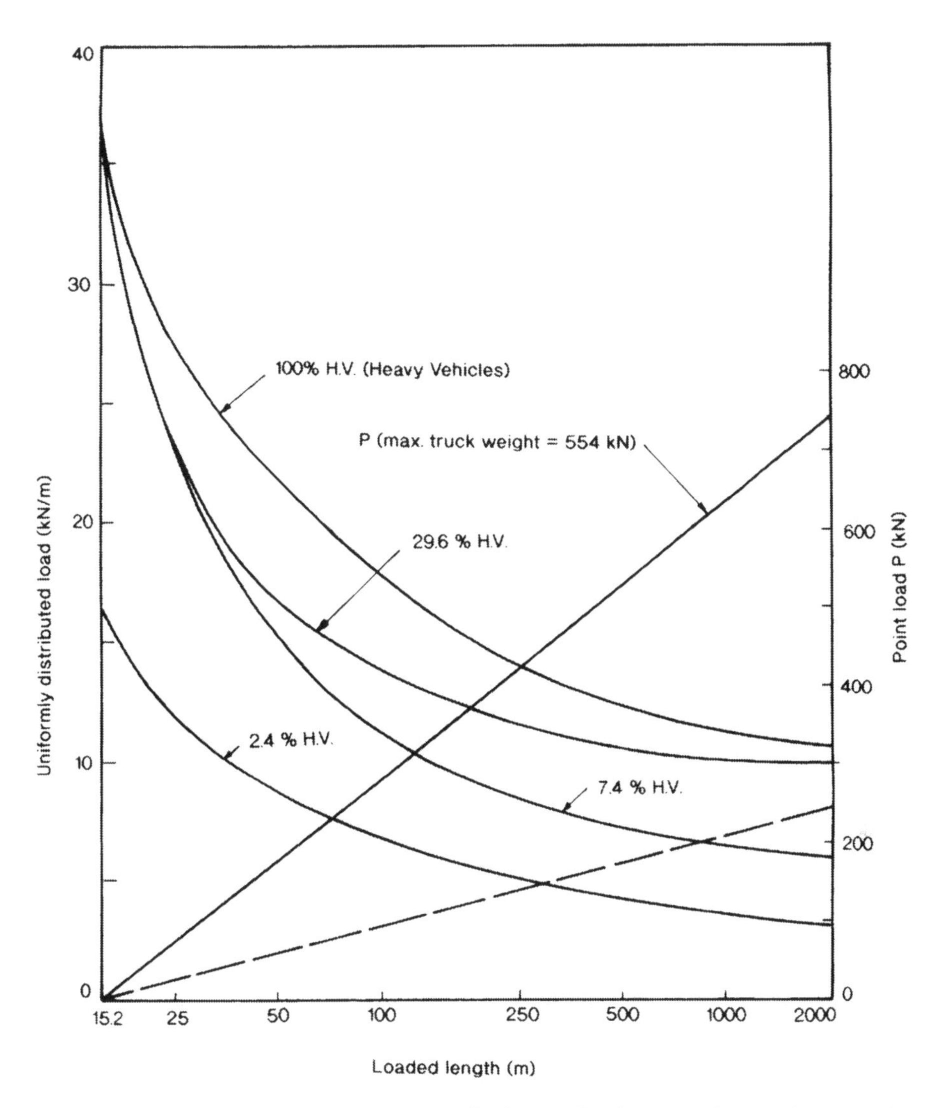

Fig. 4.5 ASCE Loading for long-span highway bridges (Redrawn from Buckland 1981)

design truck or the tandem is based on one-half of the axle load (i.e. 72.5 or 55 kN), factored up to include both the dynamic load allowance and load factor, applied to two tyres. The width (parallel to the axle) is set as 510 mm, and the length (parallel to the direction of movement) is ℓ, where

$$\ell \text{ (mm)} = 2.28 \times \text{the factored load.}$$

For example, for a load of 72.5 kN, with a dynamic load allowance of 1.33, and a load factor of 1.75, ℓ becomes 169 mm.

The design lane load is 9.3 kN/m. The older lane load was 9.34 kN/m, but applied in conjunction with a concentrated (or knife-edge) load of 80 kN for bending moments, or 116 kN for shear. For comparison, the total new design truck load became 325 kN, and the total tandem load is 220 kN. The new loading is specified to occupy a lane width of 3.6 m, although this may be reduced to as little as 3.0 m for a two-lane roadway width of 6.0 m. The older width was basically 3.05 m.

The 1994 edition includes ratios of bending moments and shears produced by the new loading to those from the old, plotted against span length. At a span of 5 m, the increases in bending moment are of the order of 10%, and these grow to about 70% at a span of 45 m. The increases in design shear are a little less, from about zero at 5 m to 50% at 30 m.

The new 1994 AASHTO dynamic load allowance, $1 + I$, is the same for all spans, being 1.15 for fatigue and fracture, and 1.33 for all other limit states. However, a factor of 1.75 is specified for the design of deck joints. The requirement for each limit state is

$$\eta \Sigma \gamma Q \leq \phi R_n$$

where

Q is a load effect, and γ is its load factor,
R_n is the nominal resistance,
ϕ is a resistance factor, or capacity reduction factor, and
η is a factor, between 0.95 and 1.0, based on ductility, redundancy and operational importance.

Typical values of the load factors for a strength limit state are:

$$1.25\ DL + 1.5\ SDL + 1.75\ LL$$

where

DL is the dead load of the structure itself,
SDL is superimposed dead load, including wearing surface and utilities (such as water pipes), and
LL is live load, with the load factor applied also to the dynamic load allowance.

Typical values of ϕ, for members in bending, are 1.0 for steel and prestressed concrete of linear-elastic, and 0.9 for reinforced concrete.

In general, the code favours the use of linear-elastic methods for the analysis of the structure to give elemental forces and moments, but then calculates the strength of these elements from their inelastic behaviour. It recognises 'an obvious inconsistency' in this, but states that this existed also in previous editions of the AASHTO specifications, and in the codes of other nations. However, it permits inelastic analysis, with 'a preferred design failure mechanism and its attendant hinge locations', this is subject to some warnings, notably, that it must be known that the relevant components of the structure will behave in a ductile fashion, with a response that 'has generally been observed to provide for large deformations as a means of warning of structural distress'.

4.4 Ontario and Canadian highway bridge design codes

The first edition of the *Ontario Highway Bridge Design Code* (Ontario Ministry of Transportation 1979) was a pioneer in the use of the limit states design philosophy for bridges, and influenced the development of many other design codes, such as the American AASHTO Specifications. The chairman of the Code Development Committee was P.F. Csagoly. An early report by Csagoly and Dorton (1973) described work that formed the original basis of its design load, and another by the same authors in 1977 described the development of the code itself. Some of this work is summarised in the Commentary on the 1983 second edition of the code. Although issued initially on a trial basis, it has been used for the design of many bridges in Ontario since 1979. One of its features was the manner in which the design loads were linked to legal limits applied to trucks in Ontario, but it seems evident that the form of these legal limits had an effect that spread far beyond that state, since one cannot drive across Canada without using the roads of Ontario. That methodology was discussed in O'Connor (1980) and (1981).

The Ontario workers defined an equivalent base length, B_M, as 'an imaginary finite length on which the total weight of a given sequential set of concentrated loads is uniformly distributed such that this uniformly distributed load would cause force effects in a supporting structure not deviating unreasonably from those caused by the sequence itself'. The 'sequential set of concentrated loads' could be taken either as a complete vehicle, or as any subset of adjacent loads. If a vehicle has n axles, then there are factorial n ($n!$) subsets of loads taken 1, 2, 3 ... n at a time. For a vehicle with 5 axles, for example, the number of subsets is 15, made up of 5 single axles, 4 pairs of axles, 3 groups of 3, 2 groups of 4 and one of 5. Each subset has a weight, W, and an equivalent base length, B_M. These may be plotted in (W, B_M) space. Generally it is only the upper bound of these points that is important, and this can be referred to as the equivalent base length signature of the vehicle. The method of calculating B_M will be discussed in Chapter 5.

Figure 4.6(a) is taken from Csagoly and Dorton (1978a, 1978b) and shows the results of a 1967 census, with the (W, B_M) space divided into segments, and the number in each segment indicating the number of occurrences of that combination. The curve shown in the figure has the equation

$$W(\text{kips}) = 20 + 2.07B_M - 0.0071B_M^2 \ (B_M \text{ in feet}) \qquad (1 \text{ kip} = 1000 \text{ lbf})$$

In SI units this becomes

$$W(\text{kN}) = 89 + 30B_M - 0.34B_M^2 \ (B_M \text{ in m})$$

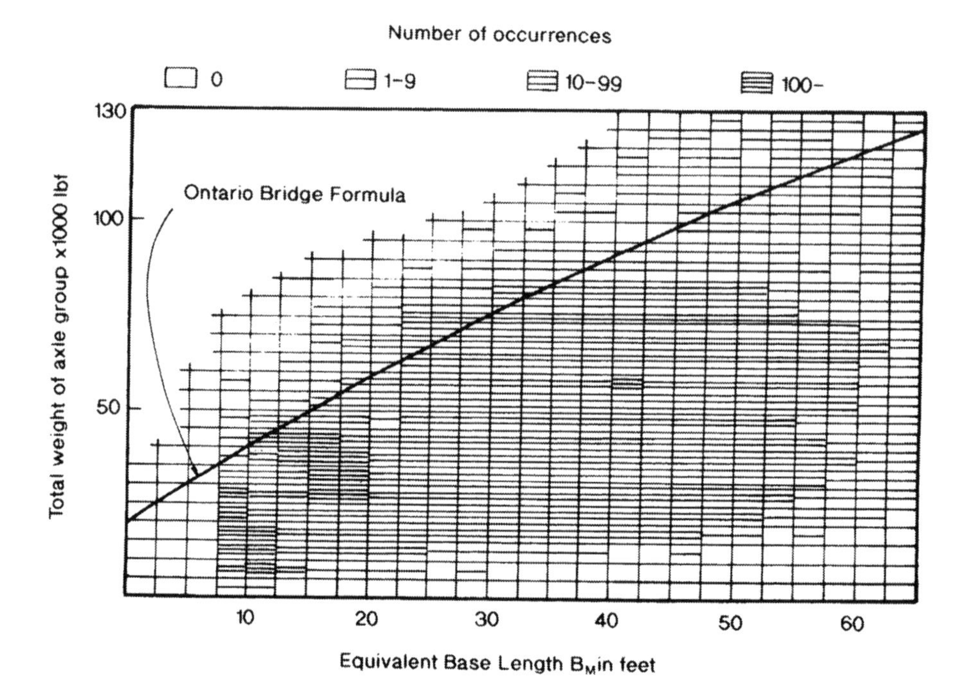

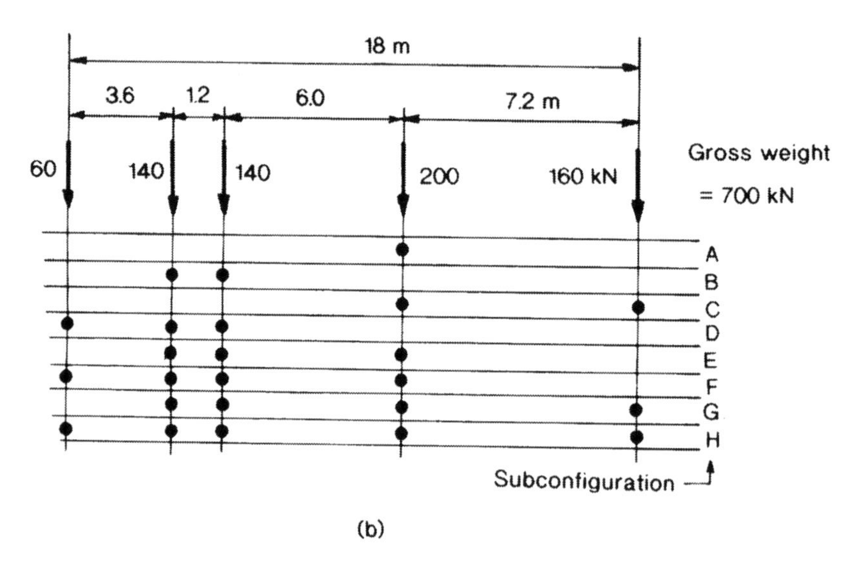

Fig. 4.6 *(continued)*

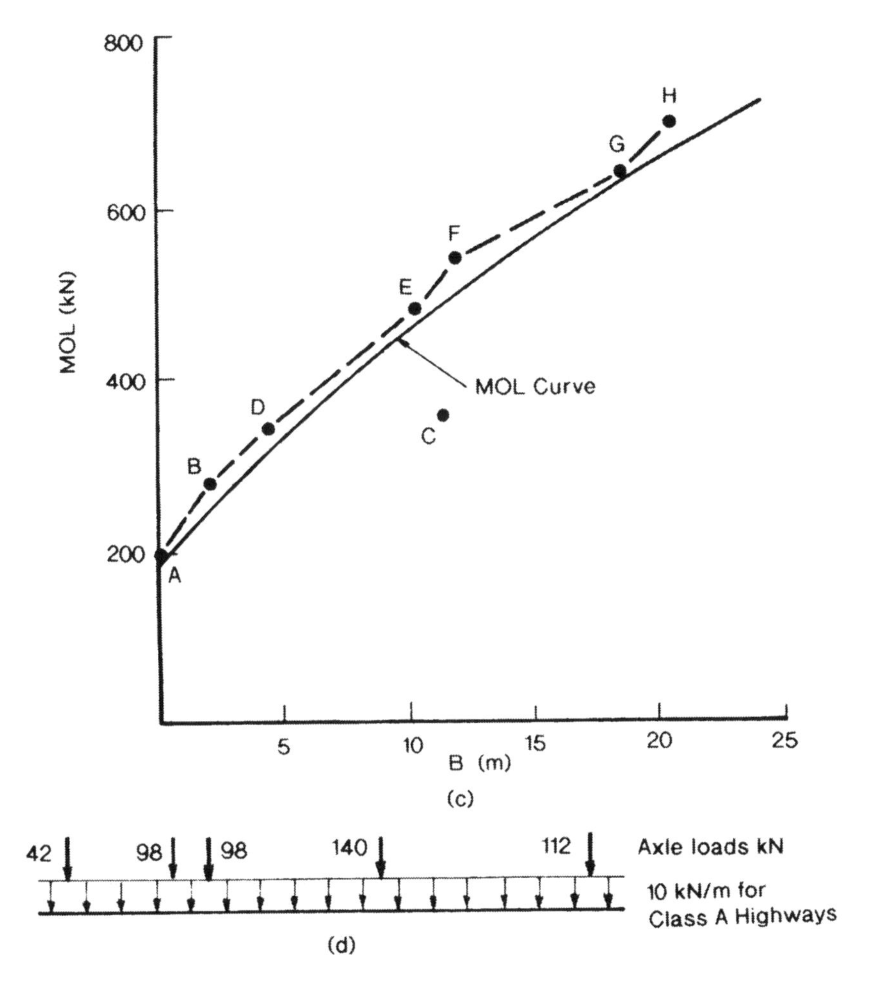

Fig. 4.6 Ontario Highway Bridge Design Code: (a) Histogram in (W, B_M) space; (b) OHBD Design Truck; (c) Signature of OHBD Truck; (d) OHBD Lane Load (Redrawn from Csagoly and Dorton 1978a and b)

This was later revised to

$$W(\text{kN}) = 9.806\,(10.0 + 3.0B_M - 0.0325B_M^2)$$
$$= 98.1 + 29.4B_M - 0.319B_M^2$$

This curve, which lies some distance below the upper bound of the survey data, was called the Ontario Bridge Formula (OBF).

A subsequent 1971 survey was used to establish a virtual upper bound to vehicles in Ontario, later revised to

$$W(\text{kN}) = 9.806\,(20.0 + 3.0B_M - 0.0325B_M^2)$$

called the Maximum Observed Load (MOL).

The OBF was used to define legal vehicle weights in Ontario, while the MOL curve became the basis of the Ontario Highway Bridge Design (OHBD) Truck, as shown in Fig. 4.6(b). There was also an OHBD Lane Load (Fig. 4.6(d)) that consisted of 70% of the OHBD Truck, together with a distributed load of 10 kN/m (for Class A Highways). Both occupied a width of 3.0 m. For components loaded by more than one axle the dynamic load allowance varied with the first natural frequency of vibration (f) in Herz, being 1.3 for $f \leq 1.0$, 1.45 for f between 2.5 and 4.5, 1.3 for $f \geq 6.0$, with linear transitions over the ranges 1.0–2.5 Hz, and 4.5–6.0 Hz.

The second edition of the Ontario Code was issued in 1983 with no change in the primary design loads (Dorton and Bakht 1984), but with the dynamic load allowances changed to 1.2, 1.4 and 1.25 for the frequency ranges quoted above. The loads were reviewed again (Nowak 1994; Nowak and Grouni 1994) for the 1991 third edition and slightly increased. Figure 4.7 shows the 1991 OHBD Truck, the only change being that the axle loads 2 and 3 are each increased to 160 kN from the earlier 140 kN. The Lane Load again consists of 70% of this truck, together with a distributed load of 10 kN/m. The form of the specification for dynamic load allowance was also changed, to depend on the number of axles considered in the design lane, to 1.40 for one axle, 1.30 for two, and 1.25 for three or more, with a value of 1.1 for the uniformly distributed portion of the Lane Load. Typical load factors for collapse in the 1983 code were:

$$(1.1\text{–}1.2)\ DL + 1.5\ SDL + 1.4\ LL \text{ (including impact)}$$

These values were not changed in the 1991 code. The methods used by Nowak to justify these values will be presented in Chapter 5 (see earlier references).

The code CSA-S6, *Design of Highway Bridges* of the Canadian Standards Association used in its earlier forms, loading models based on the AASHTO HS20 truck semi-trailer combination. By 1978 (Agarwal and Cheung 1987) the levels of the nominal axle load had been increased from the AASHTO 35.6, 142 and 142 kN to 50, 200 and 200 kN in the Canadian MS250 loading. The essential difference between the CSA loadings and those of Ontario was that the former were derived from legal limits agreed to by all provinces, whereas the OHBD Truck came from the Maximum Observed Load (MOL) in Ontario. In 1988, a Memorandum of Understanding on Vehicle Weights and Dimensions (MOU) was signed by all provinces, and this led to the CS-W loading, where W represents a total vehicle weight in kN; a value of 600 kN for W was recommended for the design of bridges on inter-provincial trucking routes across Canada. Figure 4.8 shows this loading. For ultimate limit states a load factor of 1.60 was applied, somewhat larger than the factor of 1.4 used in OHBD.

The 1988 MOU was amended in 1991 and used for the basis of a new Canadian Highway Bridge Design Code (CHBDC), due for issue in November, 2000, with the CL-600 increased slightly to CL-625 (see Cl.C3.8.3 of the Commentary of this code for references to the MOU). The load factor applied to this load was also increased (to 1.70) and the resulting factored live load is slightly above the 1991 OHBD Code Truck (personal communication, R.A. Dorton). Another important feature of this code is the removal of the 150 m upper bound

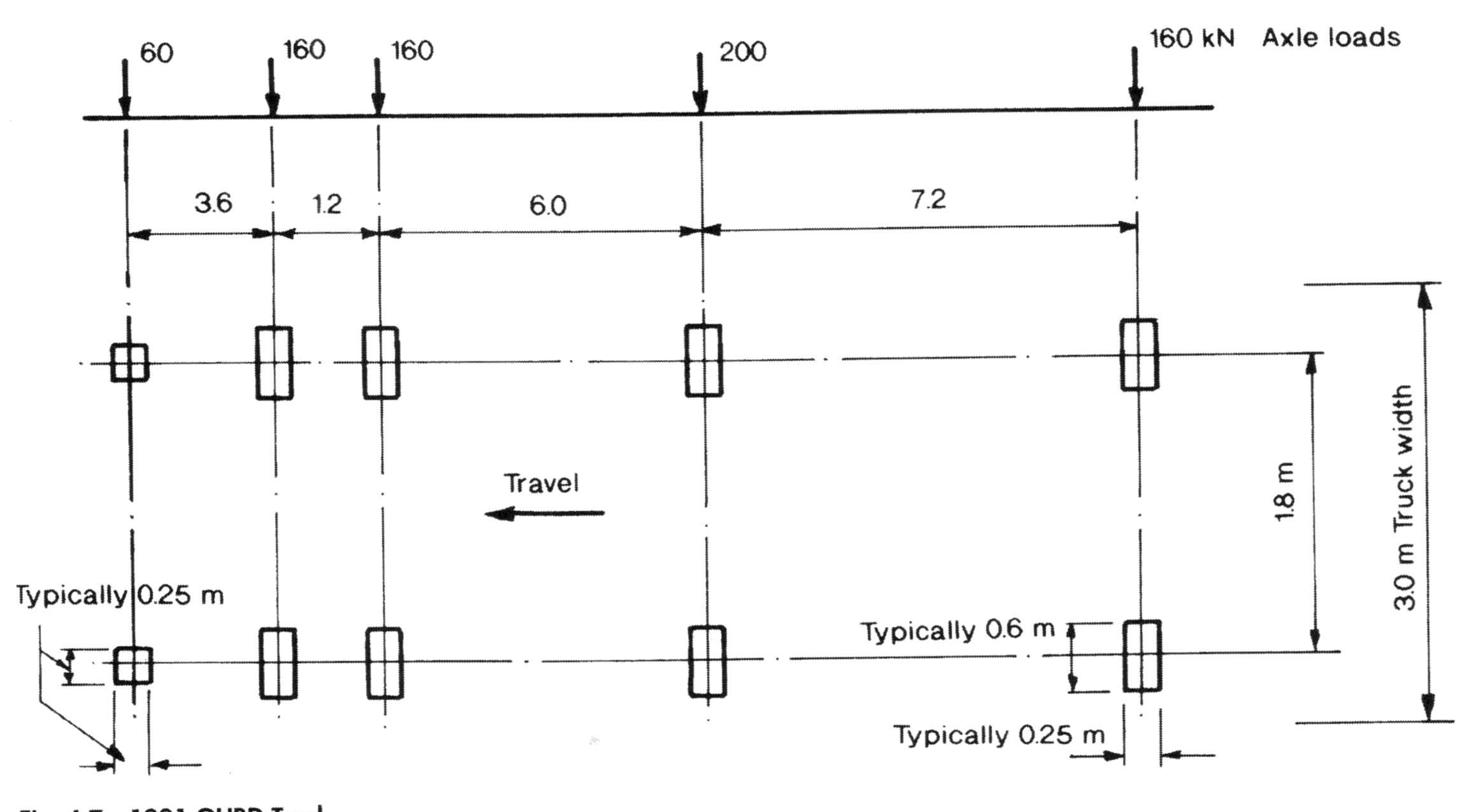

Fig. 4.7 1991 OHBD Truck

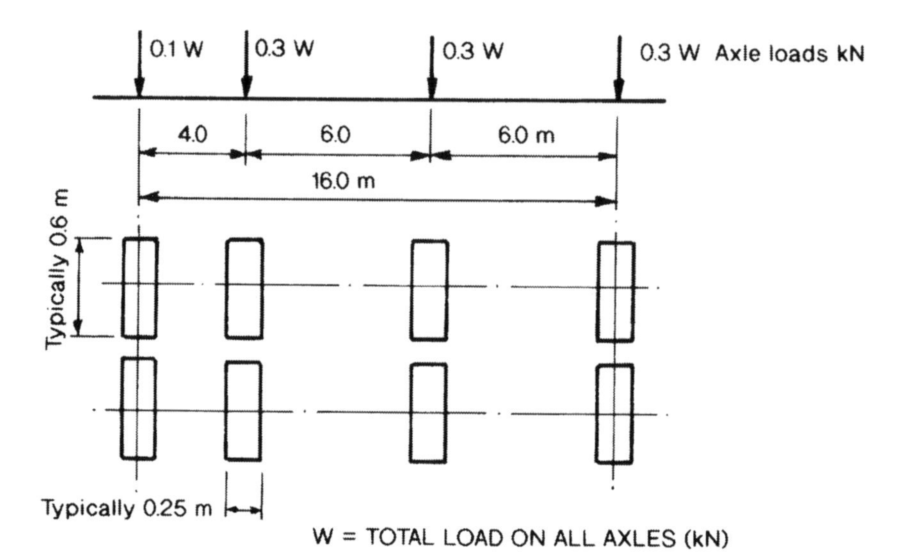

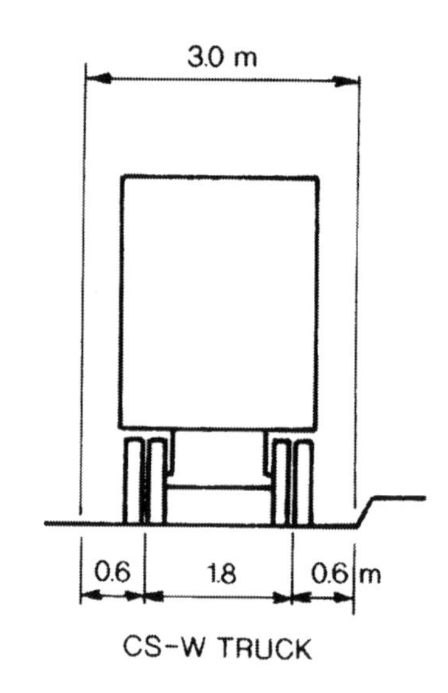

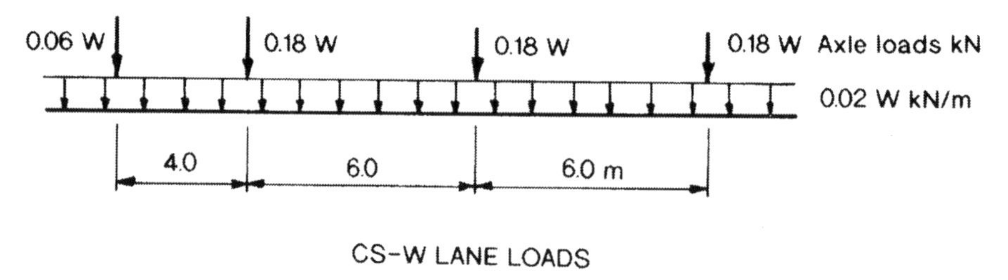

Fig. 4.8 Canadian CS-W loading

for span length. The CL-625 Lane Load consists of a CL-625 truck, with each axle reduced to 80% of its normal value, together with a uniformly distributed load of 9 kN/m, on a lane that is 3.0 m wide. The Code Commentary describes how this loading was related to earlier work by Buckland and others (Buckland et al. 1980 and Buckland 1981), that led to the loading shown in Fig. 4.5, recommended by an American Society of Civil Engineers Committee, and referred to informally as the ASCE loading. This ASCE loading consisted of a point load and a uniformly distributed load, and was based on observations of trucks using the Second Vancouver Narrows bridge. Of these, 7.4% were classed as heavy vehicles, i.e. buses and trucks over 5.4 t (12,000 lbm). Three other cases were considered:

(1) 2.4% heavy vehicles, and masses up to 18.1 t (40,000 lbm);
(2) 29.6% heavy vehicles, with masses up to 54.4 t; and
(3) 100% heavy vehicles, with masses up to 54.4 t.

All four cases are shown in Fig. 4.5; the actual loading, with only 2.4% heavy vehicles produces the lowest equivalent loads. The new Canadian lane loading corresponds in its effects closely to the ASCE (30%) load, being about 4% conservative at a 2000 m loaded length. However, the ASCE (30%) loading is itself believed to be conservative, 'by an unknown amount not exceeding 50%'.

This 1999 CHBDC code also includes an interesting reference to the risk of failure:

> Long span bridges are, by definition, likely to have serious consequences if they collapse. An importance factor is therefore desirable; or, putting it another way, the required target value of β should be increased. One way of accomplishing this is to use the formula in CSA-S408 [1981; another Canadian Code], which gives
>
> $$P_f = TAK/(W\sqrt{n})$$
>
> where
>
> P_f is the target probability of failure,
> T is the return period in years,
> A is an activity factor,
> K is a coefficient,
> W is a warning factor, and
> n is the number of people at risk

The value of $n = 10$ was adopted, but it was pointed out that, for a long span bridge, say 2000 m, n might be as large as 1000. In this case, the required value of P_f would reduce by $1/\sqrt{100}$, i.e. by 1/10. The corresponding β factor might change from 3.5 to 4.0. 'This may be accomplished without changing load factors if the design loading is increased at long loaded lengths by about 8%.' The implicit suggestion is that the proposed CL-625 Lane Load provides a sufficient reserve of strength.

4.5 Some European codes for highway bridges

The Organisation for Economic Co-operation and Development (OECD) was set up in 1960, and by 1980 included 24 member countries made up of 19 from Europe, with Australia, Canada, New Zealand, Japan, and the United States. In 1980, an OECD road research group published an *Evaluation of Load Carrying Capacity of Bridges*, which included a comparison of the design loads for highway bridges in twelve of the member countries. Some of its data will be used here.

The French *Cahier des Prescriptions Communes* (1973) specified two basic loading systems, multiplied by factors dependent on the class of bridge and the number of lanes loaded. Class I bridges have a carriageway width of not less than 7 m, and will be considered here; in Class II this width is between 5.5 and 7 m; and in Class III less than or equal to 5.5 m. Loading System A is a uniformly distributed load spread over the complete width of a lane, but with the intensity varied so that, within Class I, the load per metre is given by

$$w\ (\text{kN/m}) = 8.05 + 1260/(L + 12)$$

where

L is the loaded length in metres.

For a multiple lane bridge, the factors applied to this loading are 1.0, 1.0, 0.9 and 0.75, with 0.7 applied to each additional lane.

Loading system B consists of

(a) two 300 kN trucks per lane, as shown in Fig. 4.9(a); or
(b) an isolated wheel load of 100 kN, with a rectangular contact area 0.3 m long (parallel to the roadway) and 0.6 m wide; or
(c) a pair of axles, each 160 kN, 1.3 m apart, with a wheel spacing of 2.0 m (centre to centre).

Each truck occupies a width of only 2.5 m, with two trucks brought side by side so as to touch, while the width of the dual axle system is 3.0 m. The factors for multiple lanes are 1.20, 1.1, 0.95 and 0.8, with 0.7 for each additional lane. The impact factor applied to all forms of Loading System B is

$I = 0.64/(1 + 0.2L)$ for concrete structures; and
$I = 0.80/(1 + 0.2L)$ for steel and composite structures.

The dynamic amplification factor is $1 + I$. There is also a specified Heavy Transport, and a separate Heavy Military Load, for use on certain routes.

The West German Code, DIN 1072-1967, recognised three bridge classes:

Class 60 for motorways, federal and state roads;
Class 30 for city and country roads; and
Class 12 for rural roads with light traffic.

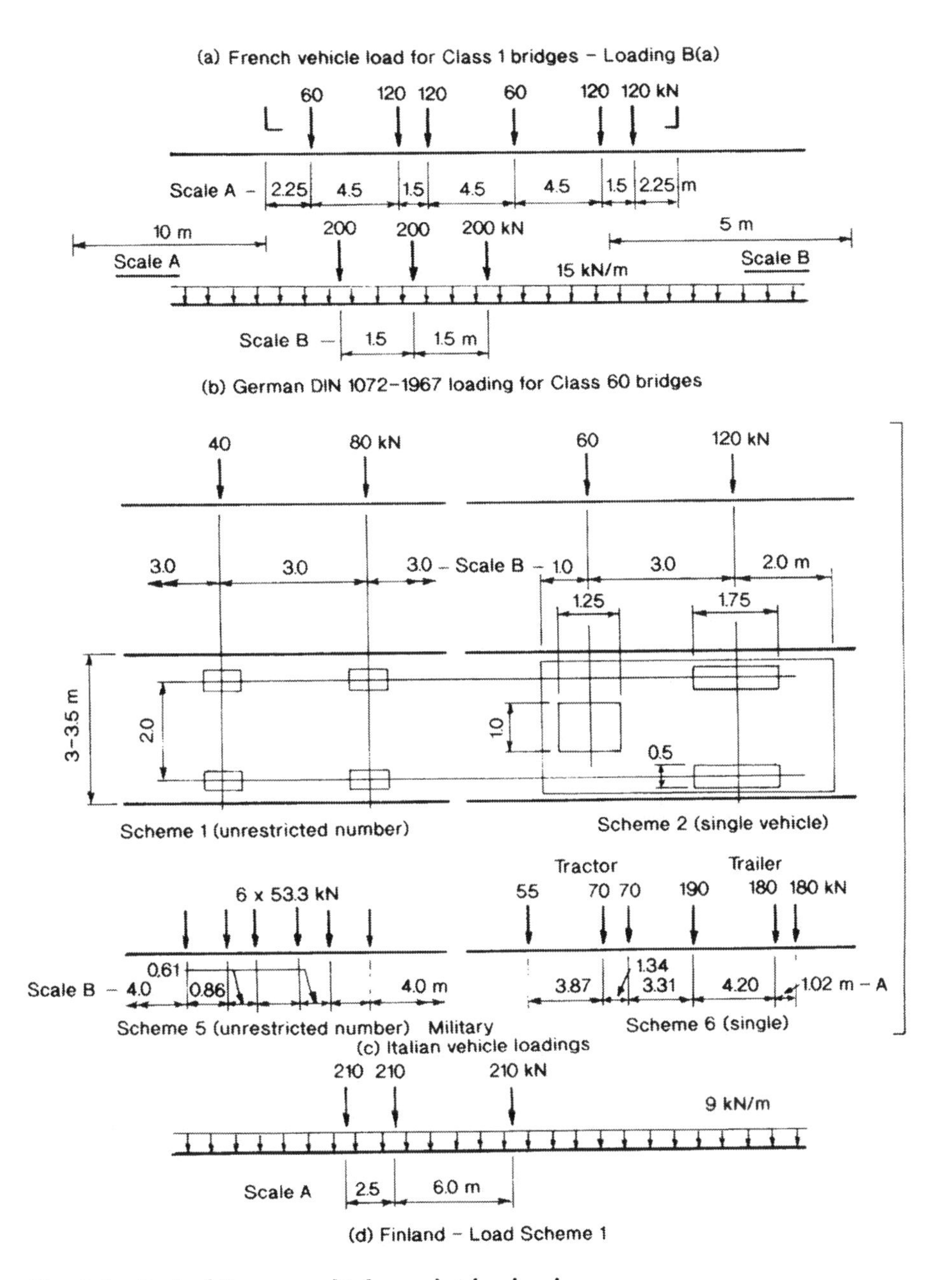

Fig. 4.9 Typical European highway bridge loads

The carriageway is divided into a 3 m wide principal traffic lane, and other lanes. For Class 60 bridges, as shown in Fig. 4.9(b), the principal lane is loaded with a group of three 200 kN axles, spaced 1.5 m apart, together with a distributed load 'in front and behind this vehicle'. It is not clear if this loading continues under the vehicle, as shown, or if it terminates some distance before and after the vehicle. It is specified as 5 kN/m^2, equivalent to 15 kN/m on a 3 m lane. The other lanes are loaded with a uniformly distributed load of 3 kN/m^2, where presumably this may be interrupted and placed for maximum effect. An impact factor, I, is applied to the loading on the principal lane only, with

$$I = 0.4 - 0.008L \text{ (m)}$$

but not less than zero. Again the dynamic amplification factor is $1 + I$.

For Class 30 bridges, the axle loads on the principal lane are one-half of those given above, but the distributed loads remain the same. It is believed these primary loads remained unaltered in the 1983 DIN Code, but that a later draft code (Chatterjee 1991) proposed that, in addition to the above loads, a second vehicle, of 300 kN (three axles at 1.5 m), be added to a second lane; and also that the Class 12 loading be abandoned.

The Spanish 1972 code specified loads similar to the 1967 DIN code, but with a distributed load of 4 kN/m^2 applied to all lanes, instead of the 5 and 3 kN/m^2 of the DIN code.

In 1979, there was a draft Belgian code in preparation which, for normal traffic, specified a uniformly distributed load of 3.5 kN/m^2 over the whole roadway and, in each lane, a pair of axles spaced 1.5 m apart, each applying 150 kN. Impact was as specified in the French code.

Italy was another country which distinguished between routes that carried both civil and military traffic (Class I), and those with civil vehicles only (Class II). Figure 4.9(c) shows the Scheme 1 and Scheme 2 civil loadings: the first had an unrestricted train of 120 kN vehicles; the second was a single machine of 180 kN. The three specified military loadings were much heavier. Two are shown in Fig. 4.9(c): the first an unrestricted train of six-axle vehicles, each totalling 320 kN, spread over an effective length of 7.8 m; and the second, a single tractor-trailer combination of 745 kN, with a length between extreme axles of 13.74 m. The specified lane width was 3 to 3.5 m, with an impact factor I given by

$$I = \frac{(100 - L)^2}{100(250 - L)}$$

but zero for spans in excess of 100 m. Typical values of I are 0.34, for $L = 10$ m; 0.28 at 20 m; and 0.08 at 60 m. These loads are from Code 384, dated 1962.

The last of these European examples is from Finland (Code 1978). Figure 4.9(d) shows its Load Scheme 1: a three-axle vehicle, consisting of three 210 kN axles, applied in addition to a distributed load of 9 kN/m, the whole occupying a lane width of 3.0 m. In addition, it was required that the bridge be checked for a single 260 kN axle load, with a loaded area for each 130 kN wheel 0.6 m wide (parallel to the axle) and 0.2 m long. There were also two specified heavy load vehicles, each applied singly.

The European bridge design codes reviewed in this OECD (1980) report all used allowable stress design. To compare these design loads, the report calculated equivalent values of a distributed load, q (kN/m^2), and a single point load, P, that produced the same maximum bending moments in simply-supported spans of chosen length (OECD 1980: 43f; 47ff; 118ff; 122ff). To do this, they obtained values of M_{MOBILE}, the moments produced by the loads themselves. They then multiplied these by the dynamic amplification factor $(1 + I)$, and another factor ν defined, for steel, as

$$\nu = \frac{\sigma \text{ yield point}}{\sigma \text{ allowable}}$$

That is, ν is the inverse of a factor of safety. Their reason for doing this is, simply, that one can adjust the design process either by changing the design loads, or by changing ν. The fact that they were able to do this confirms that these were allowable stress design codes.

Their graphs of q and P were produced for two- three- and four-lane bridges, and allow a convenient comparison of the design loads. A similar comparison, but based on their M_{MOBILE}, will be presented at the end of this chapter. Their comparison showed that the relative severity of these loadings changed with span. For example, the Italian unrestricted train of 120 kN trucks, with each occupying only 6 m, was a severe loading for the longer spans (40–100 m), whereas their loading, viewed as a whole, was light for the smaller spans (10–20 m). They also plotted the American HS20-1944 loading and showed it to be the lightest over most of the range – from about 15 or 20 m (depending on the number of lanes) to 100 m.

A more recent European code is the Swiss SIA160-1989. It uses limit states format and is based on the following loads. Load model 1 is a single two-axle group: with an axle spacing of 1.3 m, and a transverse wheel spacing of 1.8 m, on four wheel groups each taken as either 0.4 m square, or as circular, with a diameter of 0.45 m. It occupies a width of 2.2 m and may be placed anywhere within a 3 m lane. The total load is 75 kN, but to this is applied a dynamic coefficient of 1.8, to give a total of 135 kN. The lane occupied by this axle group is loaded also with a uniformly distributed load (Load model 2) of 5.0 kN/m^2; this represents a line of lorries travelling slowly, and already includes impact. Other lanes are loaded with Load model 3, a uniformly distributed load representing a mixture of stationary car and lorry traffic. This is 3.5 kN/m^2 for a carriageway width less than or equal to 9 m; 3.0 kN/m^2 for 9 to 13 m; and 2.5 kN/m^2 for widths in excess of 13 m. There is also a standard abnormal vehicle that may travel on designated routes. Typical load factors are

$$1.3\ DL + 1.3\ SDL + 1.5\ LL$$

where, as before,

DL is the self-weight of the structure,
SDL is the self-weight of non-structural elements, and
LL is the live load.

4.6 Eurocode for highway bridges

The OECD (1980) study used in the previous section drew attention to the modern spread of European road transport across national boundaries, and reached the following conclusions.

> There is an urgent need for more active and closer co-operation between Member countries on the problem of evaluating load carrying capacity of existing bridges. Co-operation is needed in the field of transport policies and regulations ... existing codes in most Member countries need further improvement.

One outcome of this has been the production of Eurocode I by the European Committee for Standardisation (CEN). This document, to which has been given the number ENV 1991, has many Parts; the two that are relevant here are ENV 1991-1 (1994) *Part 1: Basis of Design*, and ENV 1991-3 (1995) *Part 3: Traffic Actions on Bridges* (where the dates are given in brackets – it is coincidental that the number 1991 looks like a recent date). It is important to note that these Eurocode Parts are in the form of European Prestandards issued for provisional application, with the intention that they be reviewed after a period of 3 years. These reviews are, at the time of writing, still underway, and in any case it seems likely that each country will qualify the clauses of the Eurocode by issuing its own preamble, for safety levels within a country remain a national obligation. It should be noted also that ENV 1991 includes structures other than bridges.

The loading code ENV 1991-3 (1995) specifies two Load Models for normal highway bridge traffic, as shown in Fig. 4.10. Load Model 1 consists of a double axle (called a tandem system or TS), together with a uniformly distributed load, and is intended to cover 'most of the effects of the traffic of lorries and cars'. It is necessary, first, to identify notional lanes. The normal basic lane width is 3 m, with the exception that carriageways of widths 5.4 to 6 m are assumed to carry two lanes. Generally, a carriageway is divided into an integral number of 3 m lanes, which may be located transversely so as to achieve the worst effect. Of these lanes, the one causing the most unfavourable effect is called Lane 1, that with the second most unfavourable effect is Lane 2, and so on. These lanes need not correspond to the intended lane markings on the bridge; indeed, a demountable central safety barrier is ignored in locating these lanes. Space not occupied by the lanes is called remaining area. Then Lane 1 is loaded with the 300 kN tandem system shown in Fig. 4.10, together with a distributed load of 9 kN/m^2 (or 27 kN for a 3 m lane). Lane 2 is loaded with a 200 kN tandem system and 2.5 kN/m^2, and Lane 3 with a 100 kN tandem system and 2.5 kN/m^2. The remaining area is loaded with 2.5 kN/m^2. Only one tandem system is used per lane, and the distributed loads are applied only in the unfavourable areas. For general effects the tandem systems are assumed to travel along the axes of the notional lanes, but for local effects this may be varied.

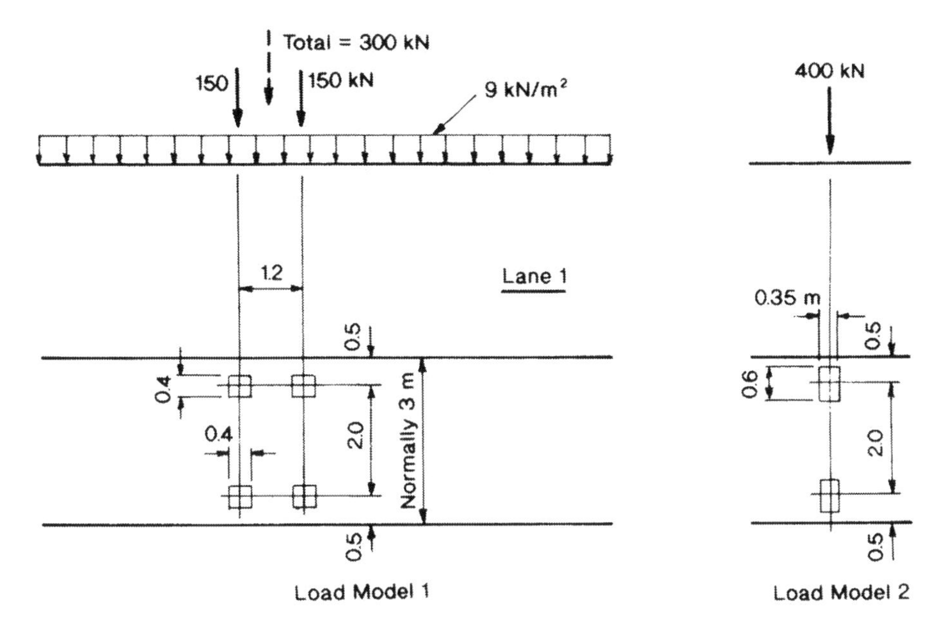

Fig. 4.10 Eurocode ENV 1991–3 (1995) highway bridge loading

Load Model 2 (see Fig. 4.10) consists of a single-axle load intended to cover the dynamic effects of normal traffic on very short structural elements. The total load on the axle is 400 kN, or 200 kN per wheel, where the assumed contact area is as shown.

These two load models are intended to represent the most severe traffic, now or expected, on the main routes of European countries. For many bridges it will be necessary to apply an adjustment factor to reduce the specified loads, but it is suggested that for bridges without signs restricting vehicle weights, this factor should be not less than 0.8. Dynamic amplification is included in the specified loads (for road bridges). The code also specifies a Load Model 3, which is an assembly of axle loads used to represent abnormal vehicles, permitted to travel only on specified routes.

Typical partial load factors for ultimate limit states are: for the self-weight of structural and non-structural elements and permanent actions caused by the ground, ground-water and free water, 1.35 if unfavourable, 1.0 if favourable; and for traffic actions, 1.35 if unfavourable, zero if favourable.

The document ENV 1991-1 suggests a design working life of 100 years for bridges, compared with 50 years for building structures. In Annex A (informative) it suggests a target reliability index, β, of 3.8 for ultimate limit states within the life of the structure, but goes on to emphasise that this value is 'intended primarily as a tool for developing consistent design rules, rather than giving a description of the structural failure frequency'. As shown in Section 3.3, if both the load effect, L, and resistance, R, are normally distributed, then the function $R - L$ is also normally distributed. Failure occurs if $R - L$ is negative. The function, β, expresses a distance below the mean of the $R - L$ distribution curve, as a multiple of the standard deviation of this curve. For β equal to 3.8, events with $R - L$ negative have a probability of occurrence of 0.72×10^{-4}, or 1 in 13800. However, the Eurocode generally uses the normal distribution only for load effects, and the lognormal or Weibull distributions for resistances and model uncertainties.

ENV 1991-1 also comments on the method of structural analysis stating that, whereas for serviceability limit states linear modelling may be used, in general 'modelling for static actions should normally be based on an appropriate choice of the force-deformation relationships', and that the effect of displacements may need to be considered.

4.7 Australian codes for highway bridges

Australia has three levels of government: (1) the local authorities or shires, (2) the states, and (3) the Commonwealth of Australia, inaugurated as a federation of states in 1901. Under the constitution, primary responsibility for roads (and railways) rests with the states, but these have vested authority for a National Highway system in the Commonwealth (Lay 1984 and 1985), and made local roads a responsibility of the shires. In 1934 a Conference of State Road Authorities was formed by the states, which became the National Association of Australian State Road Authorities (NAASRA) in 1959, and issued in 1953 the first edition of an Australian Highway Bridge Design Specification.

Before 1953, there was considerable diversity in design loads. In 1923, for example, the Queensland Main Roads Board issued a 'Diagram of Live Loads' that specified use of a standard tractor followed by an indefinite system of trailers. All were dual-axle vehicles; the tractor with axle loads of 24.9 and 49.8 kN spaced 3.05 m apart; then 3.05 m to the first trailer, with axle loads of 24.9 kN spaced 2.9 m; then 3.2 m to the first axle of the next trailer; and so on. To this was added impact, and an indication of the intended factor of safety is shown by an allowable tensile stress in steel of 110 MPa. The allowable compressive stress in concrete was 3.4–4.1 MPa.

Early editions of the NAASRA Highway Bridge Design Specifications were based on the American AASHO and AASHTO codes (see Section 4.3), and used the HS20-1944 truck and lane loads, together with the lighter American truck and truck semi-trailer combinations. Other code provisions were also similar; for example, the basic allowable stress in steel in the 1970 NAASRA code was 138 MPa. At about that time, the maximum legal load on a dual-axle group varied between states, but was typically 130 kN (NAASRA 1968). In 1973, a Metric Addendum was added which replaced the original 4- and 16-US ton axle loads by 36 and 144 kN, rather than the exact equivalents of 35.6 and 142.3 kN. In 1976, this truck semi-trailer combination (total mass 33.04 tonnes) was replaced by the T44 truck shown in Fig. 4.11, with a mass of 44 tonnes, corresponding to a further major increase of 33%. However, it can be seen that the form of the vehicle is still the same. Associated with the T44 Truck Loading was the L44 Lane Loading, consisting of a 150 kN knife-edge load,

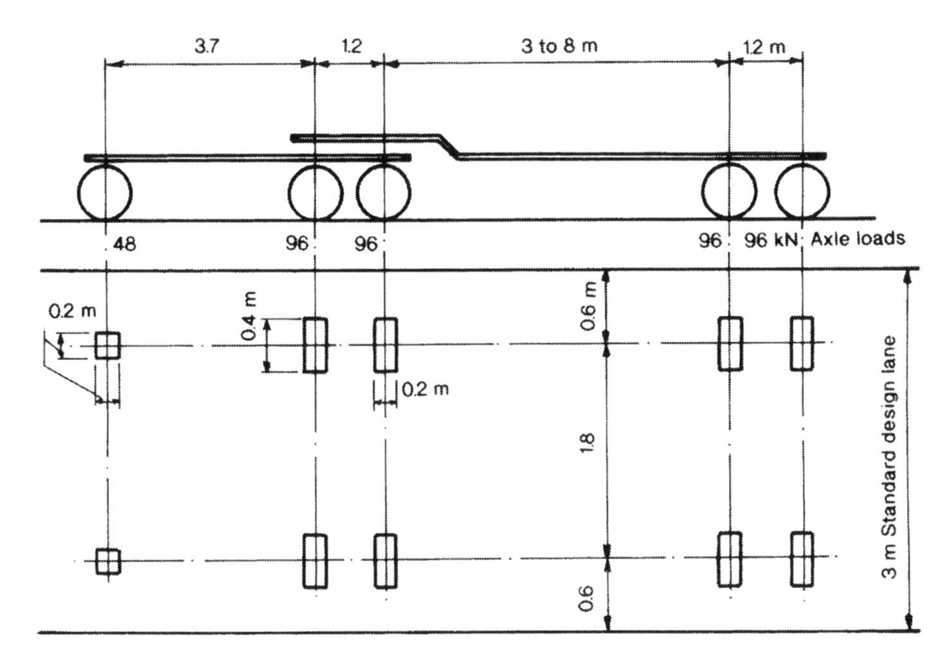

Fig. 4.11 Australian T44 truck

together with a uniformly distributed load of 12.5 kN/m applied to a 3 m wide lane, valid up to a span of 150 m.

In 1989, NAASRA was replaced by Austroads, which in 1992 issued the first Australian limit states design code. The development of design loads for the code had been described earlier by Fenwick (1985). It was decided that a design ultimate load should be one which has a 5% chance of being exceeded during the 100-year design life, where this effectively represents a suitable design load multiplied by its load factor. Figure 4.12 shows some results from load measurements on Six Mile Creek Bridge, a small bridge on a highway just outside Brisbane, chosen for the study partly because there was a nearby compulsory weigh-bridge for heavy vehicles (Pritchard and O'Connor 1984). The road bridge consisted of two separate structures for in-bound and out-bound traffic, with five simply-supported spans, of 11.4, 3 × 13.7 and 13.6 m, and a composite concrete deck on steel joists. The 11.4 m end span, carrying two inbound lanes, was selected for the study. Its end sliding bearings were lifted, cleaned and lubricated, and it was then instrumented with strain gauges to measure the total central bending moment. Calibration of the bridge was carried out under a number of test vehicles and demonstrated an acceptable, linear response. A second calibration, $2\frac{1}{2}$ years later, gave similar results. In the figure, the measured bending moments have been scaled against those produced by a single T44 truck, with an impact factor, I, of 0.30, as specified by the code, and plotted against estimated return periods, calculated as the total length of the record divided by the number of times a load level was exceeded. Over a period of 261 days, the maximum effect was about the same as the calculated effect of two

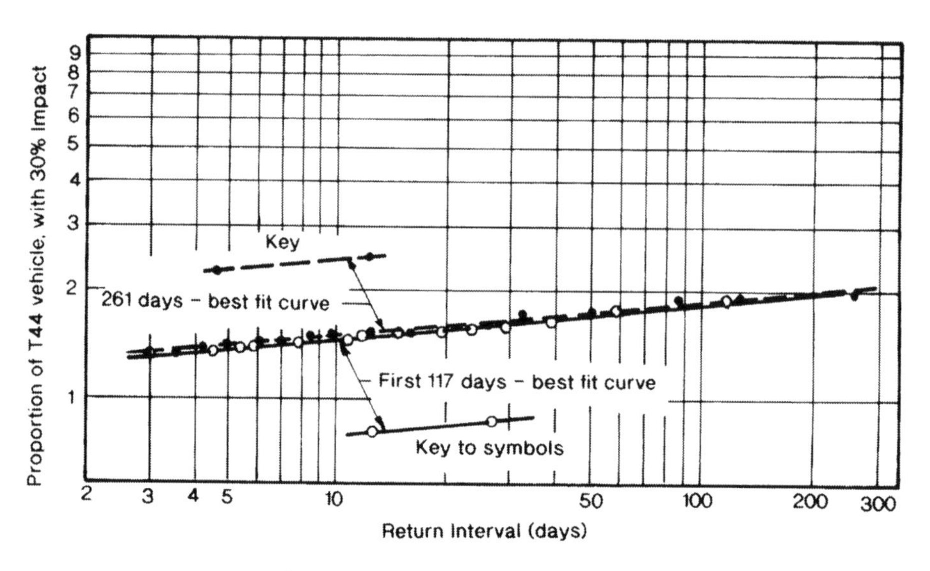

Fig. 4.12 Summary of equivalent truck weights, Six Mile Creek Bridge (Redrawn from Pritchard and O'Connor 1984)

T44 trucks (with impact); that is, about the same as the 1976 design loading for a two-lane bridge. Figure 4.12 uses log scales for both ordinates and the resulting graphs are linear, for an initial period of 117 days, and also for the extended 261-day period. If these graphs are extended to a return period of 2000 years (corresponding to a 5% probability over 100 years), the predicted vertical ordinate would be 4.07 for the 117-day data and 4.22 from 261 days. This is for a 2-lane bridge, and the recommended load factor for the new code was 2, giving a design figure of 4 times a single T44 truck with impact.

It was decided to continue with the T44 loading in the new 1992 Bridge Design Code. The L44 lane loading was also specified, together with a W7 wheel load, consisting of a 70 kN dual-tyre load, applied to an area 500 mm (parallel to the axle) × 200 mm. The dynamic load allowance was the same as in the 1979 OHBD Code (see Section 4.4). The number of traffic lanes was calculated by dividing the carriageway width by 3.1, and rounding down to the next integer. The Standard Design Lanes were then to be positioned laterally so as to produce the most adverse effect. The load modification factors then applied were 1.0 for the first lane, 0.9 for the second, 0.8 for the third, and so on, to 0.55 for 6 or more lanes. Typical load factors for an ultimate limit state were:

self-weight, steel	1.1 if unfavourable, or 0.9
superimposed dead load	2.0 or 0.7
traffic loading	2.00 plus the dynamic load allowance.

This 1992 Austroads code is now under review, and some of the work leading to this is described in Chapter 5. A proposed new M1600 Moving

Traffic Load (Heywood and Ellis 1998) consists of a uniformly distributed load of 6 kN/m together with the vehicle load shown in Fig. 4.13(a). This has four axle groups, with a variable central axle spacing, whose upper bound is undefined. The minimum length between extreme axles is 25 m, much greater than the length of the T44 truck. The total load is also much greater, being 1440 kN compared with the old 388 kN. It is also proposed that there be two additional loadings. The A160 Axle Loading (see Fig. 4.13(b)) is 160 kN shared between two equivalent wheels, of area 400 mm (transverse) × 250 mm, spaced 2 m centre to centre. The S1600 Stationary Traffic Load is intended to model the effect of a stationary queue of traffic. Although its form (see Fig. 4.13(c)) is similar to the M1600 load, the load intensities are different, with a distributed load of 24 kN/m (as compared with 8 kN/m) and an axle group load of 240 kN (compared with 360 kN). The dynamic load allowance is 0.3 for the A160 load, zero for the S1600 load, and from 0.2 to 0.4 for the M1600 load, dependent on the first natural frequency of the superstructure, as in the 1979 OHBD code and the earlier 1992 Austroads code. The number of design lanes is taken as the integral part of the width ÷ 3.1, and the standard lane width is 3.0. The allowances for

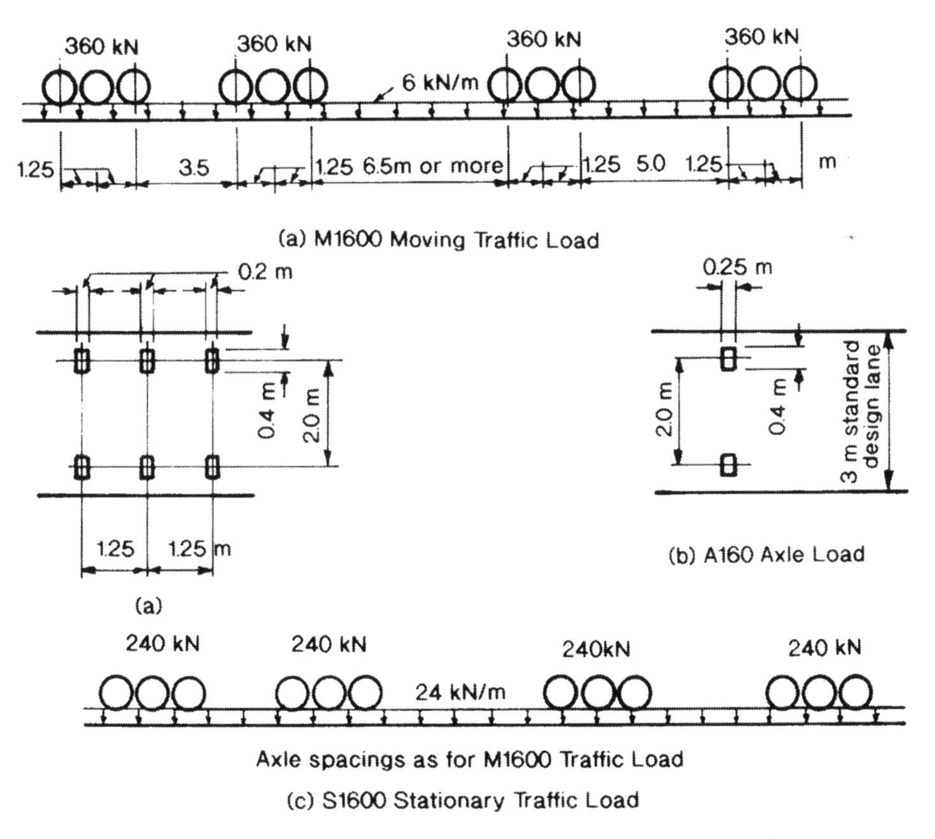

Fig. 4.13 Proposed Australian M1600 and S1600 loadings for highway bridges

multiple lanes are 0.8 for a second lane and 0.4 for three or more lanes, applied to all three loadings. For ultimate limit states, the load factor to be applied to these loadings is 1.8, less than the value of 2.0 in the previous code.

4.8 British Standards for railway bridges

The standard unit loading train specified by the old BS 153: 1923 is shown in Fig. 4.14(a). It consisted of two similar unit locomotives (at the left), each 16.8 m long, with four main driving axles, each 1.0, followed by four subsidiary axles of 0.75. The subsequent unit train of carriages or loaded railway trucks was equivalent to a distributed load of 0.1 per linear foot. A unit pair of axles, each of 1.25, spaced 1.83 m, could be applied instead of this unit train. The recommended number of units was 20, in British tons. It follows that the locomotive axle loads were 199 and 149 kN, the distributed load was 65.4 kN/m and the alternate pair of axle loads were each 249 kN. This could be replaced by the equivalent distributed load shown in Fig. 4.14(b), where it is important to note that this is the total load for a particular span; the load intensity, or load per unit length, decreases slightly with span, being 98 kN/m for a span of 15.2 m, and 74 kN/m at 76.2 m. This loading was for railway gauges not less than 1.44 m.

Dynamic interaction between a bridge and its traffic will be discussed in Chapter 7. It is, nevertheless, interesting to observe the treatment of impact in such an early bridge code (1923), due, no doubt, to a concern with the heavy out-of-balance loads that arose with the early steam locomotives. One of the earliest studies of impact was that by C.E. Inglis in 1934, which was concerned with railway bridges; its title was *A Mathematical Treatise on Vibrations in Railway Bridges*. A discussion of these early provisions for impact may be found in Scott (1931: 22ff). The initial method (1923) was to apply an impact factor,

$$I = 120/[90 + (n + 1)L/2]$$

where

L is the loaded length in feet; and
n the number of tracks supported by the member

A review made in 1928 referred to 'hammer-blow' due to a steam locomotive and provided for three loadings (expressed here in kN):

(A) 20 units of the railway loading, with a total hammer-blow of $1.99N^2$, where N is revolutions per second (or rps);
(B) 16 units plus $4.98N^2$; and
(C) 15 units plus $5.98N^2$.

This additional loading was a total load, distributed between the axles of the locomotive. The hammer-blow was then increased by a 'Dynamic Magnifier'.

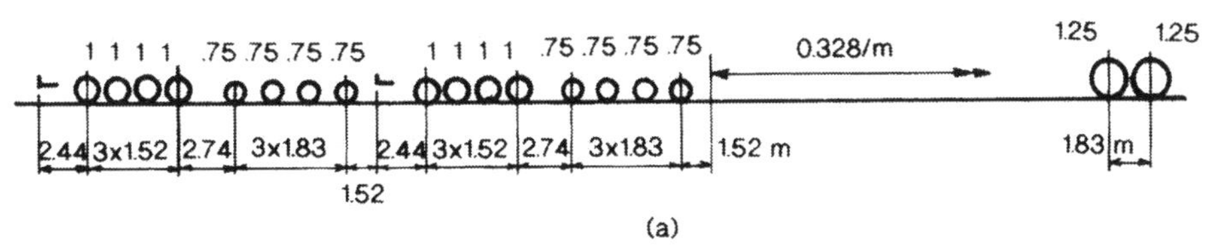

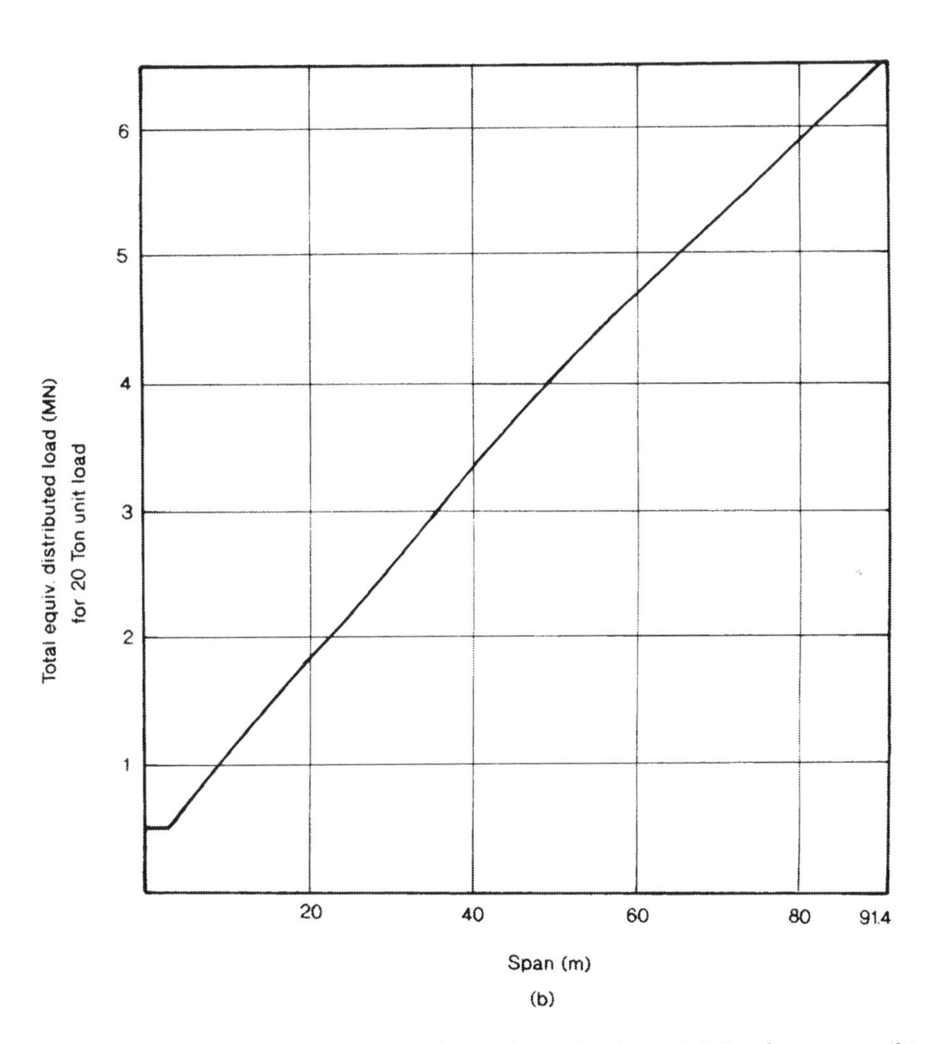

Fig. 4.14 BS 153: 1923 loading for railway bridges: (a) loading train; (b) equivalent distributed load

Scott goes on: 'In the extreme case of synchronism, the allowance for impact is 100% increase of load, but under 100 feet (30.5 m) or over 200 feet (61.0 m) span, 33⅓% increase is generally ample.' The reference to 'synchronism' shows a knowledge of the increase in dynamic stresses that can occur when the frequency of the loading corresponds to the natural frequency of the bridge. He refers to a maximum value of N of 6 rps. In addition to these loadings, there was provision for:

(a) lurching, where the load to either rail should be taken as 0.625 × the axle load;
(b) proportioning the hammer-blow so that the part applied to one rail was 1.2 in Case A, and 0.83 in Cases B and C; and
(c) the rail-joint effect, or impact due to a sudden bump at the junction between two rails.

Scott suggests that rail-joints should be avoided as far as possible on bridges but, to allow for joints and other irregularities an additional uniformly distributed load of $N^2/3$ tons ($10.9N^2$ kN) be applied to each track. It appears that this was a total load, possibly distributed along the length of the train.

A more recent railway loading is that given in BS 5400: 1978 (included in BS 5400: 1988). An appendix states that this

> has been derived by a Committee of the International Union of Railways to cover present and anticipated future loading on railways in Great Britain and on the Continent of Europe. Motive power now tends to be diesel and electric rather than steam, and this produces axle loads and arrangements for locomotives that are similar to those used for bogie freight vehicles, freight vehicles often being heavier than locomotives.

The code specifies an RU loading, to be used for bridges on main line railways with a gauge of 1.4 m and above, and an RL loading, which is a reduced loading for passenger rapid transit systems. The RU loading provides for a range of vehicles, including the locomotives and wagons shown in Fig. 4.15(a), together with a range of exceptional wagons. The locomotive axle loads are 22 or 22.2 t (216 or 218 kN), and there may be two locomotives at the head of each train. As can be seen in the figure, one of the wagons has axle loads of 25 t (245 kN); exceptional wagons may include as many as 28 axles, each of 22 t (216 kN), in two groups, with the spacing in each group as 2.25 m, and a spacing of 10 m between successive groups. For design purposes, these vehicles are replaced by the equivalent RU loading shown in Fig. 4.15(b), which shows also the RL loading.

To allow for dynamic effects, these loads are multiplied by a factor that is equivalent to the $(I + 1)$ used in other codes. Curiously, the values of this factor are based on a length, L (in metres), determined from the influence line for deflection of the element under consideration. As seen in Chapter 1, for an element of a beam, this is identical with the deflected shape of the bridge caused by a unit vertical load at the element. For a simply-supported girder, this influ-

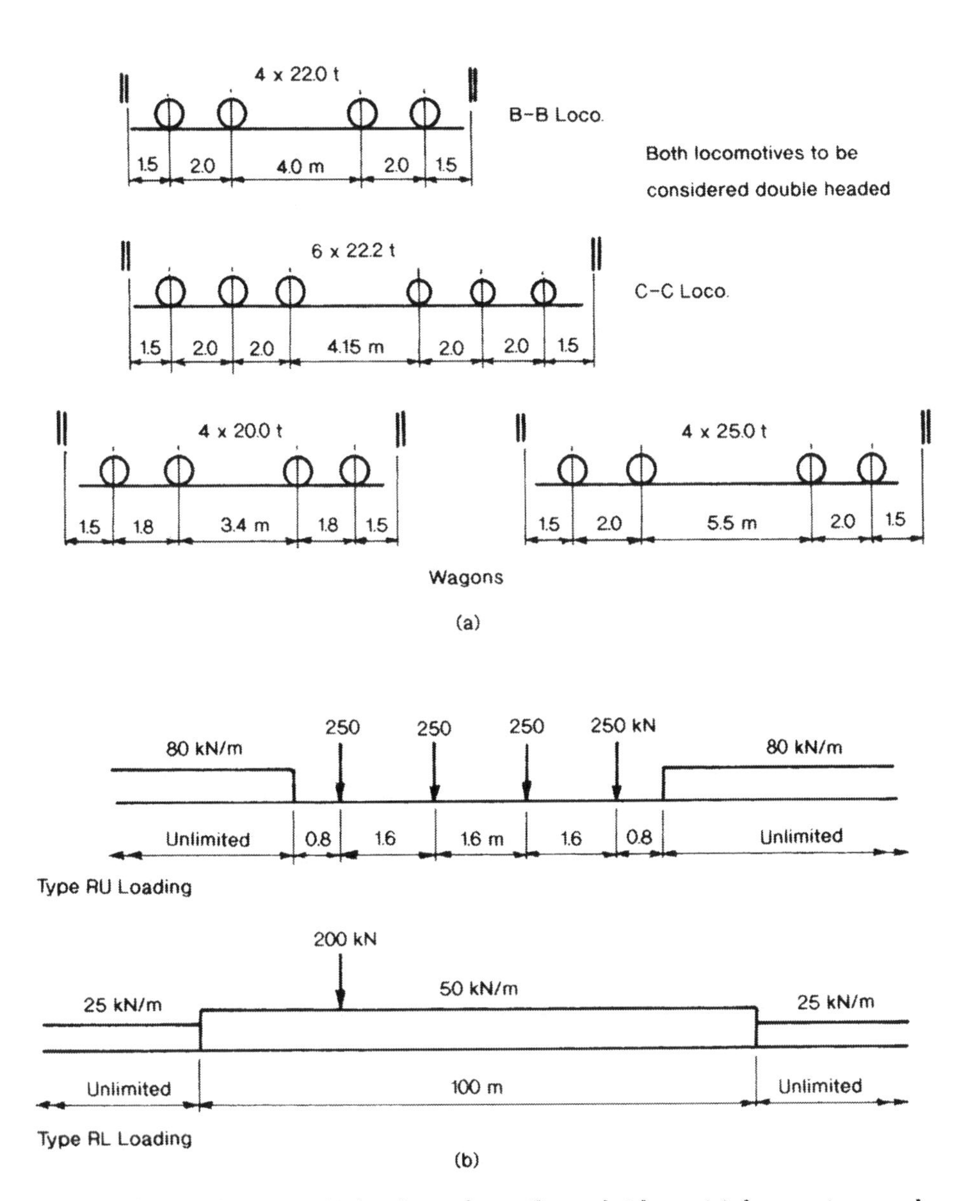

Fig. 4.15 BS 5400: 1978 loadings for railway bridges: (a) locomotives and wagons; (b) equivalent *RU* and *RL* loadings

ence line is symmetrical. For other structures this may not be the case. BS 5400: 1978 defines L as 'the length of the influence line for deflection of the element under consideration'. This is hardly correct, as the length of an influence line is the total distance over which a load may pass, or the length of the deck. It goes on, 'for unsymmetrical influence lines, L is twice the distance between the point at which the greatest ordinate occurs and the nearest end point of the influence line'. An attached table specifies L in more detail, and says that, for the main girders, it should be not less than the length of the largest span. In the case of floor members, 3 m is added to the length. Then, for bending moment (BM) and shear (S), the values of ($I + 1$) are:

(a) $L \leq 3.6$ m, $I + 1 = 2.0(BM)$; or $1.67(S)$;
(b) L between 3.6 and 67 m,

$$I + 1 = 0.73 + 2.16/\sqrt{L - 0.2} \text{ for } BM$$

or

$$I + 1 = 0.82 + 1.44/\sqrt{L - 0.2} \text{ for } S$$

(c) $L > 67$ m, $I + 1 = 1.00$ for BM or S.

There is also provision for lurching – 0.56 of the RL load shall be applied to one rail – and for a horizontal nosing force.

4.9 American Railway Engineering Association

The lack of standardisation in early codes for the design of railway bridges was mentioned earlier, and may be confirmed, for example, by looking at Skinner (1908 Part III), which quotes from 25 separate specifications used in America by various railroad companies. Another early reference of some interest is Waddell's (1908) *De Pontibus: a Pocket Book for Bridge Engineers*, which includes his attempt at a 'Compromise Standard System', and gave axle weights for ten different pairs of locomotives, up to a total mass of 170 US tons (154 tonnes) for each engine and tender, followed by equivalent distributed loads for the trucks or carriages. Waddell replaced these by charts of equivalent uniformly distributed loads to aid the design.

The American Railway Engineering and Maintenance of Way Association was formed in 1903 and issued its first *General Specifications for Steel Railroad Bridges* in 1906. The American Railway Engineering Association (AREA) loading was based on the work of Theodore Cooper (Petroski 1995: 66–121), an American bridge engineer whose career spanned the period 1872 to 1907 and who in 1894 proposed a standard system of railway loadings that still bears his name. The loading specified for American bridges by the AREA 1990 'Specifications for Steel Railway Bridges' (AREA *Manual*: Ch. 15) is the Cooper E80 load, shown in Fig. 4.16(a) in the original units of pounds and feet. There are two locomotives, followed by a train of carriages that have been replaced by a uniformly distributed load. The major axle loads for each locomotive are 4 × 80,000 lb, followed by another four axle loads, each of 52,000 lb, where this equals 80,000 × 0.65. For comparison, the old British Standard Unit Loading, shown in Fig. 4.15(a), had a ratio of 0.75, and Waddell had ratios of 0.72–0.52, reducing as the total weight increased. Various levels of the Cooper load could be specified, such as Cooper E30, 40, 45 or 50 and these retained the same axle spacings and the same ratios

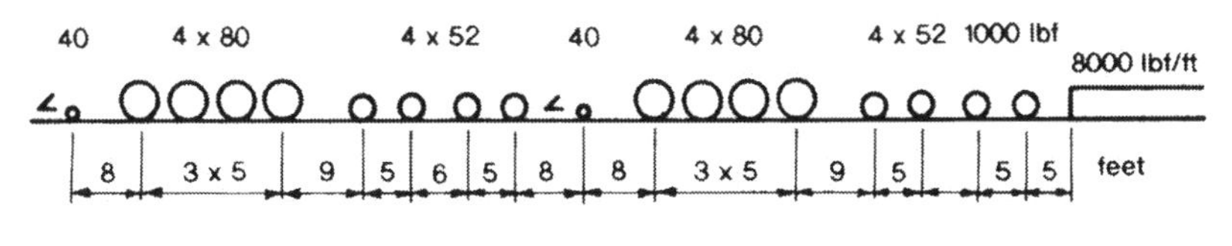

(a) Original Cooper E80 Loading

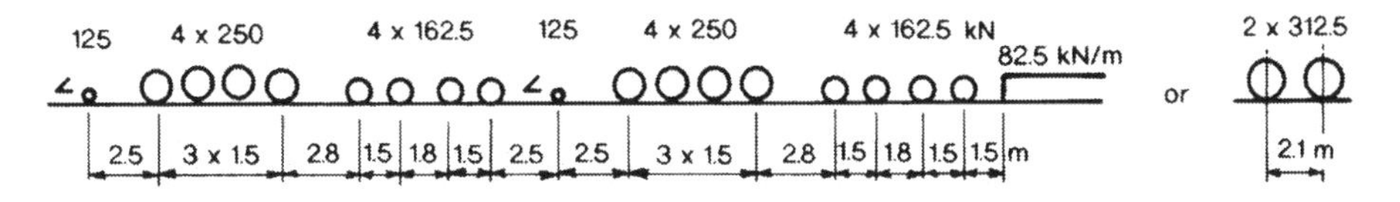

(b) Metric Cooper M250 Loading

Fig. 4.16 (a) American Cooper E80 loading for railway bridges (Imperial units); (b) Metric Cooper M250 loading (Australian and New Zealand Railway Conferences 1974)

between loads. For comparison, Fig. 4.16(b) shows the Metric Cooper M250 load specified by the 1974 *Railway Bridge Design Manual* of the Australian and New Zealand Railway Conferences, equivalent to about E56.

The AREA 1990 Specifications provided for an equal load on a second track, then one half and one quarter for succeeding tracks. The provisions for dynamic effects (called Impact) allowed (surprisingly) for '1. Rolling equipment without hammer blow (diesels, electric locomotives)', and '2. Steam locomotives with hammer blow'. In the first case, with I in per cent:

for $L < 80$ feet (24.4 m) $\quad I = 100/S + 40 - 3L^2/1600$
for $L \geq 80$ feet $\quad I = 100/S + 16 + 600/(L - 30)$

where

S is the spacing of longitudinal beams (feet), and
L is the span (feet)

this is equivalent to:

$$I = 30.5/S + 40 - 0.0202L^2 \quad (L < 24.4 \text{ m})$$

or

$$30.5/S + 16 - 183/(L - 9.1) \quad (L \geq 24.4)$$

where

S and L are in metres.

These 1990 specifications were still based on allowable stress design with, for example, an allowable stress in steel of 138 MPa. The same comment applies to the Australian and New Zealand Manual of 1974, but with this allowable stress rounded up to 140 MPa.

4.10 Eurocode for railway bridges

Eurocode I (ENV 1991-3), issued as a pre-standard in 1995, includes the railway bridge loads shown in Fig. 4.17. Load Model 71 represents the static effect of normal rail traffic, and consists of four 250 kN axle loads, preceded and followed by a distributed load of 80 kN/m. These loads may be modified for particular lines by multiplying by a factor in the range 0.75 to 1.33, but 1.0 represents 'normal rail traffic'. There are, in addition, the Load Models SW/0 and SW/2 that represent the static effect of abnormal rail traffic, but these apply only to designated lines.

The full normal load model 71 is applied concurrently to two tracks, with zero traffic load on additional tracks. A dynamic amplification factor $(1 + I)$ is specified as

$$\frac{1.44}{\sqrt{L} - 0.2} + 0.82$$

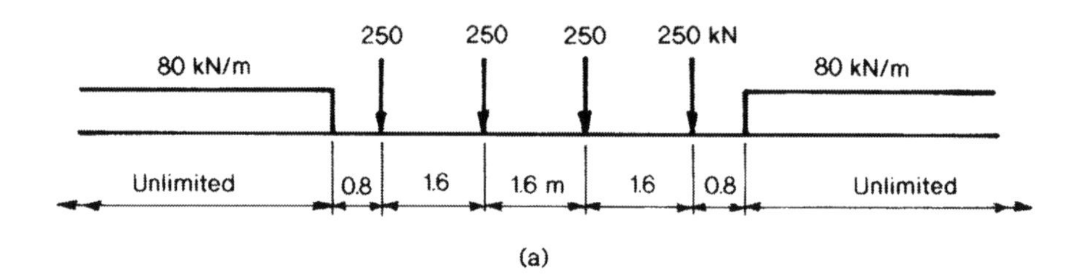

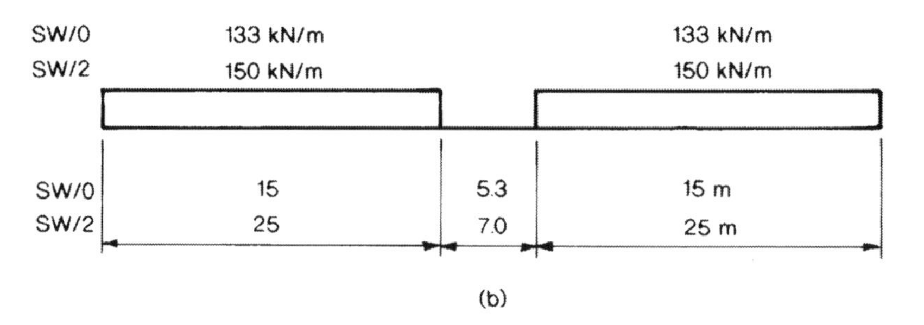

Fig. 4.17 Eurocode I (ENV 1991–33) railway bridge loads: (a) Load Model 71; (b) SW/0 and SW/2 abnormal loads

but between 1.0 and 1.67, for carefully maintained track;

$$\frac{2.16}{\sqrt{L} - 0.2} + 0.73$$

but between 1.0 and 2.0, for tracks with standard maintenance.

The length $L(m)$ is specified for particular elements. For simply-supported main girders, L is the span length. For continuous girders, the average span is multiplied by a factor 1.2 for 2 spans, 1.3 for 3, 1.4 for 4, and 1.5 for 5 or more. Lower values of $I + 1$ may be used for some types of bridges, such as arch bridges with a large cover over the arch. Higher values may be used in some cases, such as if the traffic speed exceeds 220 km/h. It should be noted that the square root sign in the denominator covers only L. The corresponding expression in BS 5400: 1978 includes the -0.2 within the square root sign. Typical load factors for an ultimate limit state are 1.35 or 1.0 for permanent loads (as for road bridges), and 1.45 or zero for the traffic loads, as compared with 1.35 and zero for road bridges, the figures quoted being for unfavourable and favourable actions.

The Swiss Code SIA160-1989 was mentioned in Section 4.5 for its rules for highway bridges. For rail bridges, it has separate clauses for standard and narrow gauge lines. For standard gauge, it gives two loadings: Load Model 1 is the same as the Eurocode Model 71; Load Model 2 is used only for designated lines carrying heavy vehicles, and consists of a uniformly distributed load of 150 kN/m, 60 m long. This second loading has the same intensity as the Eurocode Model SW2, and differs only in the length of track to which it is applied, namely, an uninterrupted length of 60 m, compared with two 25 m lengths, separated by a gap of 7 m, in the Eurocode.

4.11 Australian code for railway bridges

The Australian and New Zealand railway design manual of 1974 used the Metric Cooper M250 load, as mentioned in Section 4.9, and illustrated in Fig. 4.16(b). In 1996, a Railway Supplement was produced by the Australasian Railway Association in association with Austroads and Standards Australia, and added to the 1992 Australian Bridge Design Code.

The specified 300-A-12 Railway Traffic Loading consists of an infinite set of four axle groups, at a central spacing of 12 m, as shown in Fig. 4.18. The load to each of the four axles is 300 kN, and they are spaced 1.7, 1.1 and 1.7 m apart. Alternatively, a single axle load of 360 kN may be applied. Two adjacent tracks may be fully loaded, with modification factors of 0.85, 0.70 and 0.60 applied for further additional tracks. The dynamic amplification factor for bending moments is the same as in the Eurocode (see Section 4.10). For ultimate limit states, the load factor applied to the traffic load, including its dynamic effect, is 1.6. For ballast and track weights, the factor is 1.7.

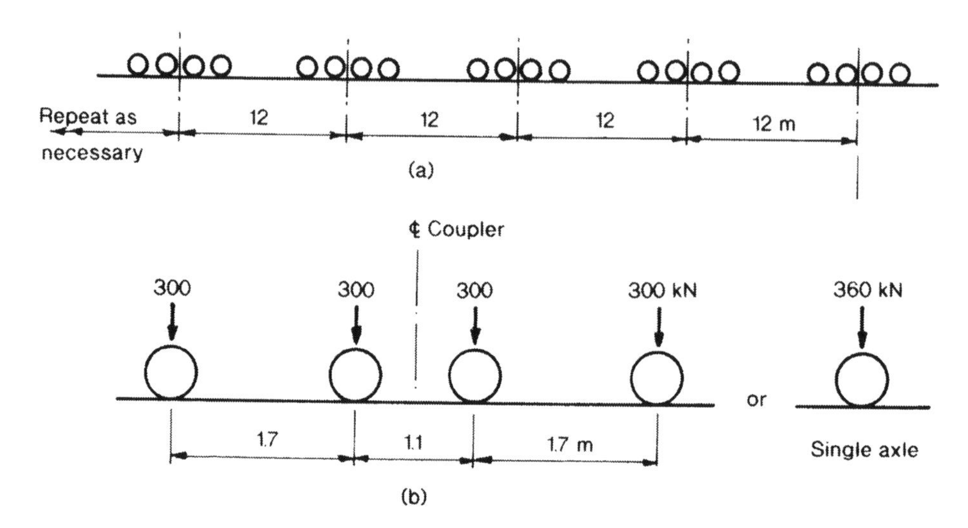

Fig. 4.18 Australian (Austroads 1996) 300-A-12 Railway Traffic Loading

This 1996 Railway Supplement was based on reports by Selby Smith and Bowmaker (August and October 1995; see also Selby Smith 1996); Appendix C of this report was concerned with the selection of load factors and utilised a special study by G.H. Brameld (July 1995). It is interesting to note some of the comments made in these studies, based partly on data obtained from some heavy haul rail lines, particularly those used in Queensland to carry coal and other minerals from major mines to the ports. To quote (Selby Smith and Bowmaker, October 1995: C1),

> Railway loading may continue at its maximum extent virtually indefinitely, with long trains of wagons.... Furthermore modern locomotives have loading characteristics which are very similar to those of the wagons they are pulling, so that the whole train may consist of loads equal to the design loading of a given bridge.

Brameld studied actual loads during a period of one month, in December 1994, for trains carrying coal from three Queensland mines. Most wagons had a nominal 90 tonne gross mass, but a few were nominally of 100 tonnes. Each wagon had two bogies, each with two axles. A train consisted of 109 wagons and 2 locomotives. During the month, 330,000 bogie loads were recorded, corresponding to about 480 wagons per day. They were found to be normally distributed, with a mean of 450 kN and a coefficient of variation of only 2%. By comparison, the mean mass of containers carried to the Port of Melbourne over a period of 12 months was only 16 tonnes (157 kN). Loads on general traffic lines were variable but also comparatively small. It was decided to use data from the unit coal trains in statistical safety analyses, and the final outcome was that a load factor on live load of 1.4 would have been sufficient. It was decided, nevertheless, to retain a factor of 1.6.

The work by Selby Smith and Bowmaker listed dimensions and standard loadings for nine locomotives and twelve wagons from four Australian rail authorities with track gauges of 1.44 and 1.07 m, and these confirm that there is little variation between locomotive and wagon axle loads, and between gauges. The heavy Queensland coal trains described earlier have a gauge of 1.07 m. These loadings were all compared with the new 300-A-12 loading using the equivalent base length concept developed by Csagoly and Dorton (1973) for the Ontario OHBD code.

4.12 South African Bridge Code

The 1983 *Bridge Code* of the South African Transport Services is included as a final example. It specified an *NR* rail loading that is similar in form to the BS 5400: 1978 *RU* loading shown in Fig. 4.15(a), but with larger loads and slightly changed dimensions. It has four 280 kN axle loads (in place of the BS 5400 250 kN loads) spaced 1.8 m (instead of 1.6 m). Before and after this loading is a uniformly distributed load of 100 kN/m (80 kN/m), separated from the extreme axles by a distance of 0.9 m (instead of 0.8 m). Dynamic amplification factors are the same as in the Eurocode. This is a limit states design code. It refers also to road bridges but states that 'road bridges must be designed in accordance with the requirements of the road authority of the province in which the bridge is situated'.

4.13 Review

It is the purpose here to review some of the highway bridge design loads described in Sections 4.2–7; railway loadings will be discussed further in Chapter 8. Figures 4.19–21 plot absolute maximum bending moment against span for simply supported longitudinal beams carrying a lane of traffic.

Figure 4.19, for Britain, shows the successive loadings: MOT: 1931, BS 153: 1958, BS 5400: 1978, and BD 37/88, together with the proposed Eurocode ENV 1991-3 of 1995. The main graph plots bending moments for spans from 10 to 150 m, while the inset graph shows spans from 0 to 15 m, at an increased scale.

Considering the larger spans first, when these graphs are compared with those for other countries (shown later) it is evident that the MOT: 1931 loading was a relatively heavy load, and many bridges were designed for its prescriptions. For much of the range of spans, BS 153: 1958 is much lighter, with the BS 5400: 1978 HA loading about the same; but the later BD37/88 has brought loads approximately back to the MOT: 1931 level. Two curves are shown for Eurocode 1: the upper curve, which lies above the other curves in the major plot, is for the first loaded lane, while the second, which is much lower, is for a second loaded lane. Most bridges have at least two lanes and, depending on the form of the structure, it is the mean of the two curves that may be significant.

The inset graphs (0–15 m) are somewhat more confused. In the range 6–15 m, the lower curves are for MOT: 1931, for BS 153: 1958, which is

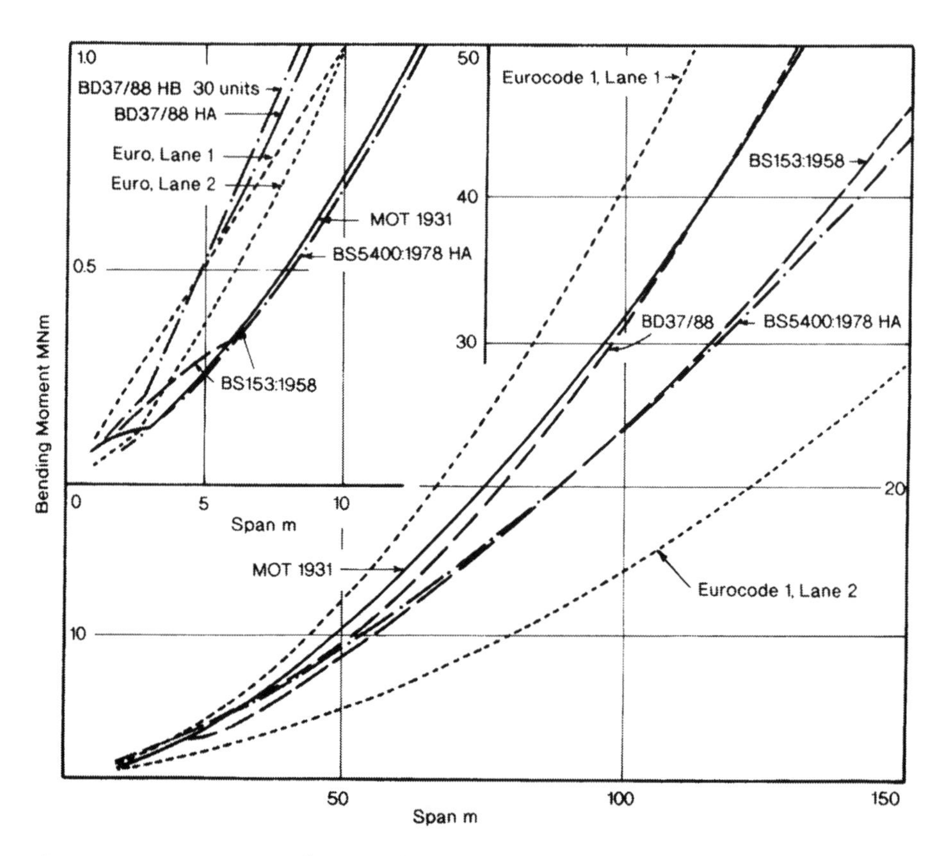

Fig. 4.19 Comparison of British highway bridge loadings (maximum bending moments in simply-supported spans)

coincident with MOT: 1931, and for the BS 5400: 1978 HA loading. The MOT: 1931 curve extends down to 3 m, with a continuation of different shape for the lowest spans. In the case of BS 153: 1958 a different and higher curve branches off at 6 m. BS 5400: 1978 also specified 25 units of the abnormal HB loading for public highways. This is not shown, but has bending moment ordinates that are in the proportion 25:30 (or ×0.83) of those shown for the BD37/88 30-unit loading. The BD37/88 HA loading and the Eurocode Load Models 1 and 2 are also shown.

Figure 4.20 compares loads used in the United States and Canada. In the main graph, the two lower curves are for the 1944 and 1994 AASHO and AASHTO HS20 loadings, with the transition from truck to lane loading at about 35–45 m, while the upper three are for OHBD 1979 and 1991 trucks, and the CSA-S6 Canadian code. The major axles of the AASHTO truck may be viewed either as single or double loads, and it is not clear where the change from one to the other should take place: the inset shows both, and also the tandem loading. In the inset, the CSA-S6 loading is also relatively low; the upper curves are for the OHBD loading, with the 1994 AASHTO loading in between.

Figure 4.21 is for Australia. The lower curves in both the major graph and

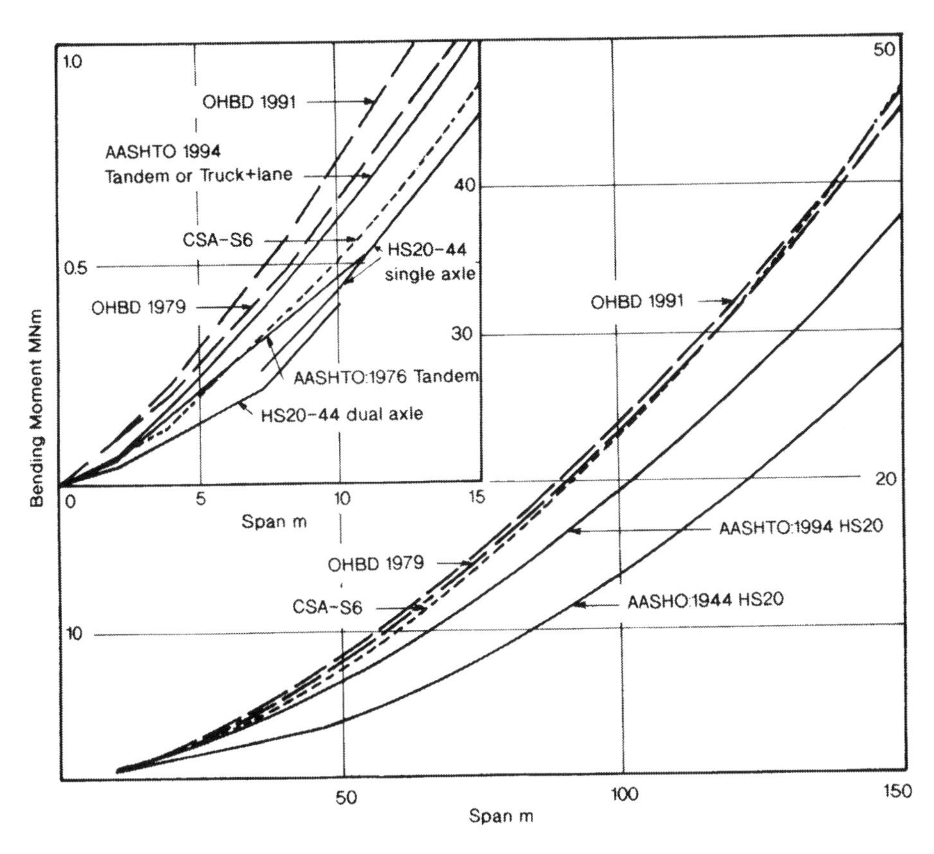

Fig. 4.20 Comparison of American and Canadian highway bridge loadings

the inset are for the AASHO HS20-44 loading, for this was used in Australia. Above it lies the similar but increased T44 loading, and then the proposed new M1600 and S1600 loadings. These are heavy loads, even compared with the first lane of Eurocode 1 and, indeed, the increase in loading, from HS20-44 to S1600 is greater than in other countries.

These graphs should be used with care, for the dynamic amplification factor and the factors for multiple lanes vary and may have a significant effect. This book is concerned with loads, and this has influenced the decision to plot loads rather than total effects.

These figures plot only one structural parameter – the absolute maximum bending moment in a simply-supported beam. Others also are important and may affect any comparisons that may be made between loads. Examples are shear, and negative bending moments above the supports of continuous beams. Nevertheless, this study suggests a number of conclusions.

1. The 1931 MOT loading represents a good example of the use of a simplified equivalent loading – here a uniformly distributed load plus a knife-edge load. There are dangers in the use of such a load: its basis is not

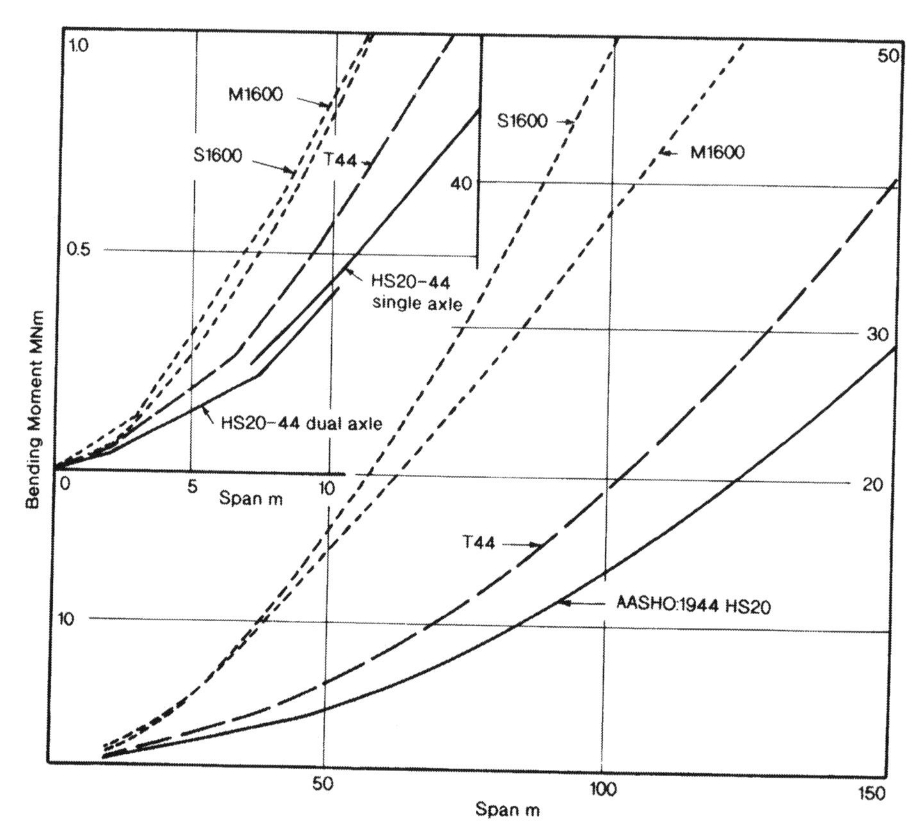

Fig. 4.21 Comparison of Australian highway bridge loadings

immediately apparent, and it may not be adequate in modelling all effects, particularly at a local level. It is, however, convenient to use.

2. The AASHO truck illustrates another basic approach, for it is apparent that it seeks to model a particular vehicle, and this may be useful when the structural form is unusual, or where local effects need to be calculated. However, although it is relatively simple for use in manual calculations, the variable back axle spacing may cause additional computations in computer use – one may need, in effect, to traverse a large number of vehicles across the bridge, each with a particular spacing, where these spacings spread across the specified range.
3. There is difficulty in modelling loads on adjacent lanes that have a reasonable joint probability of occurrence. Many codes provide for equal loads on a second lane, but Eurocode 1 is an example of one that does not. Its provisions were summarised in Section 4.6 and plotted in Fig. 4.19.
4. A further point arises from item 3. To take the Eurocode as an example, its Load Model 1 consists of a distributed load and a dual-axle group, where only one of these dual-axle groups may be applied in each lane. Certain load effects in continuous structures may require the positioning of loads on suc-

cessive spans; for example, the AASHO and AASHTO Lane Loadings allowed the use of two concentrated loads, one in each span, to cater for this effect. In using Eurocode 1, one could separate the axle groups in the two lanes, with one group on one span and the second group on another. These groups correspond to the major axle groups of separate vehicles and it is difficult to see that the joint probability of occurrence is changed by the placement of these vehicles in separate lanes. In some structures the issue may not be important, for there may be an adequate transverse distribution of these loads. In other cases, the transverse location of loads in adjoining spans may change the magnitude of primary structural actions.

5. It is evident that there has been a trend towards the increase in design loads over the years in most countries, the only notable exception being in Britain, where the 1931 MOT loading was a heavy load for its time.
6. The examples chosen here represent a great variety in the form of the design loads. There is now, however, a tendency for the use of a constant, uniformly distributed load, to model a sequence of minor vehicles, together with one or two specified major axle groups. It is important to realise that these local axle groups perform two functions:
   (a) they apply a major local load which, in conjunction with the distributed load, provides a reasonable estimate of a variety of primary load effects, and of their variation with span; and
   (b) they should correctly model local effects.

   It is not clear that the second of these objectives is always properly fulfilled. Figure 4.22 shows a single-cell box, with a diaphragm that is integral with all walls of the box. It is necessary to provide suitable longitudinal reinforcement on the top slab across the diaphragm, and the worst case may be a pair of axle groups at some spacing, ℓ. A single dual axle, as in the Eurocode Load Model 1, or a single axle as in its Load Model 2, could not provide a suitable loading.

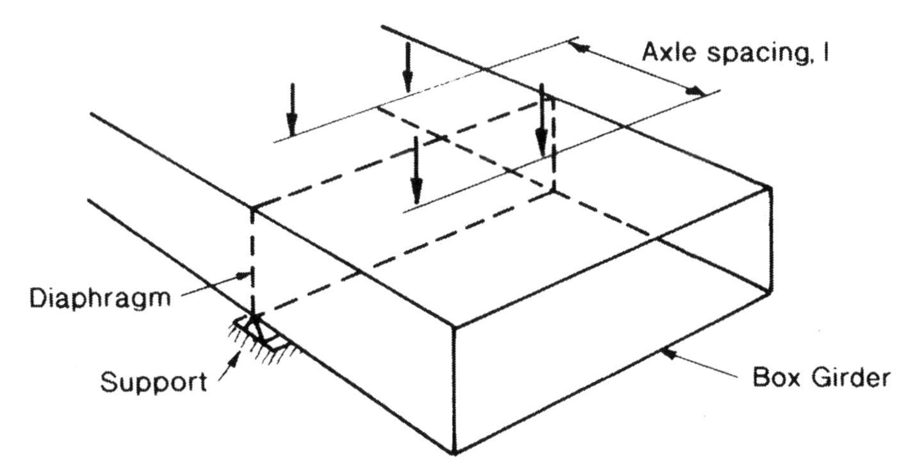

Fig. 4.22 Effects of a diaphragm on the bending moments in the deck slab of a box girder bridge

Chapter 5

Selection of primary loads and load factors

5.1 Probability models

Consider a deck of cards numbered consecutively from 1 to 100. Random selection of a card – or the use of a computer-based random number generator in this range – gives an equal probability that any number be selected, whether it be 100 or 1 or any other number, and this probability is 1 in 100 or 1%. Suppose instead that the range, 1 to 100, has units of MPa, and represents the possible 28-day compression strengths of concrete samples made to the same specification, with a required characteristic compressive strength of 50 MPa. Then these strengths will vary with a mean that will be, typically, somewhat larger than 50 MPa, possibly, say, 60 MPa; it is not difficult to see why the mean is somewhat higher, for the manufacturer will aim at a higher strength in an attempt to ensure that a sufficient number of the test results (possibly all) lie above the specified value. The distribution is said to be biased, where in this case this bias may be measured by the ratio of mean to characteristic strength. In limit states design, it is necessary to know the probabilistic distribution of these strengths, and of other resistance and load parameters.

Section 3.3 introduced the idea of probability density distributions and cumulative probability, and defined the Normal Distribution, found to describe the variability of certain functions. It also gave the mathematical description of the Lognormal Distribution shown in Fig. 3.4. This was asymmetrical, with a long upper tail giving it a positive bias. Other continuous distributions include the Gamma, Beta and Weibull distributions (Devore 1995: 167ff; Fraser 1976: 60ff), and there are many more (Sections 11.2, 12.1). It is often helpful to identify the probability distribution that best fits the available data, for this may then allow extrapolation from the data to the prediction of extreme and rare values.

One way of doing this is to use probability paper, on which cumulative distributions of the data are plotted to a non-linear vertical scale, chosen so that a specified distribution plots as a straight line. The most common example is normal probability paper; its construction may be visualised as shown in Fig. 5.1. The left-hand graph shows the normal cumulative distribution function (as in Fig. 3.1). Its ordinates are found by integrating the standard normal distribution curve. Values of these integrals (or areas) are commonly called $\phi(x)$, where values of the function ϕ are listed in many standard works (such as Devore 1995: inside front cover; Pearson and Hartley 1966, Vol.1). A cumulative probability of 0.6, for example, corresponds to a function value that is $0.25 \times$ the standard deviation above the mean. Choose any straight line AF, and carry out the projection shown to give the new vertical location of the cumulative probability of 0.6 on the vertical scale of Fig. 5.1(b). Repeat for probabilities 0.7, 0.8, 0.9 etc. Then if a normal distribution is plotted *ab initio* in Fig. 5.1(b), using its revised vertical scale, it will plot as a straight line. The consequence of this is that if the raw data is plotted on Fig. 5.1(b), and lies on a straight line, this shows that it is normally distributed. A best-fit straight line may be placed through the plotted points and extrapolated to give, for example, the value of the function that has a cumulative probability of 0.999. This would have a probability of occurrence of 0.1% and would, in fact, lie 3 standard deviations above the mean. The process is discussed by Benjamin and Cornell (1970: 453ff), who show also how this normal probability plot can be used to diagnose distributions that are non-normal.

Points 1.0, 2.0, 3.0 etc on the horizontal scale of Fig. 5.1(b) represent the

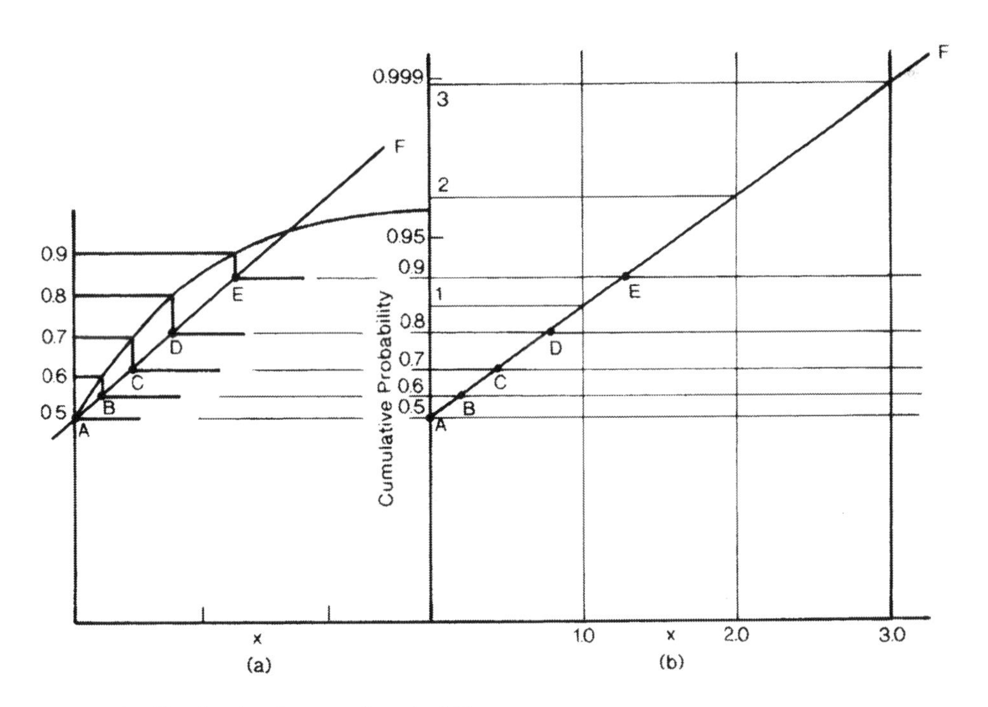

Fig. 5.1 The basis of normal probability paper

number of coefficients of variation above the mean. Projecting these points upward to line AF shows that the corresponding cumulative probabilities are equally spaced on the new, non-linear vertical scale. For this reason, the vertical ordinates on normal probability paper are commonly marked: 0 at point A, with ordinates 1, 2, 3 etc above A, and −1, −2, −3 etc below A. Instead of scaling the vertical ordinate as cumulative probability, which it does indeed represent, it is scaled as the number of coefficients of variation above the mean that would yield a particular cumulative probability, if applied to the standard normal distribution.

An example may be given, chosen as the plotting of a set of field observations of load, L.

1. Calculate the mean, standard deviation and coefficient of variation of the observed values.
2. Select a convenient load level, L. Count the number of field observations with a value equal to or less than L, and divide by the total number of observations to give the cumulative frequency, e.g. say 0.85.
3. Use a table of cumulative probabilities to obtain y, this being the number of coefficients of variation above the mean that would be required for a Standard Normal Distribution to have this cumulative probability; for example, a cumulative probability of 0.85 gives a value (y) equal to +1.036.
4. Plot (L, y).

Step 3 requires the inverse use of this table. As defined in Section 3.3, if p is the probability of occurrence of the value, x, then the normal cumulative probability, P_{ND} (equal to $\int_{-\infty}^{x} pdx$) is called $\phi(x)$. The inverse process may be defined as ϕ^{-1}, where

$$x = \phi^{-1}(P_{ND})$$

This process can be programmed for computer use. The equation to the probability density function of the Standard Normal Distribution, as stated in Section 3.3, is

$$p = \frac{1}{2\pi} e^{-x^2/2}$$

The integral of this function from $-\infty$ to x is the cumulative probability for the occurrence of values less than or equal to x. This can be evaluated numerically by observing that the integral of p to the mean value of x is 0.5. The cumulative probability given by the Standard Normal Distribution is

$$\begin{aligned} P_{ND} &= 0.5 + \int_{0}^{x} pdx \\ &= 0.5 + I \end{aligned}$$

where

I can be evaluated by numerical integration

The probability density function is symmetrical about the mean, at $x = 0$. For negative values of x, evaluate I for the absolute value of x, and subtract this from 0.5. Then $x = \phi^{-1}(P_{ND})$ may be found by interpolation or iteration.

The straight line graph on normal probability paper should pass through the points (0, 0.5) and (1.0, 0.8413), for the cumulative probability should be 0.5 at the mean, and 0.8413 at one standard deviation above the mean. Small departures from these points indicate that the values of the mean and standard deviation should be revised, with the new values obtained from the values of x at cumulative probabilities of 0.5 and 0.8413. Studies carried out for bridge codes (Nowak 1993: 19, and elsewhere) suggest that loads may be normally distributed and that resistances may be modelled as lognormal.

Loads are often referred to by their *recurrence interval*. It is important that this be properly defined. Current practice with bridge loads is based on earlier practice in hydrology for the description of flood frequencies or probabilities. The following definition is taken from Wilson 1983: 211 (see also Linsley et al. 1988: 345).

> The term *recurrence interval* (also called the *return period*), . . . is the time that, on average, elapses between two events that equal or exceed a particular level. Putting it another way, the N-year event, the event that is expected to be equalled or exceeded, on average, every N years, has a recurrence interval . . . of N years.

The point to note in this definition is 'equalled or exceeded'; where the cumulative probability being referred to is 'of exceedence', and is 1 in N-years. Nowak (1994: 40), for example, speaks of a 75-year truck, in a sample population that corresponded to 15×10^6 trucks in this period. The relevant probability of exceedence is $1/(15 \times 10^6)$. The corresponding cumulative probability of events less than or equal to this event is $1 - 1/(15 \times 10^6)$; or 0.999 999 933. Pearson and Hartley (1966) list the corresponding 'X value' as 5.27; Nowak's 'inverse normal value' is 5.26 – effectively the same. Similarly, for a 1-year truck, the cumulative frequency is $1 - 1/200{,}000$, or 0.999 995, with an inverse normal value of 4.42; and for a 1-day truck, 2.91 (Nowak rounds the number of trucks per day to 1000, giving an inverse normal value of 3.09). His normal probability plots are, therefore, as in Fig. 5.2, with the horizontal ordinates scaled against a previous design truck. He calls the vertical scale the 'inverse standard normal distribution function'. The line shown (not one of his cases) has a mean of 0.5 and a coefficient of variation of about 10% (represented by an inverse slope of 0.5/5.0). It indicates the following ratios of particular events to the mean: 1 day, 1.61; 1 year, 1.70; 75 years, 2.05.

Alternatively, the field data may be interpreted to list the maximum event in each of a succession of days; that is, the maximum daily event. Suppose that these maximum daily events are shown to be normal. Then the 100-year event

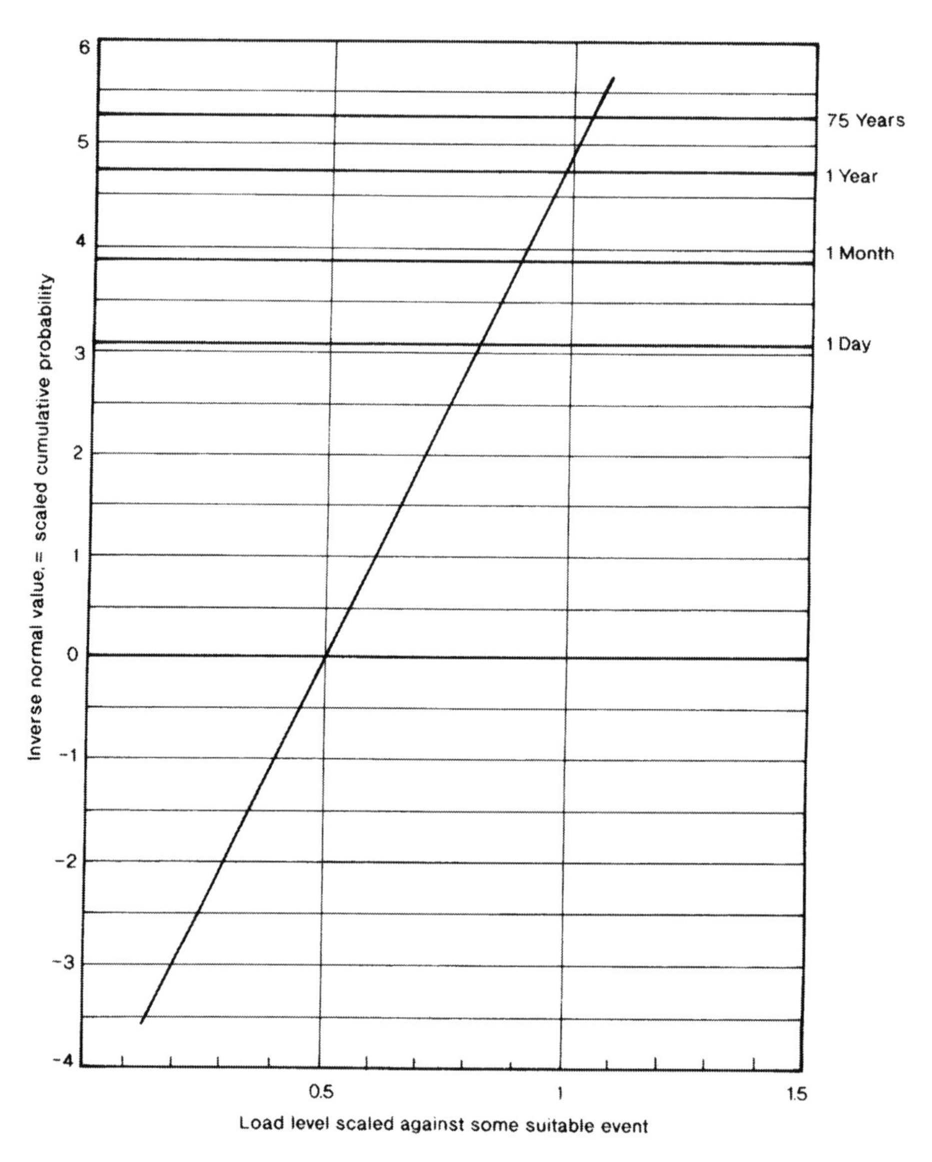

Fig. 5.2 Typical plot of field data, for loads, on normal probability paper

(to take another example) has a cumulative probability of exceedence of 1/36,500, leading to a value of the 'inverse standard normal distribution function' of 4.03.

Consider a daily maximum population with a mean of zero and a coefficient of variation of 1. Then the procedure described above gives the following values for particular recurrence intervals:

1 day	1.0
1 year	2.78
100 years	4.03

It is useful also to approach this problem in another way (Heywood 1992a: 4.12; Ransom 2000).

Let X be the extreme daily event.
Let the cumulative probability that X is less than some chosen value, x, be

$$F(x) = \int_{-\infty}^{x} p(x)dx$$

Consider 2 days. Then the probability that X is less than x on each of these 2 days is

$$F(x) \times F(x) = [F(x)]^2$$

where this is a cumulative probability

Similarly, the cumulative probability for n days is $[F(x)]^n$.

Suppose the extreme daily event has a mean of zero, and a coefficient of variation of 1.0. Then from the tables in Pearson and Hartley (1966), $F(x)$ may be calculated for any x, and then raised to some power n. Further, this cumulative probability for n days may be differentiated (here by 3-point numerical differentiation) to give the probability density curve for this n-day event. The resulting curves shown in Fig. 5.3 are for the original 1-day family of events (the Standard Normal Distribution) and for the 1-year and 100-year cases.

Consider first the 1-year case.

(a) The resulting probability density distribution has a maximum of 1.135 at $x = 2.81$.
(b) The curve is not symmetrical and it can be shown to be non-normal by plotting it on normal distribution paper.
(c) Nevertheless, it is not far from normality, and estimates of its coefficient of variation have been calculated in two ways.

The curve must contain an area of 1.0. The Standard Normal Distribution has a maximum of 0.399 and a coefficient of variation of 1.0. The 1-year curve has this maximum vertical ordinate increased by a factor of 1.135/0.399. Accordingly, its horizontal scale must be reduced by the inverse of this value, giving an estimated coefficient of variation of 0.351.

The initial curve has a semi-horizontal ordinate of 1.0 at a height of 0.2420; or at 0.6065 × the maximum ordinate. Scaling the 1-year curve at a height of 0.607 × 1.135 gives a semi-horizontal ordinate of 0.34, and this is a second approximation to the 1-year coefficient of variation.

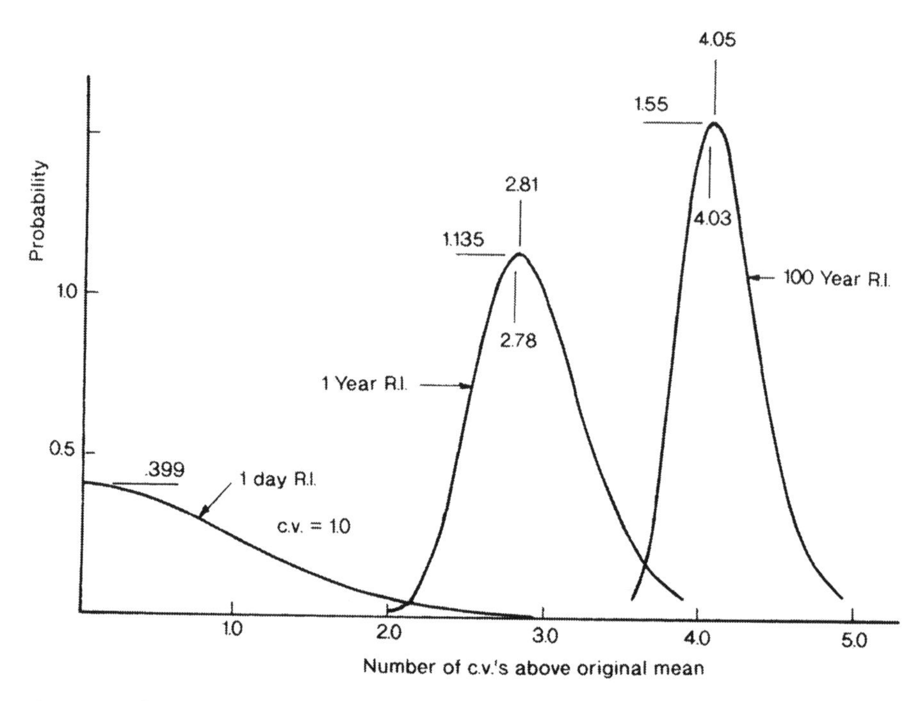

Fig. 5.3 Theoretical probability density distributions for 1-day, 1-year and 100-year events

(d) Viewed overall, therefore, the 1-year curve has a maximum probability at a value of x that is 2.81 × the original coefficient of variation above the original mean. The other analysis suggested 2.78 for the 1-year event.

The family of 1-year events has a coefficient of variation of about 0.34.

(e) Similar results can be obtained for the 100-year curve.

The maximum probability is 4.05 × c.v. above the original mean (as compared with the figure of 4.03 quoted earlier).

Its coefficient of variation, estimated as for the 1-year case, is either 0.257 or 0.24.

It is worth repeating the last result, for it is somewhat surprising. If one takes a series of 100-year samples, and for each case considers only the maximum event, then this has a coefficient of variation of the order of 0.24 – 0.257 × that of the original sample. The 100-year event still has a considerable variability.

Nowak (1993) mentioned that resistances may be modelled as lognormal (see Section 3.3 and Fig. 3.4). A function, x, is lognormal if $\ell n(x)$ has a normal distribution. It follows that, if the frequency density distribution of x is plotted, with x on a logarithmic scale, the result will be identical to a normal distribution. Similarly, if the cumulative probabilities of occurrences less than or equal to x are plotted against $\ell n(x)$ (to some suitable scale) on normal distribution paper, the points will fall on a straight line.

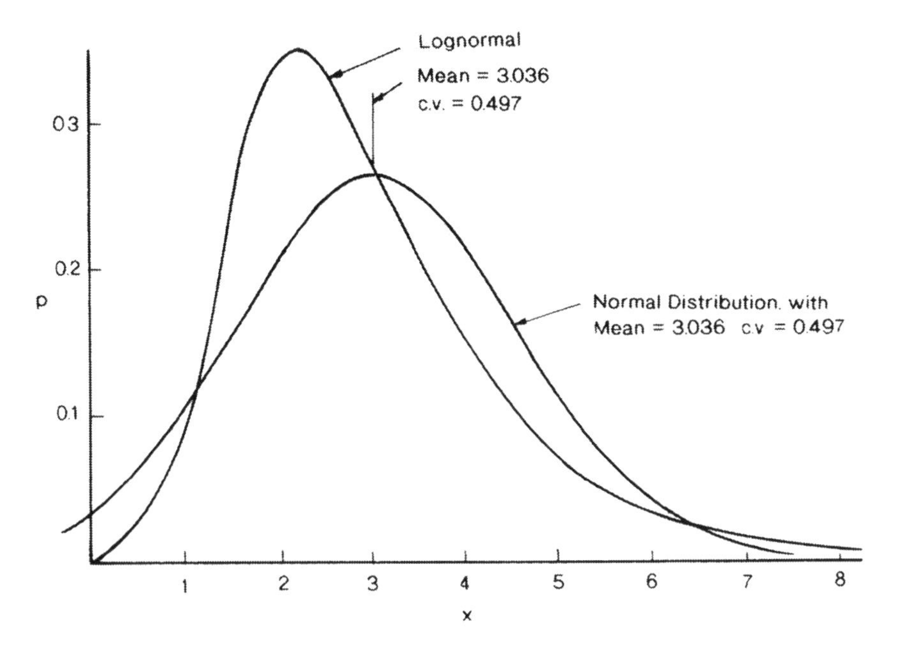

Fig. 5.4 Lognormal probability distribution

There is some value at looking again at the form of the lognormal probability density distribution. Figure 5.4 shows an example chosen so that the coefficient of variation is approximately 0.5 (actually 0.497). In the normal mathematical expression for this distribution, f is expressed in terms of the properties, μ and σ, of the function $y = \ell n(x)$. Here the values of μ and σ are 1.0 and 0.47. The resulting lognormal distribution has a mean, $\bar{x}$, equal to 3.036. Also shown on the graph is another normal distribution (not y), with a mean of 3.036 and a coefficient of variation, 0.497. The two curves could be seen as alternate representations of the same data. Some differences may be noted.

(a) The maximum of the lognormal density distribution lies to the left of that for the normal distribution.
(b) The normal distribution includes negative values.
(c) Although the areas under the two curves are the same, the lognormal distribution has a positive skew, with values at the extreme right that exceed those for the normal distribution.
(d) It is this positive skew that moves the mean (and also the median) value of the lognormal distribution to the right of its maximum.
(e) In considering the ability of the lognormal distribution to represent strength parameters, it should be remembered that test procedures during manufacture and supply should have reduced the occurrence of samples with strengths less than those of maximum frequency.

5.2 Probability of failure for non-normal distributions

Section 3.3 showed how it is possible to determine the probability of failure of a component with a resistance R under a load effect L, when both R and L are normally distributed. It was argued that, in this case, the function $(R - L)$ is normally distributed, with a mean equal to $(R_M - L_M)$ and a standard deviation σ_{R-L} equal to $\sqrt{\sigma_R^2 + \sigma_L^2}$. The reliability index, β, was defined as

$$\beta = \frac{R_M - L_M}{\sigma_{R-L}}$$

Then the probability of failure p_F was the cumulative probability for cases with $(R - L)$ negative, equal to $\phi(-\beta)$, or the area under the $(R - L)$ cumulative probability curve from minus infinity to $-\beta$. These areas can either be taken from standard reference tables or calculated by numerical integration in the manner described in the preceding section.

It is necessary here to consider the case where either R or L (or both) are not normally distributed. It is assumed, however, that the probability density distributions for both are known.

Failure occurs when $L > R$ or, at the limit, when L equals R. There are, however, many such points, for the probability density functions for R and L overlap in the manner shown in Fig. 3.3(a). Of these points, $L = R$, it is necessary to find the one which gives a maximum probability of failure. The following description is of an iterative procedure, which commences with a starting value of $L = R$, then replaces each non-normal distribution function by a normal distribution with the correct probabilities at $L = R$, and then uses these to estimate β, and to find an improved choice of $L = R$.

Consider the resistance R, for this is most likely to be normally distributed. At the chosen value, both the probability p_R and cumulative probability P_R are known. These values are now assumed to be properties of a related normal distribution, with

$$R_{ND} = R$$
$$p_{NDR} = p_R$$
$$P_{NDR} = P_R$$

where the subscripts, ND, are a reminder that these now refer to the normal distribution. Figures 5.5(a) and (b) show the standardised form of this normalised distribution, scaled to have a mean at zero, and a standard deviation of unity. Fig. 5.5(c) shows the normal distribution: its mean, μ_{NDR}, and standard deviation, σ_{NDR}, are to be found.

From the cumulative probability, P_{NDR} (Fig. 5.3(a)), find β_{R1}, either from a table of the areas of the Standard Normal curve, or by iterative numerical integration, as described in Section 5.2. For this value of β, the corresponding ordinate of the standard normal density distribution may be found (Fig. 5.5(b)). Let it be p. Then the value, p, is related to the corresponding probability, p_{NDR}, of the true normal density distribution (Fig. 5.5(c)) by the expression

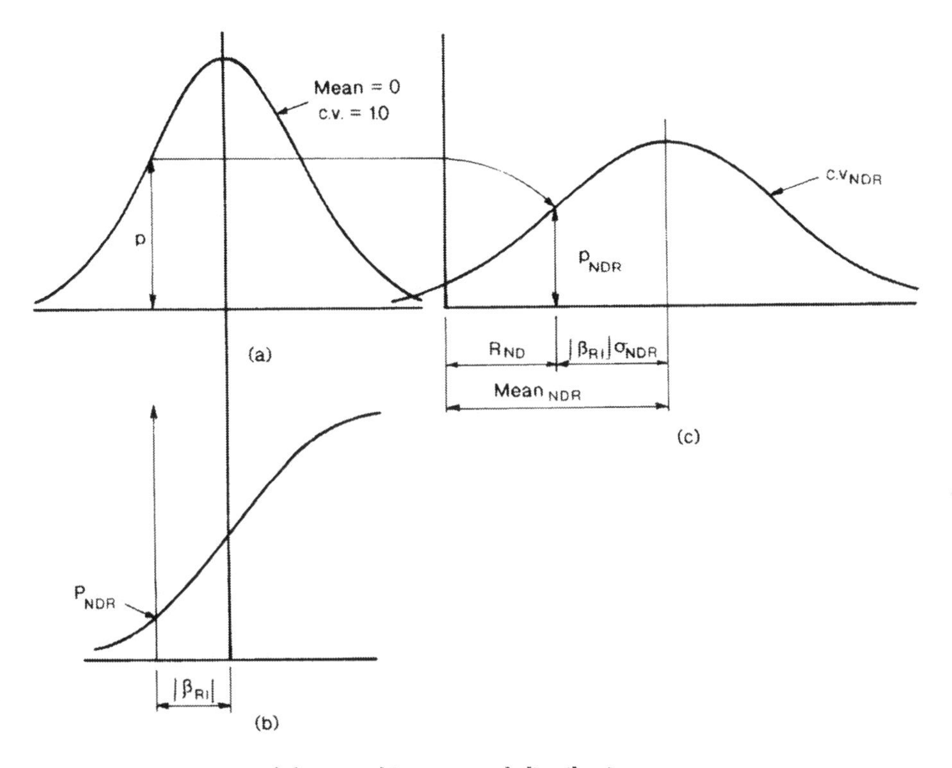

Fig. 5.5 Properties of the matching normal distribution

$$p_{NDR} = p/\sigma_{NDR}$$

Both p_{NDR} and p are now known. Then

$$\sigma_{NDR} = \frac{p}{p_{NDR}}$$

Also from Fig. 5.5(c),

$$\mu_{NDR} = R_{ND} + |\beta_{RI}|\sigma_{NDR}$$

for the case shown.

Similarly, the properties, σ_{NDL} and μ_{NDL}, of the matching normal distribution for the load, L, may be found. The corresponding function, $(R - L)$, is normally distributed, and the method described in Section 3.3 may be used to find a first approximation, β_1, to the reliability index and to the probability of failure, p_{F1}.

The present analysis can also be used to identify values of R and L for a second trial (Nowak 1993: Appendix D; Rackwitz and Fiessler 1978). The method will be described here without proof.

The standard deviation of $(R - L)$ is $\sqrt{\sigma^2_{NDR} + \sigma^2_{NDL}}$. Compute the ratios

$$\gamma_R = \sigma_{NDR}/\sqrt{\sigma^2_{NDR} + \sigma^2_{NDL}}$$
$$\gamma_L = \sigma_{NDL}/\sqrt{\sigma^2_{NDR} + \sigma^2_{NDL}} \quad \text{and then}$$
$$\beta_{R1} = \gamma_R\beta_1;\ \beta_{L1} = \gamma_L\beta_1$$

On the probability density distribution for R (Fig. 5.3(c)), β_{R1} represents a resistance R given by

$$R = \mu_{NDR} - \beta_{R1}\sigma_{NDR}$$

A similar calculation may be used for L, and should give $L = R$. These are the values for a new trial. The process is repeated. It is found that R and L converge to final values which correspond to the required maximum probability of failure.

An alternative procedure is simply to use values of R and L (with $R = L$) chosen arbitrarily over a suitable range, and for each calculate the reliability index, using the matching process described above with the analysis of Section 3.3. The correct values are those which give the maximum probability of failure.

5.3 Code calibration

The steps leading to the introduction of new design loads and other major changes in codes of practice must include a comparison with older codes and some study of the performance and strength of existing structures. To quote Galambos (T.V. *Design Codes*, 47–71, in Das 1997: 65):

> The key concept is that similar structures should have essentially the same reliability against exceeding a limit state. Connection with the experience of the past is to ensure that the reliability of the expanded set of structures in the new code is the same as that of a proven member of the set of structures designed by the old code.

This process involves the choice of load factors and is called code calibration.

As noted in Chapter 4, the 1979 Ontario Highway Bridge Design (OHBD) Code was a pioneer step in the introduction of the limit states concept, with a probabilistic basis, to bridge design. The calibration of this code was described in Nowak and Lind 1979. Although this work involved many others, notably Csagoly and Dorton, the process of code calibration in America since that time has heavily involved the work of Nowak. Papers by Nowak (1994), Nowak and Grouni (1994), Nowak and Hong (1991), Nowak, Nassif and de Frain (1993), and Tabsh and Nowak (1991), together with that by Kennedy et al. (1992) describe similar work for the review of the OHBD Code in 1991. Calibration of the AASHTO 1994 LRFD (load and resistance factor) bridge design specification was also carried out by Nowak (1993, 1995 and 1999), although a different group appear to have carried out the similar but related work for the Canadian CSA-S6 code of 1988 (see Agarwal and Cheung (1987), Allen (1992) and Buckland and Bartlett (1992)). Nowak also wrote the chapter on bridge reliability analysis and design codes for Das (1997). It has appeared appropriate therefore, to base this chapter on his report (December 1993) for the AASHTO LRFD bridge design code (reprinted as Nowak 1999) and his associated paper (1995).

It is necessary first to obtain suitable statistical descriptions of the loads where, here, the focus will be on the major primary loads and the dead loads. Nowak found the dead load to be normally distributed with, for factory-made members, the mean being 1.03 times the nominal value, with a coefficient of variation of 8%; the corresponding figures for cast-in-place members were 1.05 and 10%.

Nowak used truck survey data obtained in Ontario in 1975 for 9250 trucks, chosen on the basis that they appeared to be heavily loaded (Argarwal and Wolkowicz 1976; Harman and Davenport 1979: 497), and for these, calculated load effects scaled against one HS20-44 standard truck or lane loading, or the 1976 alternate military loading tandem (two 106.8 kN axle loads), for spans from 9.1 to 61 m. The chosen load effects were maximum bending moment and shear in simply supported spans, and maximum negative bending moment for a girder with two equal spans. These trucks were assumed to represent about two weeks of traffic. The results were plotted on normal probability paper in the form, chosen load effect against cumulative frequency of occurrence, where the chosen load effect might be, for example, the maximum positive bending moment in a 9.1 m span. These plots were judged to be sufficiently linear to justify the assumption that they were normally distributed. They were then extrapolated to give results for a chosen bridge life, where, in particular, a life of 75 years was assumed to correspond to 20×10^6 vehicles, a little larger than the number 18×10^6 obtained if the 9250 trucks passed in exactly two weeks. The quantitative significance of this extrapolation was not great; for example, from 2 weeks to 75 years corresponded to an increase in maximum positive bending moment by 1.30 for a 9.1 m span, 1.22 for a 30.5 m span, and 1.23 for a 61 m span. Expressed alternatively, the slope of the curves represents their standard deviation. Nowak gives estimates of the corresponding coefficients of variation that lie typically in the range 10 to 20%, although it must be remembered that this is for these chosen trucks. Assuming, therefore, that the extrapolations to a 75-year life (20×10^6 vehicles) are reasonable, it is important to note some of the values that were obtained, grouped here as (value, span in metres):

maximum positive bending moment	(1.72, 9.1), (2.0, 30.5), (1.82, 61)
maximum simply supported shear	(1.49, 9.1), (1.90, 30.5), (1.60, 61)
maximum negative moment	(1.77, 9.1), (1.27, 30.5), (0.92, 61).

Although this number, 20×10^6, is large, it is not infinite. If a number of tests could be carried out, each with 20×10^6 vehicles, and the maximum effects observed, then these maxima would vary to some extent. The extrapolated values quoted above represent the mean maximum values for 20×10^6 vehicles.

Their variability can also be estimated (see Section 5.1) and Nowak gives a value of 0.11 as the coefficient of variation for the maximum moment for most span lengths, for a 75-year period.

Nowak went further and estimated girder effects, assuming longitudinal primary girders, with spacings from 1.2 to 3.7 m. To do this, it was necessary to make assumptions concerning the possibility of joint or multiple presence. This was based on work by Nowak, Nassif and de Frain (1993), although Nowak notes that 'there is little data available to verify the statistical parameters for multiple presence'. Distribution factors were based on a linear finite element analysis, instead of the values S/10 for a concrete slab on steel and prestressed concrete girders, and S/12 for a similar slab on reinforced concrete girders, of the total lane loads, as specified in the current AASHTO. The resulting ratios of mean maximum 75-year girder moments to those based on the current AASHTO varied with girder spacing, from about 1.7 for a spacing of 1.2 m, to about 1.25 for a spacing of 3.7 m. There were also studies of dynamic effects, but these will not be discussed here (see Chapter 7). This truck survey data was relatively old (1975) compared with the date of issue of the new code (1994). It was assumed further that the legal load limits would not be changed and that the truck population would remain the same.

The next phase was to look at resistance data. Earlier Canadian work by Kennedy and Baker (1984) considered resistance factors for steel highway bridges, and pointed out that uncertainties in the estimates of resistance resulted from three factors:

(a) variations in material properties;
(b) changes in geometry; and
(c) variations between design procedures and test performance.

They presented statistical data for all factors, in the form of ratios of mean to nominal values and coefficients of variation, V. All material strengths and ratios of test to computed resistance were expressed as lognormal distributions with geometrical variations taken as normal. Typical results were:

yield strength of steel:

mean ÷ nominal = 1.06
$V = 0.051$

plastic section modulus for rolled sections:

mean ÷ nominal = 0.99
$V = 0.038$ (including machine error)

discretisation factor in the choice of an available section:

mean ÷ nominal = 1.059
$V = 0.039$

test to predicted ratios of plastic moment:

mean ÷ nominal = 1.02
$V = 0.045$

Combining all factors and using Monte Carlo simulation techniques (using a randomly selected family of 50,000 values, with the above properties set so as to match the assumed probability density distributions), they found, for the plastic hinge moment, the ratio of mean to nominal was 1.21, with a coefficient of variation, V, equal to 0.077.

Nowak (1993) considered the same sources of variability, and, using different techniques, estimated statistical variations in resistance for steel and composite steel and concrete sections, for reinforced concrete girders, and for a concrete slab composite with standard AASHTO precast and prestressed concrete girders.

Data for more than 100 actual bridges was used to compare minimum values of resistance (R_{min}) required by the current AASHTO code with that actually provided. These ratios (actual ÷ minimum required) varied from about 1.5 to 2 for moment in steel girder bridges, from about 1.2 to 1.8 for moment in prestressed concrete girder bridges, and about 1.7 to over 4 for shear in steel girder bridges. The number of these bridges was insufficient for an adequate study of the effect of all variables. He therefore used the current code to obtain values of R_{min} for steel and composite steel and concrete girder bridges, reinforced concrete T-beams and prestressed concrete girder bridges for spans from 9.1 to 61 m, with girder spacings from 1.2 to 3.7 m. Assuming that the resistances actually provided were set equal to these values of R_{min}, he was then able to determine probability distributions for resistance for all these cases. He then used these with the normally distributed girder load effects computed previously to obtain values for β, the reliability index, using the method described in Section 5.3. Three of his results are shown in Fig. 5.6:

(a) for moment in simply supported composite girders;
(b) for moment in simply supported prestressed concrete girders; and
(c) for shear in prestressed concrete girders.

Some observations may be made.

1. Nowak's load effects for girder bending moment and shear used his analysis for the distribution of the primary loads between girders, whereas the resistances were based on procedures in the current AASHTO code. This change in distribution coefficients has led to a marked variation of β with girder spacing.
2. The variation of β with span is small for bending moments in girders with spans greater than 18.3 m, and in shear for spans equal to or greater than 27.4 m. Presumably this results at least partly from differences in Nowak's loads, compared with the current AASHTO loads.
3. For the longer spans, values of β range from about 2.7 to 4.5 for bending,

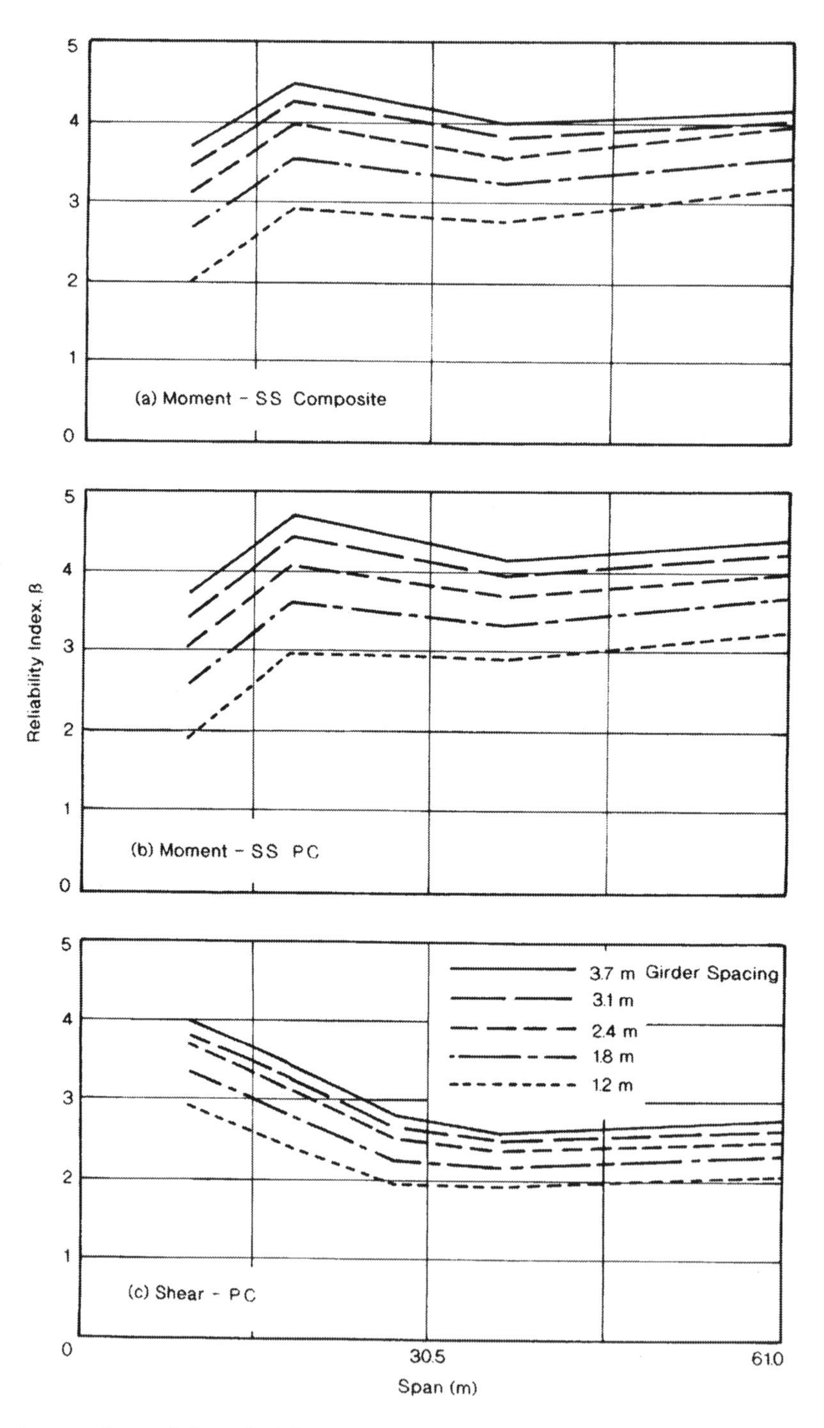

Fig. 5.6 Values of the reliability index, β, for current AASHTO code (Redrawn from Nowak 1993; used with permission)

and 1.9 to 3 in shear; values for bending in the shortest spans are somewhat less than for the longer spans, down to about 1.9 for a span of 9.1 m.

4. The effect on β of the material type can be gauged by observing the following values in bending for a span of 27.4 and a girder spacing of 3.3 m:

non-composite steel	3.76
composite steel and concrete	3.76
reinforced concrete	3.53
prestressed concrete	3.88

5. Based on this data, Nowak chose a target value of β for the new code of 3.5.

Section 4.3 presented the new AASHTO design loads, revised dynamic load allowances, and the load factors that were finally chosen, these being:

$$1.25DL + 1.5SDL + 1.75LL$$

Nowak tried a range of load factors and was guided in his selection by the rule

$$\gamma = \lambda(1 + kV)$$

where

γ is the load factor for a particular load type,
λ is the ratio of its mean to nominal value,
V is its variance, and
k is a constant for all load types

His values of γ and V were:

(a) dead load of factory-made elements:

$\gamma = 1.03$; $V = 0.03$;

(b) dead load of cast-in-place concrete:

$\gamma = 1.05$ $V = 0.10$;

(c) dead load of asphaltic wearing surface:

$\gamma = 1.00$ $V = 0.25$;

(d) live load (with impact)

$\gamma = 1.10$–1.20 $V = 0.018$.

His trial values of k were 1.5, 2.0 and 2.5; the value finally selected was 2.0. This gave a range in load factor for live load from 1.5 to 1.6. He finally proposed the more conservative value of 1.7, but this was increased to 1.75 in the code. He used these load factors with the proposed new design live load to select members with trial Resistance Factors, ϕ, of 0.95 and 1.0 for steel, composite, and prestressed concrete members in bending, 0.85 and 0.90 for reinforced concrete members in bending, and values from 0.90 to 1.0 in shear. For these members he then calculated reliability indices, β, using the procedure described previously, including his finite element values for the distribution factors.

Some of his results are shown in Fig. 5.7; they may be compared with results obtained using the current code, as shown in Fig. 5.6, and this leads to the following observations.

1. Finite element values for the distribution factors were used in both the design and checking parts of the second analysis, and this necessarily removed the dependance of β on girder spacing.
2. The variability of β values with span length, as shown in Fig. 5.6, is now largely removed, suggesting that the new form for the live load is suitable.
3. As would be expected, values of β vary with the resistance factor, ϕ, and the load factor, γ.
4. The target value for β was 3.5. For the cases shown in Fig. 5.7, this led to the selection of $\gamma = 1.6$, with $\phi = 1.0$ for bending and 0.9 for prestressed concrete in shear. This value of γ was later upgraded to 1.7 and then 1.75. His final recommended values of ϕ were 0.9 for reinforced and prestressed concrete in shear, and reinforced concrete in bending, with 1.0 for all other cases.

5.4 Selection of the primary live loads

The procedures for code calibration described in the previous section do not truly form a method for the selection of the geometry and nature of the design primary loads, except by trial and error. The selected load is based on the old AASHO load and differs, for example, from the Ontario OHBD loads of 1979 and 1991 on which Nowak also worked. This is not intended as a criticism, for it is appropriate that designers within a particular country continue with methods that are at least partly familiar; such a procedure may also provide a clearer link between new designs and those in service. Furthermore, the review of codes of practice in Chapter 4 indicated a trend towards the use of an arbitrarily chosen load that consisted of a set of local loads and a distributed load (see conclusion 6 of that chapter). It is, nevertheless, appropriate to enquire if these are procedures leading directly from truck surveys to a particular form of the primary load. This Section will describe two.

As mentioned in Section 4.4, the 1979 Ontario Highway Bridge Design Code used the concept of Equivalent Base Length to replace surveyed truck data (axle weights and spacings) by an equivalent design vehicle, and also to define

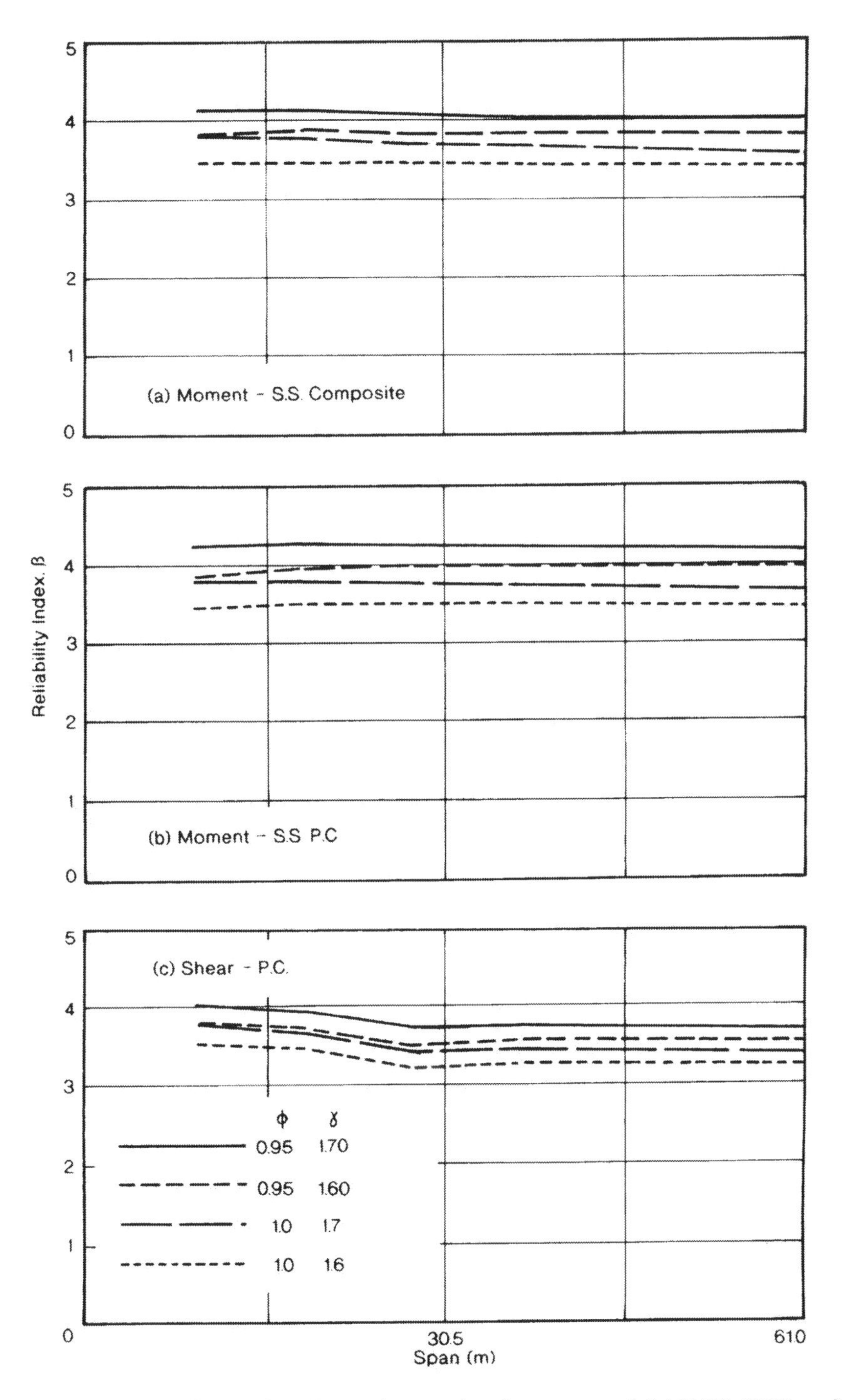

Fig. 5.7 Values of the reliability index, β, for the proposed AASHTO LRFD code (Redrawn from Nowak 1993; used with permission)

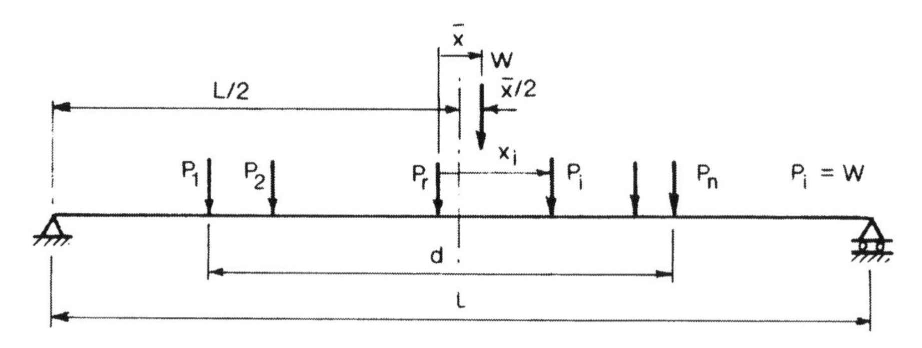

Fig. 5.8 System of concentrated loads

legal truck weights (Dorton and Csagoly 1977); the concept was later reviewed by O'Connor (1980 and 1981). The equivalent base length (B_M) was defined as 'an imaginary finite length on which the total weight of a given sequential set of concentrated loads is uniformly distributed such that this uniformly distributed load would cause force effects in a supporting structure not deviating unreasonably from those caused by the sequence itself' (O'Connor 1981: 105).

Figure 5.8 shows a typical load group. It was shown in Chapter 1 that the absolute maximum bending moment under a particular load (say P_r) of a load system as it moves across a simply supported beam occurs when the centre of the beam bisects the distance between P_r and the centre of gravity of the load system. Here,

$$\Sigma P_i = W$$

$$\bar{x} = \frac{\Sigma P_i x_i}{W}$$

The moment of the loads, P_i to P_n, about P_r

$$= \sum_{i=1}^{n} P_i x_i$$

where this may be evaluated alternatively as

$$\tfrac{1}{2}[\Sigma \mid P_i x_i \mid + \Sigma P_i x_i]$$

with both sums including all values of i. It follows that the bending moment at P_r is given by

$$\begin{aligned} M_r &= W(L + \bar{x})^2/4L - \tfrac{1}{2}\Sigma \mid P_i x_i \mid - \tfrac{1}{2}\Sigma P_i x_i \\ &= WL/4 + \tfrac{1}{2}\Sigma P_i x_i + (\Sigma P_i x_i)^2/4WL - \tfrac{1}{2}\Sigma \mid P_i x_i \mid - \tfrac{1}{2}\Sigma P_i x_i \\ &= WL/4 + (\Sigma P_i x_i)^2/4WL - \tfrac{1}{2}\Sigma \mid P_i x_i \mid \end{aligned}$$

A distributed load, W, of length B_M, will produce a maximum bending moment when its centre is located at the centre of the beam. Let B_M be such that this bending moment is M_r. Then

$$M_r = WL/4 - WB_M/8$$

Equating the two,

$$B_M = \frac{4\Sigma \mid P_i x_i \mid}{W} - \frac{2}{L}\left[\frac{\Sigma P_i x_i}{W}\right]^2$$

The span length appears only in the second term. The total length occupied by n loads is d, and their average spacing is $d/(n - 1)$. For the smaller spans, where this second term is relatively large, there are other loads not far beyond the ends of the beam. Csagoly and Dorton (1978a and b) introduced the approximation,

$$L = n\, d/(n - 1)$$

Then

$$B_M = \frac{4\Sigma \mid P_i x_i \mid}{W} - \frac{2(n-1)}{nd}\left[\frac{\Sigma P_i x_i}{W}\right]^2$$

In this expression, the distances x_i are measured from load P_r, and the bending moment used to derive the expression was under this load. It is sometimes difficult to identify the load P_r which causes the greatest bending moment. Define the central load, P_c, as the load such that the sum of the loads to the left, together with some proportion (aP_c) of P_c, with $0 < a < 1.0$, is equal to $W/2$. Consider the influence line for function F, shown in Fig. 5.9, with equal slopes rising to a maximum ordinate. Place P_c at this point. Then the total load to the left of the central point, including (aP_c), is equal to $W/2$. Similarly the load to the right is $W/2$. Suppose the load system is moved a distance s to the right. If the influence line were given the dotted extension shown, the load $W/2$ to the left would rise by an amount (ts); the load to the right would fall by an equal amount; the change in F would be zero. In fact, the load, (aP_c), will drop a distance, ($2ts$), below this dotted extension, and there will be a reduction in F equal to aP_c ($2ts$). Similarly, if the load system is moved to the left, the reduction in F will be $(1 - a)P_c(2ts)$. It follows that the maximum value of F will occur with P_c at the maximum ordinate. An ambiguity arises in defining the central load if, for a particular load in the system, $a = 0$. Then, for the next load to the right, $a = 1$. The load system can be divided into two groups of loads, each totalling $W/2$. Then if either of the neighbouring loads ($a = 0$ or $a = 1$) is placed at the maximum of the influence line, equal maximum values of F will occur. Here the central load will be defined as that which has the smaller distance from the resultant of its $W/2$ group. This central load is, for some other pointed influence lines, more likely to achieve a maximum effect than its neighbour.

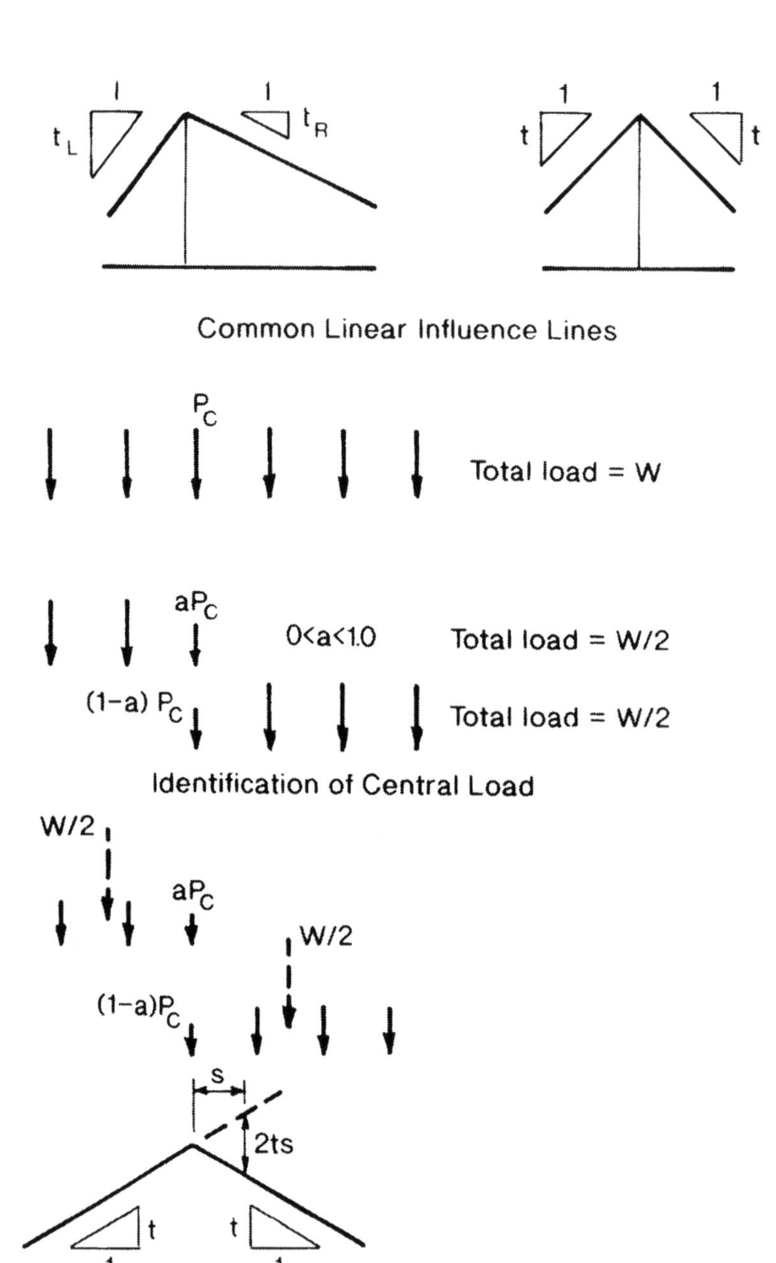

Fig. 5.9 Identification of 'the central load' (Redrawn from O'Connor 1980, 1981)

The Ontario concept is not simply that a truck may be replaced by its weight, W, distributed over a length, B_M, but that there is a base length signature in (W, B_M) space for the truck, with each subset of adjacent axle loads plotted in this space. Figure 4.6 provided an example of this process. Another example may be taken from O'Connor (1981). At that time the Australian legal limits were:

one-axle group	8.5 t (83.4 kN)
two-axle group – steering	10 t (98.1 kN)
two-axle group – nonsteering	15 t (147.1 kN)
three-axle group	18 t (176.5 kN)
maximum length	16 m
maximum total weight	varying in a specified manner from 18 t (176.5 kN) at a 3 m total length to 36 t (353 kN) at 12 m, and then constant to a total length of 16 m

These legal limits were used to generate a family of 191 trucks, with 4050 subsets of axle loads. For these, values equivalent to (W, B_M) (see below) were plotted as in Fig. 4.6(a), and this plot used to generate the upper bound line shown in Fig. 5.10(a), using the equation for B_M given above, but omitting the approximate, second term. The values shown for W and B_M are slightly rounded as a result of the intervals used in the counting procedure.

This curve can now be used to generate equivalent vehicles, three of which are shown in Fig. 5.10(b). All contain one single steering axle at the legal limit of 83.4 kN. The first two (LT1, 2) also contain a three-axle group at the legal limit, with an assumed axle spacing of 1.0 m, together with a further, two-axle group (1.2 m spacing) chosen so that the total truck weight is at the legal limit. The third (LT3) has a legal limit steering axle, a dual axle at the legal limit, and another two single axles, each of 61.3 kN, chosen so that the total load is at the legal limit. These three trucks include certain axle spacings determined so that their base length signature lies on the curve of signatures (Fig. 5.10(a)). An example may be given.

For truck LT1, consider the combination of the three-axle group with the adjacent two-axle group. For this group, $W = 269.9$ kN. One half of this value is 135 kN. It follows that the third axle is the 'central load'. The spacing, 2.39 m, between the two axle groups gives

$$B_M = 6.75 - 0.23$$

or, omitting the second, approximate term, B_M is 6.75 m, as on the curve. The example illustrates the relative size of this second term. In this case it was omitted both in the generation of the curve, and in the design of the truck LT1.

The family of legal vehicles was used to calculate maximum values for five load effects, with spans to 100 m:

(a) the central bending moment in a simply supported span, and a 3-span continuous girder;

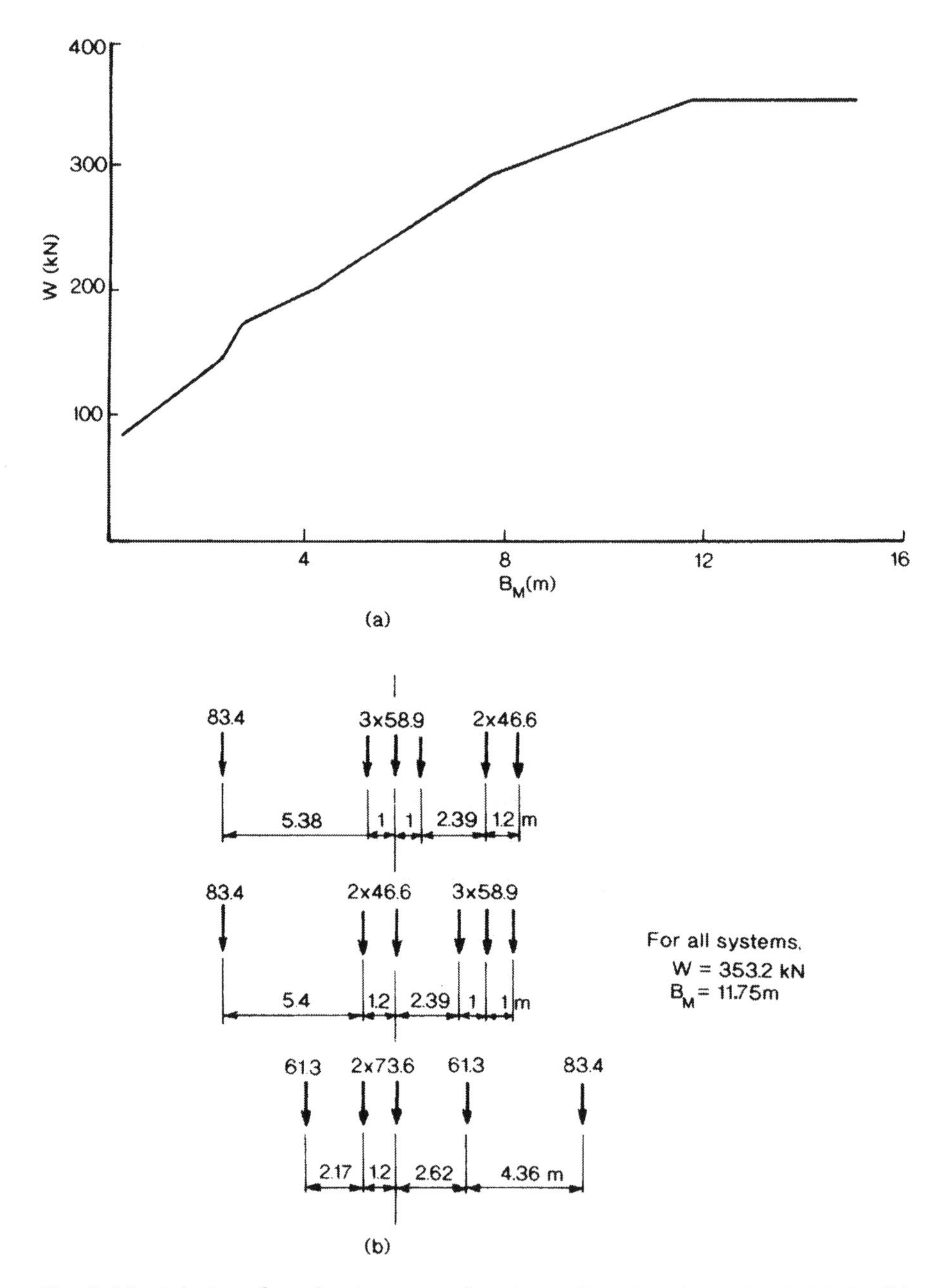

Fig. 5.10 (a) Base-length signature for Australian legal trucks (1981); (b) Equivalent vehicles

(b) the bending moment over an inner support in a 3-span cantilever bridge; and
(c) the end shears for a simple span, and a 3-span continuous girder.

The three trucks, LT1–3, were then used for the same cases. The resulting errors were typically small; for example, for truck LT1, the maximum error in the simple span bending moments was 3.2% at a span of 9 m, and less than 0.5% for spans in excess of 16 m. Errors in shear were greater, but this is not surprising. The maximum axle load for a dual axle is 73.6 kN, greater than that for a three-axle group, and this load is not present in LT1. It is difficult to select a truck that includes all possible extreme cases for the shorter spans.

O'Connor (1981) also discussed a variation to the Ontario Equivalent Base Length. The set of concentrated loads to the left of the central load, together with part of the central load, is equal to $W/2$. The total load can be replaced by two concentrated loads, each equal to $W/2$, as shown in Fig. 5.11, with b_L and b_R the distances from the central load to these resultants, and $b_L + b_R = b$. It may be shown also that 'b' precisely equals $B_M/2$, where B_M is calculated with the omission of the second term. Define the Concentrated Base Length as b, with the location parameter, x, as the lesser of b_L/b and b_R/b. Then

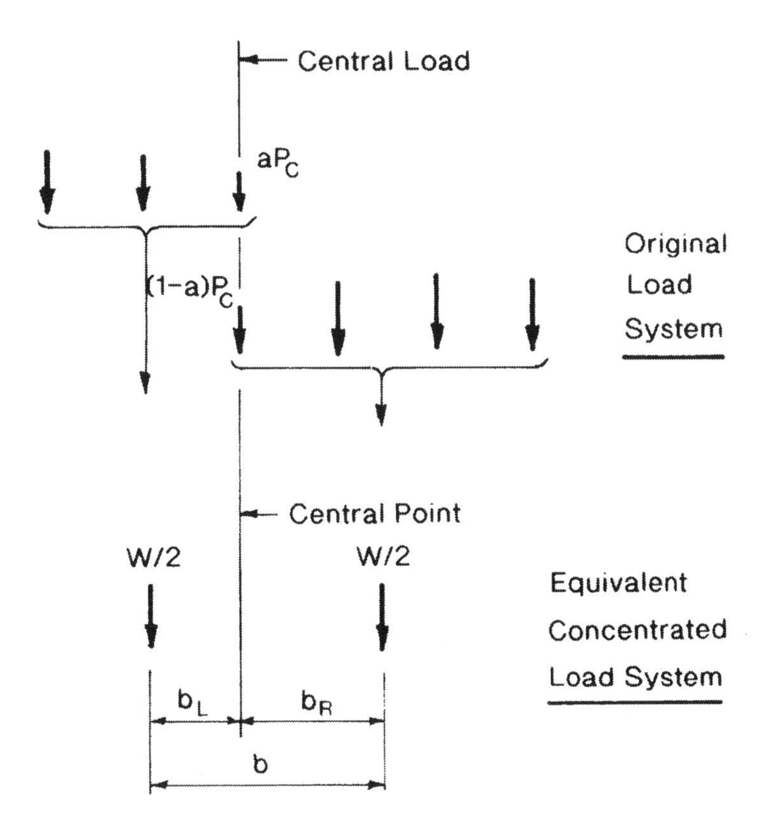

Fig. 5.11 Equivalent concentrated load system

this system of two equal concentrated loads has three parameters, W, b and x. If this pair of concentrated loads is applied to any linear influence line, it will provide the same maximum value of the function, F, as the original system, provided that the central points are placed at the same point. The study described above showed that, of the legal truck set, those near the upper bound tended to have large values of x. Values of x for the complete trucks LT1–3 are 0.43, 0.17 and 0.28. It was found also that the truck with the largest value of x, LT1, gave the best results. This parameter, x, cannot exceed 0.5. A larger value of x tends to mean that the loads are equally distributed on either side of the central load.

It is concluded that

(a) the Ontario Equivalent Base Length, B_M, is a useful tool;
(b) an alternative is the Concentrated Base Length, b;
(c) a truck with a large value of the location parameter, x, for the complete truck, appears to give the best results.

A second procedure for the selection of a design vehicle from traffic surveys was developed by Heywood (1992a). The Australian Road Research Board, in the period prior to 1990, developed CULWAY, an instrumented reinforced concrete box culvert, with an associated processing system, based on work by Peters (1986), and earlier work by Moses (1979) and others. A related study on Six Mile Creek Bridge, Queensland, was carried out in 1981–2 by Pritchard (1982; see also Pritchard and O'Connor 1984). The CULWAY system gave vehicle speed, axle spacings and axle weights for all records above a pre-set threshold value. Heywood analysed data for more than 679,000 vehicles from 31 sites in all States of Australia, for recording periods between 1989 and 1991, not continuously monitored, but totalling 7.74 years for the sum of the sites.

The nature of the traffic using these 31 sites varied. Central Australia is relatively flat and without major streams; for many routes there is no railway that could be used as an alternative means of transport for heavy or bulk loads. Here road trains are permitted, made up of a semi-trailer or rigid truck followed by two further trailers with relatively short tow bars. The heaviest road train recorded in this study had a mass of 138 t, equal to a weight of 1354 kN. These cannot be used on all roads – to quote Lay (1985: 348):

> Road trains operate in many parts of Australia, their field of legal operation (designated routes) depending on the permits granted by the relevant local authority. The area of road train operation continues to increase as the overall economic efficiency and transport flexibility associated with the units is very great.

Heywood identified 99% of Australian bridges as having a maximum span less than 40 m and decided to direct his study to spans in this range. He focused attention on six functions whose influence lines form a characteristic set: for a simply-supported span, the central bending moment and end shear; for a 3-span continuous structure with equal spans, the moments at mid-span and above the interior support, the shear at the exterior face of the interior support, and the interior reaction. He calculated values of these functions for each recorded vehicle, for spans between 5 and 40 m, at a 5 m interval.

For a simply-supported span, the maximum mid-span moment caused by a single load, P, is of the form

$$M = PL/4$$

If, instead, there is a series of axle loads, P, spaced s apart, where s is small, then the bending moment takes the form

$$M = PL^2/(8s)$$

Both are special cases of

$$M = \psi L^{\zeta}$$

with $\zeta = 1$ for a single axle, and 2 for a uniformly distributed load (see also Heywood and O'Connor 1992). Heywood found that for his sites, M could be related to L empirically by expressions of this form.

Heywood evaluated the best fit values of ψ and ζ for all sites and concluded that low values of ζ indicated sites carrying shorter vehicles. He decided to group sites according to the following rules:

$1.4 < \zeta < 1.5$	short vehicles
$1.5 < \zeta < 1.65$	medium combination vehicles
$1.65 < \zeta < 1.75$	long combination vehicles

His resulting expressions for bending moment were:

'short' sites	$M = 19.25\ L^{1.45}$
'medium' sites	$M = 13.75\ L^{1.55}$
'long' sites	$M = 10.62\ L^{1.69}$

The accuracy of these expressions can be gauged from the coefficients of variation obtained by comparing field data with predicted values. These were: short, 0.13–0.15; medium, 0.19–0.22; long, 0.16–0.21.

For each of these groups he combined data for the six chosen functions; he showed that, to an acceptable degree of approximation, it was normally distributed and, from linear plots on normal distribution paper, found the average extreme daily moments and shears (six functions) for each of the 8 spans. He then used an optimisation routine to select loadings that best modelled these load effects.

In this optimisation process, a general vehicle with k axles was modelled by its axle loads and spacings, and could be applied with or without a distributed load, with intensity allowed to vary with some power of its loaded length. The objective function, F_o, was expressed in the form

$$F_o = \sqrt{\frac{A}{m\ell}}$$

where

m was the number of functions (6),
ℓ was the number of spans (8), and

$$A = \sum \frac{(LE - LE_T)^2}{LE_T}$$

with

LE_T = the load effect from the surveyed vehicles (the target load effect), and
LE = the load effect from the trial vehicle

This sum was evaluated for a number of terms equal to $m\ell$ (or 48). Each term could be modified by a pre-set importance factor. Also each variable could be constrained to be within a specified range. If this range was zero, then the variable was fixed. Also, the maximum gross mass and/or the maximum vehicle length could be specified.

The optimisation routine used by Heywood was that given by W.H. Swann (*Constrained Optimization by Direct Search*, in Gill and Murray 1974: 191–217). Heywood checked the procedure by generating simple span bending moments and shears, 5 to 40 m, for the Australian T44 truck shown in Fig. 4.11, based on the American AASHO HS20-1944, converted into metric units and increased by about 33%. He then used his routine to check if this input vehicle would be recovered by the optimisation process, commencing with a seed vehicle of similar configuration, with axle group loads set at the current Australian legal limits (58.8, 161.8, 161.8 kN), and axle spacings of 4.0, 1.2, 5.0 and 1.2 m. This seed vehicle gave an error, as measured by the objective function of 21.3%. After 378 iterations, this was reduced to zero. The resulting vehicle differed only slightly from the T44 truck, with axle group weights of 34.2, 199.5 and 198.9 kN (T44: 48, 192, 192 kN), and axle spacings of 3.34, 1.34, 3.07 and 1.39 m (T44: 3.7, 1.2, 3.0, 1.2 m). The total weight was 432.5 kN (T44: 430 kN) and the total length, 9.14 m (T44: 9.1 m). A further trial was carried out with the dual axle spacing set at 1.2 m and gave the T44 truck. This demonstration showed (a) that the system worked; and (b) that there may be more than one equivalent truck. When used to find a suitable three-axle vehicle, it yielded a result that was significantly different from the simplified T44 truck, with axle loads of 145.8, 141 and 146.9 kN, and axle spacings of 3.55 and 3.58 m. The resulting objective function was 0.73%. This again is small, and confirms that a range of vehicles may give suitable results.

Heywood proceeded to apply this method to model the Australian CULWAY data. Only a few of his conclusions can be presented here. He considered, first, the combination of those sites identified with short and medium vehicles. His best result was obtained for a vehicle with an associated distributed load, as shown in Fig. 5.12(a), giving a value for the objective function, F_o, of 0.98%, with a maximum error in one of the load effects of 1.7%. For comparison, Fig. 5.12(b) shows a single dual axle with a distributed load, generated by

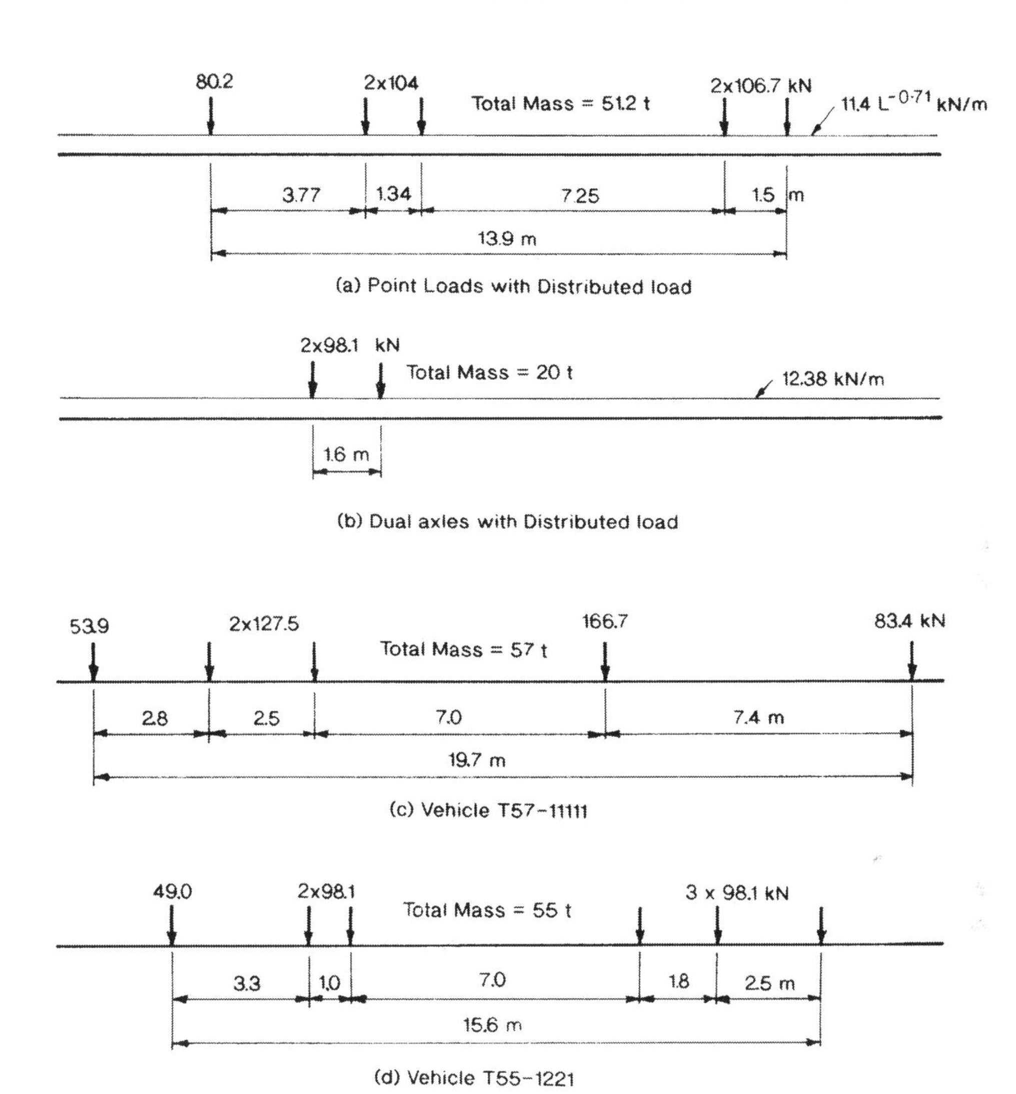

Fig. 5.12 Vehicles chosen by Heywood (1992a) to model Australian CULWAY data for short and medium vehicle sites

the program as having an intensity that did not vary with the span. Here F_o was 6.39%, with a maximum error of 10.3%. When the program was run without a distributed load, the optimum vehicle was that shown in Fig. 5.12(c), where the designation, T57-11111, indicates that it has a total mass of 57 t, with five single axles. It gave F_o as 2.01%, with a maximum error of 4.4%. It bears some similarity to the Ontario OHBD truck shown in Fig. 4.6(b). One of the benefits of the Heywood method is that it allows the progressive modification of the chosen vehicle, possibly with some reduction in accuracy. His survey of truck loads gave the values in Table 5.1 for average extreme daily axle loads.

Table 5.1

Axle group	Load per axle (kN)	
	Short–medium sites	*Long sites*
Single steer	76	67
Single non-steer	108	95
Tandem steer	61	59
Tandem non-steer	102	103
Tri-axle groups	81	80
Quad-axle groups	43	34

For tandems, the legally acceptable range of axle spacings is 1.0–2.0, and 1.0–1.6 for tri-axle groups. The minimum span used by Heywood in this study was 5 m. To ensure a vehicle that suitably modelled local effects (spans less than 5 m), he then constrained the axle loads to lie within 10% of their average extreme daily values, yielding the truck shown in Fig. 5.12(d), with F_o equal to 2.35%, and a maximum error of 6.1%. Expressed in tonnes, the loads on the successive axle groups are 5.0, 20.0, 20.0 and 10.0 t, giving a total mass of 55 t. He calls it the T55-1221 vehicle and identifies it as his representative vehicle for short–medium sites; that is, for sites with normal traffic.

A similar study was carried out for the long-vehicle sites. The best value for F_o was 1.87%, for a vehicle generally similar to the T90-12223 vehicle shown in Fig. 5.13. This latter vehicle had its axle loads constrained to within 10% of the observed average extreme daily values, and gave F_o equal to 2.45%, with a maximum error of 5.65%. It is similar in many respects to the T55-1221 vehicle selected for the short–medium sites.

These results have been presented to demonstrate a method for the selection of a design vehicle from truck-survey data. Heywood called them 'representative live load models', but, in the absence of further work on the loads actually present on bridges of longer span, and on future legal load limits, thought it premature to recommend them for use. However, it is important to note some further conclusions from his survey, for it is either the largest, or one of the largest, such surveys ever carried out.

It is useful, first, to see the nature of the probability distributions from this data. Figure 5.14 shows two plots on normal distribution paper of mid-span moments for a simply-supported span (L), scaled against $L^{1.5}$. The first, (a), is

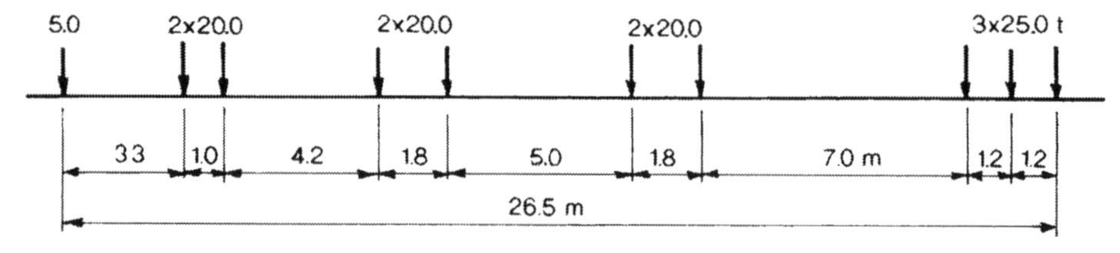

Fig. 5.13 Vehicle chosen by Heywood (1992a) to model Australian CULWAY data for long vehicle sites

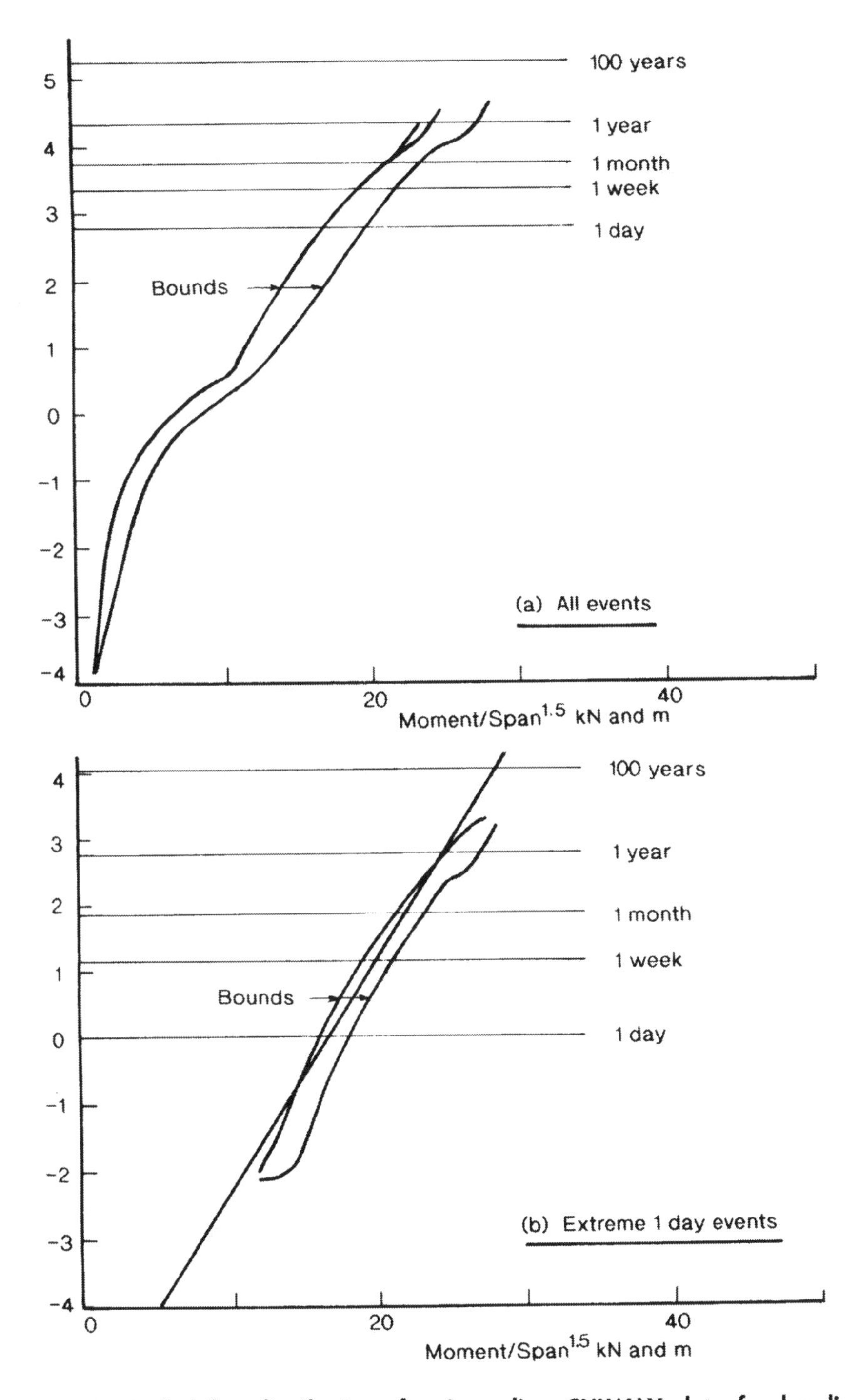

Fig. 5.14 Probability distributions for Australian CULWAY data for bending moments (Redrawn from Heywood 1992a; used with permission)

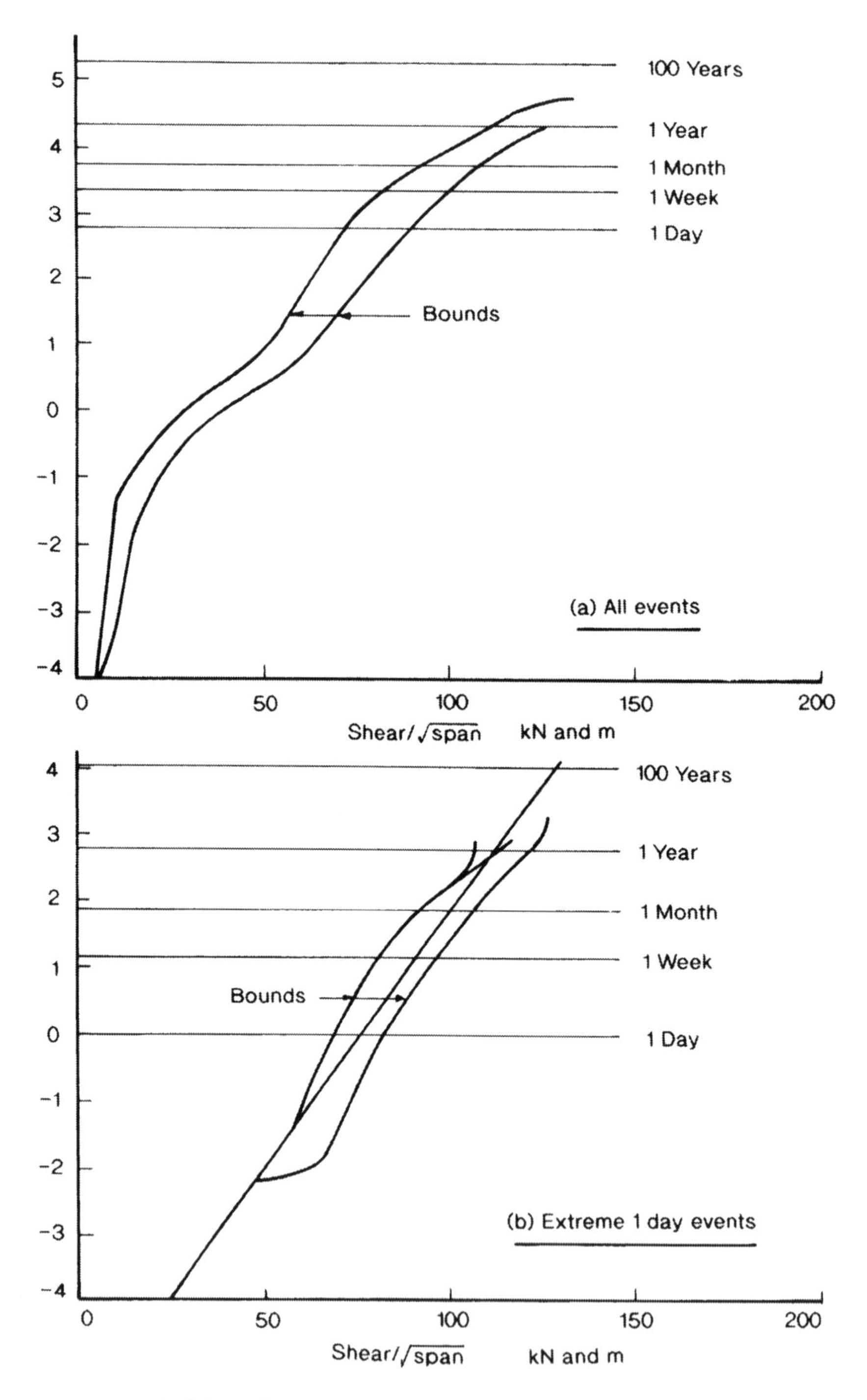

Fig. 5.15 Probability distributions for Australian CULWAY data for shears (Redrawn from Heywood 1992a; used with permission)

the cumulative distribution for all events; the second, (b), is the cumulative distribution for maximum daily events – the second truncates the lower tail and, for this reason, gives a more linear plot. The use of $M/L^{1.5}$ brings the results for the nine spans (5–40 m in intervals of 5 m) closer together; the curves for the eight spans, 10 to 40 m, are close to coincident, and the second graph shows the best fit line. The relative values of bending moment for various periods are:

1 day	1.0
1 month	1.33
1 year	1.48
100 years	1.70

Similarly, the cumulative probabilities for simple span shear are shown in Fig. 5.15, here scaled against $L^{0.5}$. Again, a best-fit straight line is shown; relative values of shear for the four periods are as for moment. It can be seen that in neither case are the curves truly linear, and this must imply some uncertainty in the extrapolation to 100 years (see also Heywood and O'Connor 1992).

Figure 5.16 compares load effects calculated (a) from a single T44 truck, and (b) for a single T44 truck and lane loading, with the average extreme daily events computed from the survey data at the short–medium sites (a positive error indicates that the T44 value is high). The effects shown on the graphs are:

1*M*	one span, central bending moment
1*V*	one span, end shear
3*MNP*	three continuous spans, negative moment over pier
3*VEP*	three continuous spans, shear at exterior pier face
3*MNM*	three continuous spans, negative moment at centre
3*RP*	three continuous spans, reaction at pier.

It can be seen that a single T44 truck often gives a value that is low – up to 10%. The combined T44 truck and lane load gives larger design effects, particularly for the three-span bridges and the longer spans, with errors between about −10% and +40%. If the maximum 1-year truck were used, this would be 1.48 by the average maximum 1-day event, and the predicted load effects would generally be too low. It should be remembered also that for many years prior to 1973 (Section 4.7) the old AASHO HS20-1944 loading was used, this being about 0.75 of the T44 loading. To quote Heywood (1992a: 6–39), 'On this evidence it is expected that there should be many bridges suffering distress. However, failure due to overload remains rare in Australia'. Some of the reasons for this will be discussed later.

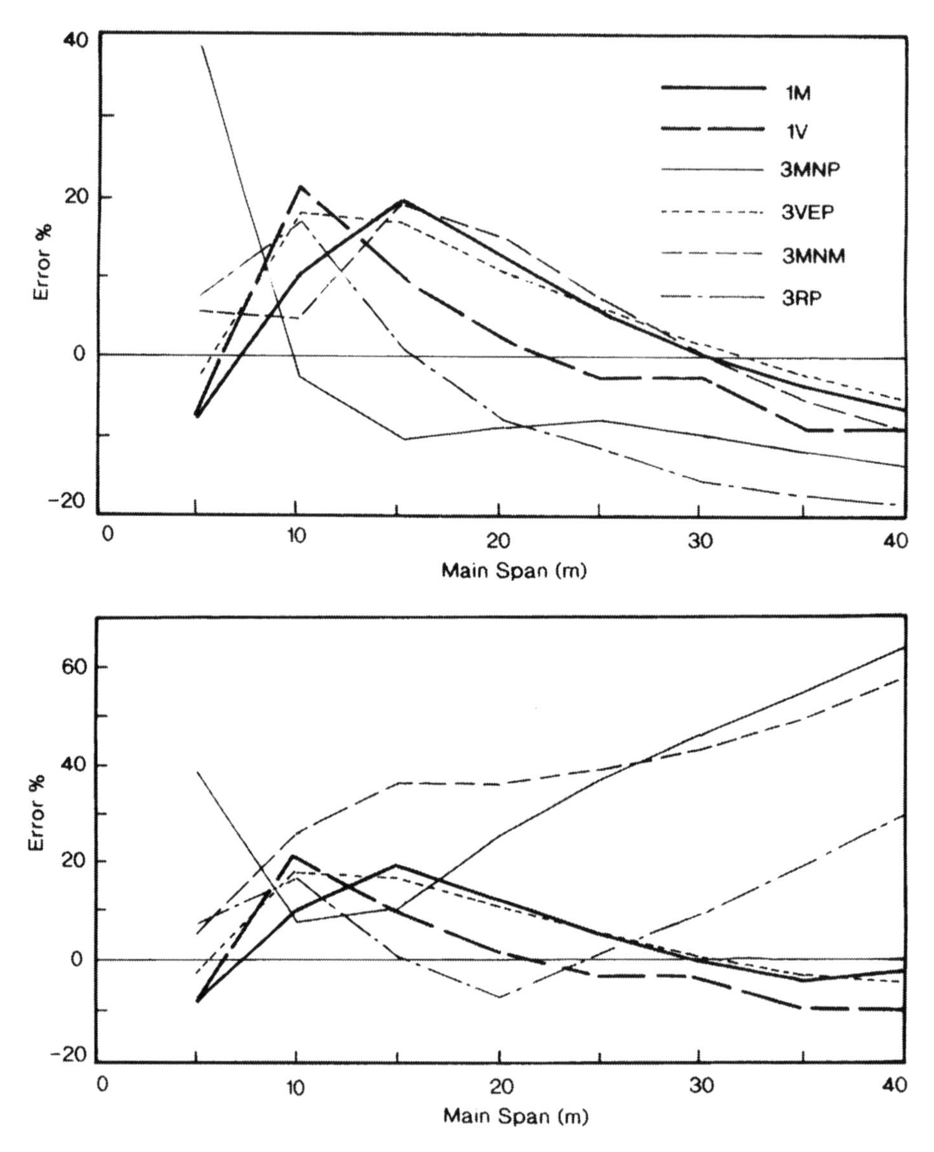

Fig. 5.16 Comparison of load effects from T44 loadings and Australian CULWAY data (Redrawn from Heywood 1992a; used with permission: see text for definition of terms)

5.5 Legal limits and costs

There is a necessary link between the primary live loads specified for bridge design and the legal limits imposed on vehicle weights and geometries. This link is different for railways as compared with roads.

The American Railway Engineering Association specifications for 1990 required the application of the Cooper E80 load, shown in Fig. 4.16 in the original units of pounds and feet. Also shown in that figure is the Metric Cooper M250 load specified in the 1974 manual of the Australian and New Zealand Railway Conferences, equivalent to about Cooper E56. Both model a train made up of two locomotives drawing a long series of wagons, with these replaced by a uniformly distributed load. It is not difficult to show that the larger loads are applied by the locomotives. In the second case, for example, the four major locomotive axle loads are each 250 kN, at a spacing of 1.5 m. The average loading is 167 kN/m. For comparison, the equivalent wagon load is 82.5 kN/m. In such a case, the specified locomotive loading is intended to be equivalent to the weights of actual locomotives in service on that railway system. The question of accidental or deliberate overloading arises only in the case of the relatively smaller wagon loads.

By comparison, Appendix D of the British standard BD37/88 (see Section 4.2) concludes that axle loads and arrangements for modern locomotives tend to be similar to those used for bogie freight vehicles (or wagons), freight vehicles often being heavier than locomotives, and this was illustrated in Section 4.11 by reference to a recent study of Australian coal trains. Here, 30,000 bogie loads had a mean of 450 kN with a coefficient of variation of only 2%. This mean is equivalent to a mass of 45.9 t. Most of the wagons were nominally of 90 t, with a few of 100 t, in both cases with two bogies per wagon. The point made here is that the variability of the recorded loads is small, either because of controls exercised at the point of loading, or because of the restriction on load automatically applied by the volume of the container. The only real source of variation was the moisture content of the coal, leading to a possible change in density. Economy in transport costs dictated the full usage of the transport vehicles and this resulted also in an automatic barrier against accidental overload. Although special purpose heavy or maintenance vehicles could be used on some lines, their weights should also be subject to careful control.

With highway loadings the case is different. Although each country will have legal restrictions on vehicle weights and geometries, and methods in place to police these limitations, yet in practice these controls are often avoided. A case in point arises from Heywood's (1992a) study of Australian road traffic loads, described in the preceding section. That section listed measured values of average extreme daily axle loads for various axle groups, for two classes of sites. The corresponding legal limits for the complete axle groups are listed below.

Single axle – steer	6.0 t	(58.8 kN)
Tandem – steer	11.0 t	(53.9 kN per axle)
Tandem – nonsteer	16.5 t	(80.9 kN per axle)
Tri-axle group	20.0 t	(65.4 kN per axle)

The ratios of measured values to legal limits (short–medium, long sites) were (1.29, 1.14), (1.13, 1.09), (1.26, 1.27) and (1.24, 1.22) for these four axle configurations.

At the time of this survey, Australian legal limits varied a little from State to State, as each applied a so-called 'bridge formula' to limit the total mass, M(t), as a function of the distance, L(m), between extreme axles, with an upper bound in the range 38 to 42.5 t; two examples are

Queensland	$M \ngtr 3.0L + 8.0$
Northern Territory	$M \ngtr 3.6L + 7.2$.

These formulae were generally applied also to the truck of a truck-semitrailer combination.

The presence of overload is recognised by most codes, which specify a design vehicle somewhat heavier than the legal limit. This provision is appropriate, but it should be observed that its influence on bridge design is more apparent than real, as the primary loads are multiplied by load factors, whose level may be adjusted.

Some other examples of legal limits for highway vehicles may be given. Chatterjee (1991: 64, 5) quotes an agreement reached in 1986 for a single regulation throughout all countries of the European Community. For Great Britain, this meant 'an increase in the maximum axle weight from 10.5 to 11.5 t, and in the maximum vehicle weight from 38 to 40 t'.

An earlier OECD report (1979) summarised legal limits in twelve European countries as at 1 January 1977. The maximum permitted length for a two-axle vehicle varied from 10 to 12.4 m; for an articulated vehicle, from 15 to 16.5 m; and for a road train, from 18 to 24 m. The maximum masses per axle group were:

single axle	10 to 13 t
dual axle	16 to 21 t

and for complete vehicles:

with 2 axles	16 to 20 t
articulated, with 5 axles	generally 32.5 to 44 t (28 t in Switzerland)
road trains	generally 38 to 50 t (32.5 t in the United Kingdom)

Two extensive reviews of American truck loads were published in 1989 and 1990, the first entitled *Providing Access for Large Trucks*, and the second, *Truck Weight Limits: Issues and Options* (Humphrey et al. 1989; Cohen and Hoel 1990). The following summary of American legal limits is taken from the second. In 1946, AASHO recommended single axle limits of 8.16 t (18,000 lbm) and, for a tandem axle, 14.52 t (32,000 lbm), with a maximum total vehicle mass of 33.24 t (73,280 lbm), provided that the extreme axle spacing was not less than 17.4 m (57 ft). For shorter lengths, L, the limiting gross mass, M, was given by

$$M = 10.25\,(L + 24) - 3L^2 \quad \text{(lbm and ft units); or}$$
$$M(\text{t}) = 1.525\,(L + 7.32) - 0.01465L^2 \quad \text{(for } L \text{ in m)}$$

The intention of this constraint on total mass was to limit overstressing of bridges, and the formula was referred to as a bridge formula (Formula A).

These AASHO rules were adopted in 1956 for Interstate highways, and then increased in 1974. The new limits for axle groups were 9.07 t (20,000 lbm) for single axles, and 15.4 t (34,000 lbm) for tandems. The new limit on total mass was 36.29 t (80,000 lbm), with a new bridge formula (Formula B):

$$M = 500\,[LN/(N-1) + 12N + 36]\ \text{(lbm and ft units); or}$$
$$M(\text{t}) = 0.2268\,[3.281LN/(N-1) + 12N + 36]\ \text{(for } L \text{ in m)}$$

The 1990 study retained these limits on axle and total mass but proposed a new bridge formula, called the TT1 HS-20 formula:

$$M = 1000\,(2L + 26) \text{ for } L \leq 24 \text{ (lbm and ft units)}$$
$$M = 1000\,(L/2 + 62) \text{ for } L > 24 \text{ (lbm and ft units); or}$$
$$M(\text{t}) = 0.4536\,(6.562L + 26) \text{ for } L \leq 7.32 \text{ m}$$
$$M(\text{t}) = 0.4536\,(1.64L + 62) \text{ for } L > 7.32 \text{ m}$$

Heavier loads were permitted on Special Permit Vehicles.

In Canada, the CL-W design truck was based on a Memorandum of Understanding (MOU) between the Canadian Provinces reached originally in 1988, and then amended in 1991. This design truck was originally CL-600, with a total weight of 600 kN, or mass of 61.2 t; it was then increased to the CL-625 (625 kN, 63.7 t). It will be recalled that this design vehicle was intended to model legal limits, whereas the Ontario OHBD Truck was based on the Maximum Observed Loads (MOL) in the Province of Ontario, somewhat higher than the legal limits. Some of the MOU legal limits are:

single steer axle	55 kN (5.61 t)
other single axles	91 kN (9.28 t)
tandem axle group	170 kN (17.34 t)

The maximum gross weight is 225 kN (22.9 t) for a 'straight truck', rising to 625 kN (63.7 t) for what was called a B-train, made up of a tractor vehicle with a single steer axle and a dual or tandem drive axle group, followed by a trailer with a tri-axle and a tandem. The B-train has a minimum length of 19 m.

It is important to observe that, whereas the OECD (1979) study of European legal limits in 1977 specified *maximum* values for the lengths of vehicles, the later references require certain *minimum* lengths, either defined specifically, or by the so-called bridge formulae. These newer limits reflect the effects of these loads on bridges and express a desire to increase total vehicle mass without placing undue stress on an existing stock of bridges. The classic strategy of this kind was that adopted in Ontario in connection with the development of the 1979 OHBD Truck (see Section 4.4). The concept of Equivalent Base Length (Dorton and Csagoly 1977) was used to plot data from the current legal limits for trucks in Ontario and gave rise to the following formula:

$$W\,(\text{lbf} \times 10^3) = 20 + 2.07B_M - 0.0071B_M^2\ (B_M \text{ in feet})$$

This was later revised to

$$W(\text{kN}) = 9.806\ (10 + 3.0B_M - 0.0325B_M^2)\ (B_M \text{ in m})$$

It was called the Ontario Bridge Formula (OBF) and used to define a new flexible system of legal limits. With the earlier form, the maximum legal limit for vehicle length was 65 feet (19.81 m); the maximum value of B_M corresponding to this was 80 feet (24.4 m) giving W equal to 140,000 lbf, or 63.5 t (623 kN). The later 1991 Canadian Memorandum of Understanding gave a figure of 63.7 t (625 kN), as quoted earlier. The Ontario MOL curve, from which the OHBD Truck was derived, lay 22,500 lb (10.2 t or 100 kN) above the OBF curve. As stated by Dorton and Csagoly (1977), this difference was a measure of the effectiveness of enforcement of legal limits at that time. The OBF was initially used as a formula for the specification of legal limits in Ontario. It was found difficult to obtain court convictions for overloaded vehicles, and so it was later replaced by a series of tables. The upper bound for vehicle weight of 63.5 t may be compared with the corresponding figures for other countries quoted earlier; for example, the range from 38 to 50 t for European countries in 1977. The Ontario practice was a step towards obtaining heavier vehicle loads without penalty to existing bridges, for by distributing the load in an appropriate manner, over a suitable length, the total vehicle weight could be increased. It appears that this provision had a significant effect on trends in vehicle design, such as the use of a separate trailer with a long, rigid towbar.

There are significant cost benefits in the use of heavier vehicles, provided that these can be accommodated safely on national roads. Four choices are available:

(a) existing load limits may be retained, so as to safeguard existing bridges;
(b) load limits may be increased if it is perceived that older design procedures have resulted in bridges with a sufficient reserve of strength;
(c) a device such as the Ontario Bridge Formula may be used to increase the total legal load without risk to existing bridges; or
(d) it may be judged that the economic benefits achieved in national transport by the use of heavier vehicles may justify the construction of new bridges to a higher standard, accompanied by a programme for the strengthening of existing bridges.

In this connection, it is necessary to have some idea of the costs involved; it should be noted that some of these will vary greatly from country to country. The two American studies quoted previously (Humphrey et al. 1989, Cohen and Hoel 1990) investigated the effect of changes to legal limits in the United States, and of the rules governing the access of the heavier vehicles to local roads. Detailed safety and cost studies were carried out; and as a result the maximum gross vehicle mass was held at 36.29 t (80,000 lbm), the legal axle loads were kept unchanged, but the bridge formula was altered, as described earlier. Other options were also examined, including use of the 1988 Canadian Memorandum of Understanding rules, which permitted the total truck weight to rise as high as 61.2 or 63.7 t (135,000–140,000 lb) for long vehicles. The new bridge formula gave the following changes in total costs ($US billions/year; + indicates an increase, − a decrease; Cohen and Hoel 1990: 13):

total transport	−5.1
pavement maintenance	+0.1
bridge strengthening or replacement	+0.4
total	−4.6

If the Canadian MOU limits were adopted, the corresponding changes would be −11.7, +0.5, +2.4, −8.8. Both show that the savings in general transport costs greatly outweighed the outlays on pavements and bridges. These figures apply only to National Highways, or to roads and bridges funded under the Federal Aid programme. The report by Humphrey et al. (1989: 189, 242–3) lists 182,710 miles (294,000 km) as the National Network Mileage, or 209,310 miles (337,000 km) if other roads under the Federal Aid system are included. The corresponding bridge numbers are given in Table 5.1.

A similar major study has been carried out in Australia (Austroads 1998) in connection with the proposal to adopt the M1600 and S1600 design vehicle loads shown in Fig. 4.13, where the first of these consists of four tri-axle groups, each with a total weight of 360 kN (equivalent to a mass of 36.7 t), together with a uniformly distributed load, one of the heaviest of proposed or recently adopted bridge design loads (see Figs. 4.19–21). Work by Pearson (included in this report) suggests that the use of longer vehicles with payloads increased from 26.5 t (the present value) to 70 t would reduce Australian transport costs by a little over $A1 billion per year; that the incremental benefits were greatest in the range 28.5 to 40 t, and small beyond about 65 t. Australia currently permits the use of a range of vehicle types (Fig. 5.17) on inland routes, with a limiting mass of 20 t applied to a three-axle group, where this may be increased in the future to 40 t. The total vehicle masses are listed in Table 5.2 (Gordon and Boully 1998).

The Articulated Vehicle is a semi-trailer, like the HS-20 truck. The B-Double has a further axle group, supporting a platform that is pivoted to the preceding semi-trailer. The B-Triple has a third platform, supported in the same way from the preceding B-Double. On the other hand, the road train is a semi-trailer followed by trailers, each of which has two axle groups, consisting typically of a dual – followed by a triple-axle group.

Table 5.2

Designed to	*Bridges*
HS-20 or greater	100,370
HS-15 and H-20	43,369
H-15 or less	67,366
other (generally < HS-20)	33,910
Total	245,015

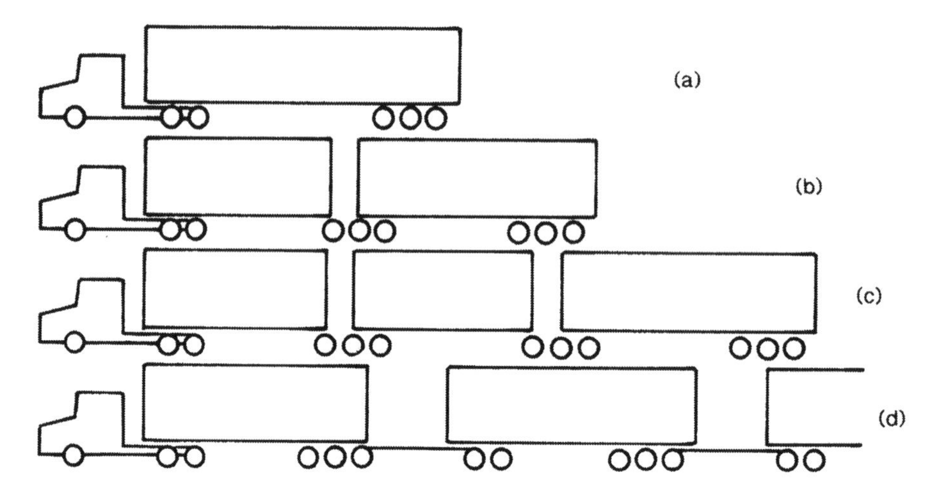

Fig. 5.17 Vehicle types permitted on some Australian roads

Table 5.3

Vehicle type	*Fig. 5.17*	*Current maximum legal length (m)*	*Total vehicle mass (t), with 20 t or 40 t per axle group*	
Articulated	(a)	19.0	42.5	74
B-Double	(b)	25.0	62.5	114
B-Triple	(c)	33.0	82.5	154
Road train	(d)	53.5	115.5	208

A study was made of future bridge costs, with three alternative vehicles – articulated, B-Double and B-Triple – each with 40 t axle groups, as listed above. The resulting bridges were compared with those designed to existing standards (T44 or more), using an appropriate mixture of bridge spans and types. The increases in annual costs were $A1.8 million (0.8%), $A4.2 million (1.9%), and $A6.4 million (2.9%). As with the American study, these increases are relatively small. Similarly, estimates were made of increases to pavement maintenance costs and the costs necessary to upgrade existing bridges (R. Gordon in Austroads 1998). This estimate of upgrading varied from $A2.45 billion to $A3.85 billion, or from 2.2 to 3.5 years of the saving in general transport cost. Although considerable, this cost is not excessive.

Perhaps of more general significance are the strategies recommended for the implementation of these proposals. The new design load would be introduced immediately for the design of new bridges. The mass limits on vehicles, on the other hand, would be implemented in stages, with increases in years 6, 14 and 20. It is rather too easy to do cost estimates to show the economic desirability of increases in design load and legal limits; it is rather more difficult to ensure the safety of existing bridges and the feasibility of their replacement. This is particularly so in the case of larger bridges with long spans, and the matter

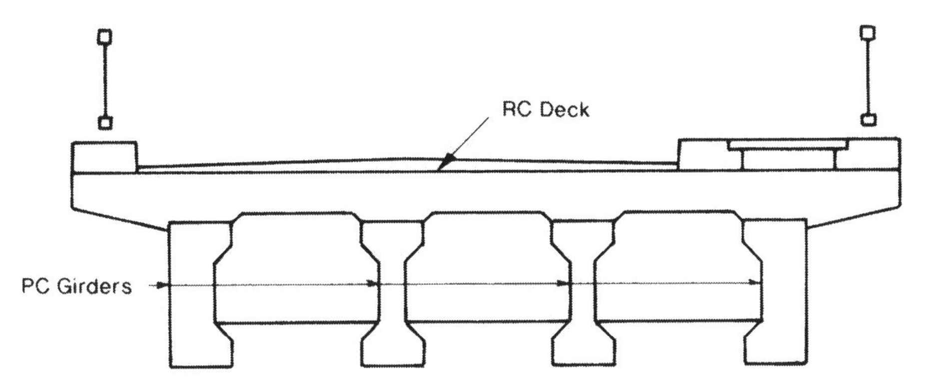

Fig. 5.18 Typical cross-section of small span bridge in study by Gabel (1982) of minimum-cost design

will be referred to again in Section 5.6. This may be of less significance in Australia than in other countries, for the Australian routes carrying heavy loads tend to be inland routes with fewer waterways and space to relocate bridges where this is required.

The effect of traffic loads on bridge design may be illustrated by two examples. Gabel (1982) developed a program for the minimum-cost design of bridges of small span, with cross-sections similar to that shown in Fig. 5.18, using prestressed concrete girders of standard shape. This program was based on the 1976 NAASRA Code, which in turn was derived from the American AASHO Code. The loading was the HS-20 truck or lane load, with provision for an abnormal vehicle load. The spans considered ranged from 15 to 32 m, the latter figure representing the maximum span that could be achieved with the specified girder geometries. The design procedures incorporated analyses for the distribution of the primary loads between the main girders, with proper allowance for the effect of diaphragms. A 40% increase in live loading was found to require only a 7 to 12% increase in the total superstructure cost.

Another study (O'Connor 1971b) considered prestressed concrete box girders of much larger span, from 61 to 305 m (200 to 1000 feet) and produced optimum designs for bridges continuous over an infinite number of equal spans. Although the study considered also girders of constant depth, the cases quoted here have variable depth with a soffit that is circular in elevation. The form of the bridges is shown in Fig. 5.19, with three lanes on an independent structure that forms one half of a divided highway. They were designed to the 1965 AASHO Code, for the HS-20 loading, and the Australian Code CA35 for Prestressed Concrete (1963). The self-weights of some optimum designs are shown in Fig. 5.20, in the form of weight, w, per unit deck area (kN/m^2). The equivalent straight line shown on the figure has the equation,

$$w\ (\text{kN/m}^2) = 7.2 + 0.07L\ (L \text{ in m})$$

The AASHO lane loading consists of a concentrated load (for bending moment) of 107 kN on each span, together with a uniformly distributed load of

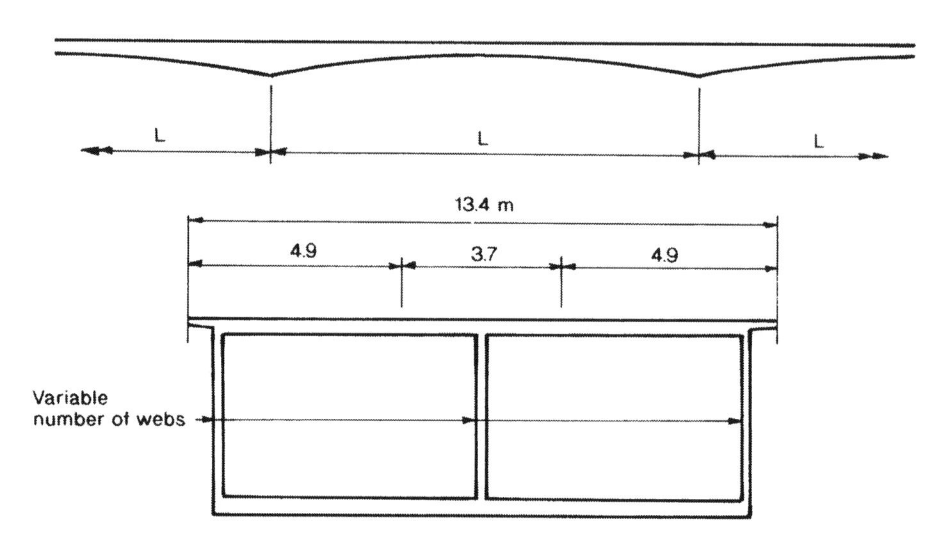

Fig. 5.19 Bridge form considered by O'Connor (1971b) in study of minimum-cost design

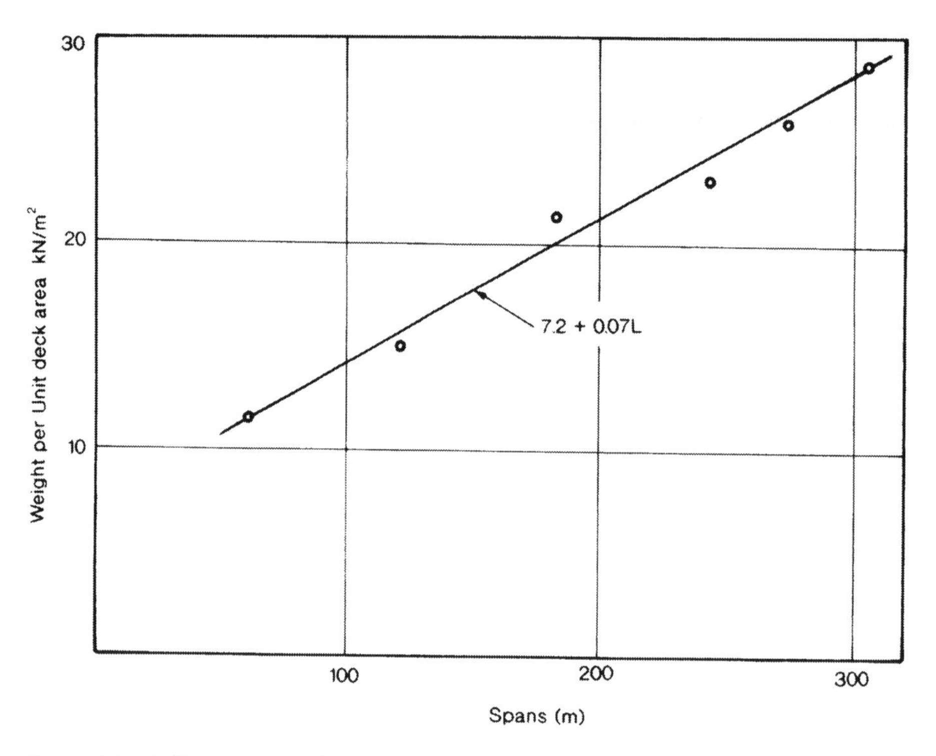

Fig. 5.20 Self-weight of long-span, computer-designed, prestressed concrete bridges (O'Connor 1971b)

9.34 kN/m for each lane. The design has assumed three lanes. If the distributed live load on these three lanes is averaged over the 13.4 m total width, with the specified reduction of 90% for multiple loading, the resulting load is 1.88 kN/m^2. For comparison, the self-weights, w, given by the above expression, and the ratios, $v = 1.88/w$, are:

100 m	$w = 14.2$ kN/m^2	$v = 0.132$
200 m	$w = 21.2$ kN/m^2	$v = 0.089$
300 m	$w = 28.2$ kN/m^2	$v = 0.067$

For these spans, the effect of the concentrated load is small. For example, a load of 107 kN averaged over the total deck area is equivalent to only 0.08, 0.04 and 0.026 kN/m^2 for these spans.

The costs of the optimum designs were plotted against span length, and took the form (revised to metric units),

$$\text{cost } (\$/\text{m}^2) = A\ (1 + 0.0125L)$$

where A is a constant.

The validity of the procedure can be judged by observing that the optimum design for a span of 182.9 m (600 feet) had a depth of $0.07L$ at the supports and one half of this at mid-span. The unit costs used in the study were derived from those for the Captain Cook Bridge, a bridge with the same maximum span then under construction in Brisbane. It had depths of $0.05L$ and $0.025L$. The program was re-run with these depths and gave a cost only 7% above the optimum. The choice of the shallower depth was justifiable on other grounds.

When these factors are put together – the relative size of the live load compared with the self-weight, and the relative insensitivity of cost to span (and other causes of increased bending moment and shear) – one can judge that increases in the design live load would have only a small effect on total cost. It is of some interest to observe that this linear increase of cost with span reproduced a result observed in other studies (O'Connor 1971b).

5.6 Bridge life and acceptable risk

There is a possible ambiguity in the use of the term 'life'. Consider the case of a building. Here the basic concept of life is the time it is expected to remain in service, until it is replaced for reasons that include the following:

(a) its form may no longer be appropriate;
(b) it may be functionally inefficient;
(c) the use for which it was designed may have disappeared; and
(d) it may now constitute an uneconomic use of its site.

It is, then, desired that the building should not suffer structural failure during this functional or economic life and this gives rise to a structural use of the word 'life' – as the time until the building fails (or collapses), or our best estimate of this time. It follows that a life chosen on non-structural grounds becomes the *target* life for the structural design.

This idea of a functional or economic life is less easily defined in the case of a bridge, although some become obsolete because of width or alignment considerations or other factors, or may indeed fall out of use. If the concept of bridge life is fully embraced, strategies for bridge replacement should be developed. More truly, the public perception of major bridges, particularly those that become a distinctive city feature, such as the Sydney Harbour Bridge, is that they are there forever, and maintenance procedures are put into place in an endeavour to ensure that this is so. A major bridge becomes the focus of major traffic routes and, if it is a city bridge, the adjoining land is typically heavily developed. Both combine to create a future difficulty. A true dedication to the concept of a limited bridge life would require that, from the outset, land is retained for a replacement bridge with a suitable geometry for its approaches.

It is appropriate to reflect on the true significance of some elements of the current approach to probabilistic limit states design. The Eurocode ENV 1991-1 (1994) suggests a target bridge life of 100 years, with a target reliability index, β, of 3.8. The calibration procedures for the 1994 AASHTO code chose a value, $\beta = 3.5$, and, implicitly, a bridge life of 75 years. The corresponding annual risks of failure are 0.724×10^{-6} and 3.1×10^{-6}. As shown in Section 5.3, the AASHTO values were based on a desire to keep the risk of failure of bridges designed to the new code about the same as those from the preceding code, and to make this risk more consistent.

It is suggested that this method of code calibration is entirely appropriate. To start, *ab initio*, to determine an acceptable risk in cases involving the possible loss of human life is difficult, although two studies may be quoted. Menzies (Menzies, J.B. 'Bridge failures, hazards and societal risks', 36–41, in Das 1997: 36–41) refers to the *fatal accident rate* (FAR) as the risk of death from 100 million hours of exposure to some particular activity, where the basic value of FAR for home risks is about 1. He states that 'in the United Kingdom there are about 4000 deaths each year on the roads in a population of 56 million which equates to a FAR of about 20'. This implies an exposure to road death of 1 hour per person per day, as can be shown as follows.

A FAR of 20 implies a risk of 20 deaths in 100×10^6 hours of individual exposure, equivalent to 0.274×10^6 years. For a population of 56×10^6 individuals, one would expect 20 deaths in 0.274/56 years, or 4088 per year. He subdivides the causes of these 4088 road deaths by quoting the following relative figures for various modes of travel:

by motor cycle	300
by pedal cycle	60
walking beside a road	20
travel by car	15
travel by bus	1

For 100,000 highway bridges in the United Kingdom, he quotes a recorded failure rate of about one in every 1 or 2 years, from all causes. If such factors as scour and flooding are deleted, a possible collapse rate 'due to structural deficiency or overloading' is about one in every five years, with a loss of life of about one in 10 years. There is some uncertainty in estimating the corresponding exposure time. If it were taken as 1 hour per day (as before) then the corresponding value of FAR is much less than 1 (actually about 0.5×10^{-3}). Menzies suggests that this historical failure rate has been acceptable. Accordingly he quotes the following risks per bridge as acceptable:

annual risk of accidental death due to structural bridge failure	1 in 10^6
annual risk of structural bridge failure	1 in 0.5×10^6
risk of structural collapse in a 100-year bridge life	1 in 5000

The target annual risks of failure for the Eurocode ENV 1991-1 (1994) were quoted above as 0.724×10^{-6} and 3.1×10^{-6}, or, in inverse form, 1 in 1.38×10^6, and 1 in 0.323×10^6. These straddle the value of 0.5 in 10^6 suggested by Menzies, although the differences are not great.

Another major study was done by Allen (1992), although this was concerned with the assessment of existing bridges, or bridge evaluation, rather than new bridge design.

It is useful to reflect again on some of the data found by Nowak (1993 and 1995). Figure 5.6 presented some of his values for β, calculated for survey truck data, using bridges that incorporated the minimum strength permitted by the old code. For the cases shown (moments in composite steel girders, moments in prestressed concrete girders, and shears in prestressed concrete girders) the minimum value of β was 1.9, for bending in a prestressed concrete girder of 9.1 m span. The corresponding probability of failure in 75 years was 0.0287 (Devore 1995, Fraser 1976), or 1 in 2610 years. Section 5.5 quoted from the American Transportation Research Board report written by Humphrey (1989: 189), which listed a total of 245,015 bridges on Federal Aid Highways. A probability of failure of 1 in 2610 years (i.e. 1/0.0287 in 75 years) would imply that 94 bridges should fail because of primary overloads each year. This is, however, an overestimate, as only a few of Nowak's girders yielded a β value of 1.9. On the other hand, more than half the bridges were designed for loads less than HS-20. Nowak's β values for shear were typically of the order of magnitude shown in Fig. 5.6(c), with most less than 3.0. If the calculation is repeated for β equal to 3.0, this gives a probability of 0.00135 in 75 years, or 4.4 bridge failures per year. Even this figure may be an overestimate, for Nowak also records that a study of actual bridges gave strengths ranging from 1.2 to 2.0 times the minimum required values. These rough calculations still suggest, however, that one would expect more than four cases of failure through overloading each year; there is no evidence that this number is occurring, and there is a need for improved data on the probability of failure as demonstrated by bridges in service.

5.7 Reserves of strength

The calibration procedures built into modern limit states codes should ensure that part of the reserve of strength held by existing bridges is fully utilised, such as that between the occurrence of first yield in a steel bridge and the development of the plastic hinge moment. There may be additional reserves of strength which include the following.

1. A bridge may behave in a way which differs from that assumed in its original design. For example, the moving bearings of a bridge may seize up, preventing the lower flange from expanding under the action of the primary loads (O'Connor and Pritchard 1982, Ransom 1999).
2. Yield may modify behaviour in two ways. In a beam continuous over a number of spans, true collapse can occur only when a suitable mechanism is formed, with a longitudinal redistribution of bending moments. In most bridges there is the possibility also of a transverse redistribution. Yield in a longitudinal beam may transfer load to its neighbours and collapse may involve the failure of a number of parallel members. Even a bridge consisting of a single primary member, such as a multi-cellular box girder, will undergo change in behaviour during collapse.
3. Figure 2.20 showed that the yield strength of a particular steel increased with the strain rate. Soroushian and Choi (1987) have reported similar increases, not only for structural steel, but also for steel reinforcement. Soroushian, Choi and Alhamad (1986) have described increases in the strength of concrete caused by increases in the strain rate. It follows that elevations in strength may occur under the dynamic components of the primary loads.
4. As mentioned in Section 2.3, the collapse of a bridge is essentially a dynamic phenomenon. The passage of a load large enough to cause static collapse may instead leave the bridge in place, but with some permanent deformation. On the other hand, dynamic factors can cause premature collapse, as illustrated in Section 6.1.

These factors mean that bridge failure may be evidenced not so much by a sudden collapse but by a growth in damage caused by successive large events. This damage may lie in the development of permanent deflections, in the onset of cracking in concrete structures, or in the growth of fatigue cracks in steel. Regular inspection by an experienced bridge engineer should form part of any bridge maintenance programme. These remarks lead also to two final conclusions.

(a) Further work needs to be done on the mechanisms of bridge failure.
(b) There is also a need for improved failure statistics, where these should take note of progressive failure as well as total collapse.

5.8 Bridge evaluation and proof testing

Bridge evaluation and *assessment* are terms commonly applied to the examination of existing bridges to determine their condition and load-bearing capacity. Such an examination may lead to a decision that a bridge needs to be *rated*; that is, that a limit must be imposed on the weights of vehicles allowed to use the bridge. These topics are discussed in many codes. Some of this discussion lies outside the scope of this book, and the treatment here will be limited to the *proof testing* of bridges.

A proof load may be defined as a load applied to an existing structure to determine or 'prove' its load-bearing capacity. The areas of deck loaded to produce critical effects in bridge components will vary. The proof loading of a bridge may, therefore, consist of a programme of tests with the superimposed (or applied) loads moved carefully from one location to another. The most common proof loads are stationary and applied incrementally; that is, there will be stationary vehicles or load locations, with loads added to these vehicles or locations in increments. This allows the effects of the loading to be assessed as it is increased, so as to give some warning of impending collapse. For example, a deflection may be plotted against the applied load. The initial response may be linear, but as the load is increased it may be found that the slope of the load-deflection curve increases, where this may be taken as a warning that there should be no more increases in load.

The design of a proof testing programme for a particular bridge requires, therefore, the selection of particular locations for the loads, together with some decision as to the manner in which the loads are applied and increased. There will also be decisions concerning the observations to be taken during loading, and these will commonly include deflection and strain readings at well-chosen locations.

Two important matters arise and these will be discussed here:

1. There is a risk of collapse during the proof test.
2. The outcome of the test will commonly include an assessment that the structure may safely carry a particular level of service load. The knowledge gained from the proof test may be used to reassess the probability of failure in a manner that needs to be determined.

A useful discussion of the problem may be found in the paper by Moses, Lebet and Bez (1994). Consider the case of a simply-supported girder bridge of moderate span. The loading may take the form of a particular truck, or a particular set of trucks placed side by side in adjacent lanes, where the longitudinal location of these loads to produce maximum girder bending moments may be determined by the method described in Chapter 1. The number of trucks in adjacent lanes and their transverse location will need to be determined. In many cases, girder bending moments will vary with this transverse location, and this is one reason why a number of tests may be required. Let it be supposed, however, that a suitable transverse location has been determined. Then the resulting bending moment in the girder of interest may be calculated as a function of the level of the applied load. The corresponding service bending moments will vary. Suppose further that the statistical distribution of these bending moments has been assessed in the manner described in Sections 5.1 and 5.3. Then this loading may be compared with resistance or strength as in Section 3.3 and Fig. 3.2, to give an estimate of the probability of failure. The decision to carry out the proof test has, presumably, resulted from an initial observation that this probability of failure is unacceptable for current levels of the service loads.

In the present context, the first benefit that results from prior statistical estimates of resistance and strength is that they allow for an assessment to be made of the probability of failure during the proof test, and the initial selection of the magnitude of the proof load. If perfect control were exercised over the loading, then the load effect, L, is known. Failure will occur for any value of the resistance, R, less than L. The probability of failure is represented, therefore, by the area of the probability density distribution for resistance below the level $R = L$; or by the cumulative probability that $R \leq L$. The actual risk is somewhat less, for two reasons.

(a) Suppose the load level is being increased from L_{-1}, the prior test load, to L. Then it is known that the resistance, R, is not less than L_{-1}. Part of the probability density distribution for the resistance R may be deleted (but with some adjustment, as will be seen below).

(b) The observations being made during the test provide a safeguard against failure, for, at any time, the test may be terminated if the evidence indicates that collapse is imminent.

The second benefit is similar to that mentioned in (a). Let L_{PL} be the maximum level of a particular load effect produced during the programme of proof tests. Then it is reasonable to assume after the test that the corresponding resistance is not less than L_{PL}. The probability density distribution for R may, therefore, be truncated as shown in Fig. 5.21, where this truncation has two logical stages. Initially, all probability of resistances less than L_{PL} may be deleted, leaving only the right hand portion of the initial curve ($R > L_{PL}$). However, any probability density distribution must have a total area of unity. The initial, truncated distribution for R needs to be revised. The simplest way of doing this is to adjust the scale of its vertical ordinates.

Let A_{RPL} equal the area of the initial probability density distribution to the left of the point $R = L_{PL}$. Then the corresponding initial area to the right is $(1 - A_{RPL})$. The area to the right may, therefore, be brought back to zero if each vertical ordinate is divided by $(1 - A_{RPL})$. Section 3.3 presented a numerical method for the evaluation of the convolution integral that yielded the probability of failure when R and L were both normally distributed. It is not difficult to adjust this for the present case. The initial, untruncated, probability of failure may be evaluated, and divided by $(1 - A_{RPL})$, and then the contribution from that part of the resistance curve with $R < L_{PL}$ may be evaluated and deleted.

It should be observed, however, that this calculation, although not unreasonable, may involve a significant assumption. The original probability density

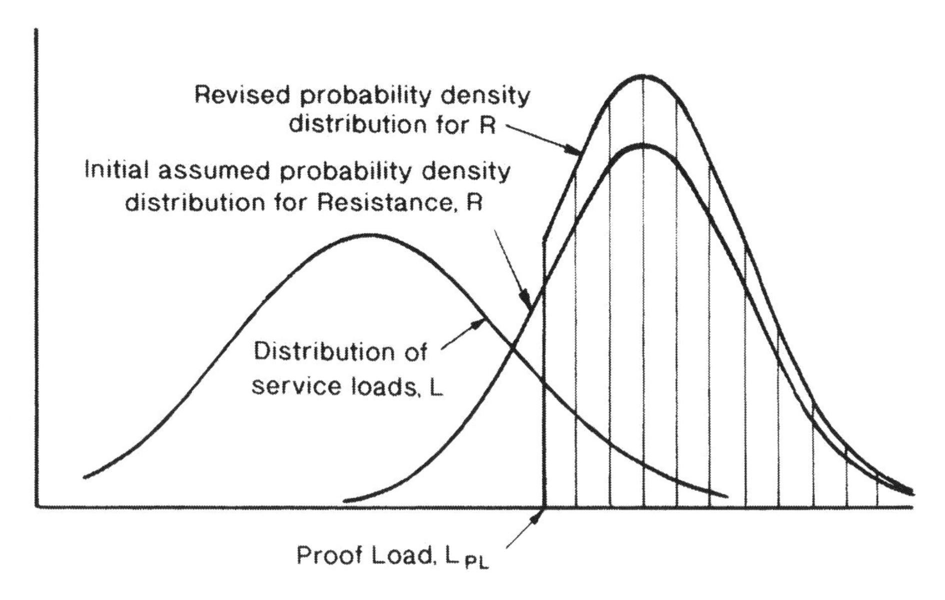

Fig. 5.21 Proof load testing

distributions for R and L were, presumably, based on observed data. In some cases, a proof test may indicate increases in R, or reductions in L, that were not present in these observations. A not uncommon effect is that seizure of bridge bearings reduces bending stresses and indicates a reserve of strength beyond that observed by the designer. Before relying too heavily on this reserve of strength, it would be desirable to have some data on its statistical distribution. That is, the modification described above after truncation of the resistance curve, may not be fully justified. Despite the precautionary warning, however, it does seem reasonable to assume that the estimated probability of failure, following a proof test, will be reduced.

A further note should be added. The objectives of a programme of proof tests should include not only an intention to determine the safety of existing bridges, but also a desire to improve the quality of future theoretical predictions. An observation that some experimental reserve of strength is present is not really sufficient. Rather, one would hope that the source of this reserve of strength is identified and built into future design and assessment calculations.

5.9 Serviceability loads

A bridge may fail in ways other than collapse, and for this reason all modern bridge codes specify loads for two classes of limit state: serviceability and collapse.

Part 1 of Eurocode 1 (ENV 1991-1) deals with structures generally, including buildings as well as bridges, and includes the following statements:

> 3.1 Limit states are states beyond which the structure no longer satisfies the design performance requirements. In general, a distinction is made between ultimate limit states and serviceability limit states.
> 3.3 Serviceability limit states correspond to conditions beyond which specified service requirements for a structure or structural element are no longer met. The serviceability requirements concern:
> - the function of the construction works (bridges in the present case) or parts of them;
> - the comfort of people;
> - the appearance.

The 1992 Austroads Bridge Design Code goes further:

> The serviceability limit states to be considered include:
> (a) permanent deformation of foundation material or a major load carrying element, of sufficient magnitude that the structure is unfit for use, or such that the public would become concerned as to the safety of the structure;
> (b) permanent damage due to corrosion, cracking or fatigue which significantly reduces structural strength or useful service life;
> (c) vibration leading to structural damage or public concern;
> (d) flooding of the road network and surrounding land and scour damage to the channel bed, banks and road embankments.

Some of these items are not concerned with traffic loads, such as item (d), which speaks of scour and flooding. Other common serviceability problems in bridges relate to such matters as the performance of deck expansion joints, deck drainage and the disposal of the water that may pass through the deck joints. None are concerned chiefly with major traffic loads; rather, all have to do with the quality of the design details. Again, there are matters such as the development of a bump on the road surface as traffic enters the bridge – uncomfortable, but also leading to increased dynamic effects in the loads applied by vehicles to the bridge. Again, temperature changes may be at the root of certain serviceability problems.

However, some are related directly to traffic loads, including (b) and (c). The bending behaviour of reinforced and prestressed concrete members divides into two important stages, before and after cracking. The ultimate limit state generally occurs after cracking, and the tensile strength of the concrete may be ignored. On the other hand, the durability of the concrete depends partly on the presence of cracks. It is appropriate, therefore, for codes to specify loads that may be used to check this limit state. The same applies to dynamic effects that cause perceptible vertical accelerations at frequencies that may lead to discomfort or public concern (see Chapter 7).

Codes of practice must, therefore, specify two levels for the primary design loads: one to check serviceability, and the other, collapse. It was noted elsewhere that, as far as collapse is concerned, codes commonly specify both a load and a load factor, where it is the multiple of these that is relevant. The need to specify serviceability loads adds another dimension. Common practice typically specifies a basic design load that is either a little less than, or about the correct order for, serviceability checks, together with a load factor that enables this to be increased for ultimate limit states. The need for a serviceability load causes the specified loads and load factors to have an appropriate, rather than an arbitrary relationship to one another.

In the Australian context, Fenwick (1985) was quoted earlier as speaking of the ultimate load as that which has a 5% chance of occurrence within the 100-year life of a bridge. It should be noted that this is not a statement that says that one should identify such a truck, allocate to it a weight and coefficient of variation, and use these to achieve a desired level of reliability. Rather, this statement should be interpreted as suggesting that a reliability analysis would finally yield load effects (specified loads × load factors) that approximate to this description. Similarly, Fenwick speaks of two possible 'definitions' of the serviceability load:

(a) a load with a 95% chance of occurrence in the design life; or
(b) one which as a 5% chance of occurrence per year.

The second of these definitions was chosen as the preferred basis for selection of serviceability loads for Australian bridges. It indicates that such vehicles may be expected to occur a number of times (notionally 5) within the 100-year life of a bridge. The load actually chosen was the T44 truck (Section 4.7) with a serviceability load factor of 1.00, and to this was then applied a dynamic load allowance.

Serviceability loads are required for various kinds of behaviour. It is impossible, therefore, to derive them generally in the same way as that used in Section 3.3 for the ultimate limit state. It is difficult also to summarise code provisions, for they apply to a range of limit states. Some of these matters, however, such as fatigue loads and deflection limits for comfort, will be considered later (see Chapters 6 and 7).

Chapter 6

Multiple presence, progressive failure and fatigue

6.1 Multiple presence

Much of the work done on the selection of primary design loads for bridges has been concerned with the identification of a single heavy vehicle, for example, a truck with a specified return interval. This work needs to be extended in two ways: (a) all modern bridges carry at least two lanes of traffic, and truck loads in parallel lanes need to be considered; (b) in long span bridges, there are also significant effects from the presence of succeeding vehicles in each lane.

This need to allow for multiple presence was evident in the discussion of codes of practice in Chapter 4. The old Ministry of Transport loading, and the primary design loads specified in BS 5400 and BD37/88 all model trains of vehicles in one or two lanes (Section 4.2). Successive American AASHTO codes have been based on a single heavy truck per lane, but then apply a similar truck load in adjacent lanes, with reduction factors for the loads in lanes beyond Lane 2 (Section 4.3). The draft Eurocode applies a reduced load in the second lane (Section 4.6).

It is appropriate, therefore, to consider the probability of occurrence of multiple vehicle loads. Before doing this, one must observe that a variety of cases arise.

1. Load effects in some bridge components, such as deck details, may be due to local loads, such as a single axle or wheel load.
2. In some cases, the application of loads in a second lane will not change the design, for example, in closely-spaced parallel girder systems, or in the case of longitudinal girder units with limited shear transfer between units.
3. The lanes of a two-lane bridge commonly carry vehicles moving in opposite directions.
4. On the other hand, a divided highway may be carried by separate parallel bridges, with all loads on one bridge moving in the same direction.

5. The traffic on a city bridge may be stationary, with a mix of vehicles filling all lanes at a minimum longitudinal spacing.
6. The traffic on country bridges may vary widely. In some cases, the probability of one heavy vehicle passing or overtaking another is low. In other cases, heavy trucks may be bunched together, occupying more than a single lane. An example occurs on motorways, such as in the United Kingdom, where there are commonly three lanes in each direction, but of these, heavy trucks may be allowed to use only the outer two. Cars can readily pass, using the third lane. Trucks wishing to overtake may find it difficult to do so, with the result that an uninterrupted succession of heavy vehicles may occupy the outside lane. Further, when trucks do overtake, their speed differential is small; overtaking takes some time, and a large group of trucks may often be found filling the outer two lanes.
7. Military traffic also may completely fill a lane.

It is difficult, therefore, to give general conclusions concerning multiple presence. Some discussion of methods of analysis, however, may be useful.

Earlier studies include those by Asplund (1955), Ivy et al. (1954) and Murakami et al. (1972). Ivy et al. gives two typical graphs of observed vehicle weights along a 550 m length of the lower deck of the San Francisco-Oakland Bay Bridge (with a main span of 704 m). The AASHTO code (Section 4.3) at that time recommended the HS20-44 truck and lane loads, where the latter consisted of an 80 kN line load for bending moments, and a uniformly distributed load of 9.34 kN/m; the code introduction suggested that this be used only for spans less than 152 m. The traffic patterns observed by Ivy et al. indicated that these design loads were conservative for spans in excess of 120 m. They recommended that the line load be reduced to 40 kN for spans from 183 to 244 m, and then to zero for larger spans, with the concurrent distributed load taken as 8.75 kN/m for spans from 305 to 366 m, and then 8.17 kN/m for longer spans.

The work carried out by Buckland (1981), Buckland et al. (1980) began as a study of loads on the Lion's Gate suspension bridge in Vancouver (main span 473 m). It relies on the observation that maximum load effects occurred only when the traffic was stationary and 'bumper to bumper'. For this reason, their final recommendations (see Section 4.3 and Fig. 4.5) were for design loads without impact. Their work was in two phases.

The first phase was called a simulation program. Random sequences of traffic were generated and fed on to the bridge, in one or more lanes. In this traffic, cars were all assumed to have the same weight and length, and the same treatment was applied to buses. Trucks were given a probability distribution of weight and on this basis were divided into six (or fewer) groups. Each group was given a length probability distribution. Each traffic sequence was modelled, therefore, as a series of finite uniformly distributed loads of varying length and weight intensity. The total traffic volume for each hour of the day, the mean vehicle speed and the relative numbers of cars, buses and trucks in each lane were specified as input data. The vehicle spacing consisted of a specified, constant value for zero speed, with an additional spacing for moving traffic, taken as proportional to the traffic velocity.

These streams of moving traffic were interrupted by stoppages caused by mechanical breakdowns or accidents involving one, two or three lanes, where the numbers of these events were based on observation, together with a statistically varied time to clear each stoppage. During a stoppage, therefore, a certain number of vehicles would become stationary on some part of a bridge or its approaches, and in one or more lanes. This loading was then scanned, for lengths from 15.2 to 1951 m (50, 100, 200, 400, 800, 1600, 3200 and 6400 feet) to identify groups of vehicles that would produce maximum total load, maximum central bending moment and maximum end shear in a simply-supported span of this length. In doing so, the length of 15.2 m was considered first. Each lane of traffic, whether stationary or moving, was scanned to find the group of vehicles, 15.2 m long, that had a maximum weight. The total train of vehicles in each lane was then moved longitudinally so that these maximum-weight 15.2 m groups were side by side. In this process, the only difference between a 'moving lane' and a stationary lane lay in the vehicle spacing, although the speed of the moving traffic was adjusted, depending upon traffic direction, if it was moving past a stationary lane.

The chosen input data meant that a certain number of stoppages would occur during a 3-month period. For each of the lengths (15.2–1951 m) the extreme values of total load (W), simple span bending moment (M) and simple span shear (S) were stored. The process was repeated for many such periods, enabling a statistical description of these values and an extrapolation to give extreme values for a chosen return period, actually 5 years.

Buckland and his co-workers then replaced these three extreme values by two, in the following way. Consider a simply-supported span, L, under the action of a moving point load, P, and a uniformly distributed load of intensity, w. For this loading, the central bending moment, M, maximum end shear, S, and total load, W, are given by:

$$M = PL/4 + wL^2/8$$
$$S = P + wL/2$$
$$W = P + wL$$

The observed data corresponds to M, S and W. Only the second and third of these equations were used to derive P and W. Then

$$W - S = wL/2, \text{ or}$$
$$w = 2(W - S)/L$$

Also,

$$P = 2S - W$$

Buckland argues that, for his type of data, the values of P and w derived in this fashion yield only small inconsistencies in M. His final recommendations (see Fig. 4.5) are based on the following sample data (more complete data may be found in his 1980 paper):

- number of stoppages per year amount to 800, of which 560 were mechanical breakdowns, 120 were single lane accidents, 104 involved two lanes and 16 involved three lanes;
- in 90% of cases, the time to clear an event was 15 minutes following mechanical failure, or 23 minutes following an accident;
- cars were assumed to be 4.9 m long and weigh 1588 kg; the corresponding figures for buses were 12.2 m and 13,610 kg;
- of the trucks, 60% weighed 16,330 kg or less; 99% were 48,990 kg or less; none weighed more than 54,400 kg;
- of trucks in the range 5,400 to 27,200 kg, 90% had lengths equal to or less than 18.3 m; of those with weights between 27,200 kg and 54,400 kg, 95% had lengths less than 29 m;
- the space between stationary vehicles was 1.5 m, increasing by a vehicle length for every 16 kph of velocity;
- the total daily traffic volume in two directions on six lanes was 85,900 cars, 350 buses and 6,510 trucks; corresponding to a total of 92,760 vehicles, of which 7.4% were heavy vehicles (buses and trucks).

The equivalent loads shown in Fig. 4.5 have a point load that increases with loaded length. It is, however, still less than the total distributed load, where this ratio increases from zero to about 0.26 at 80 m, and then reduces to 0.155 at 488 m, and 0.066 at 1950 m.

Buckland et al. (1980) also used what they call an Analytical Program, but this will not be described here. They did not subdivide their final results by lane. Further work on the longitudinal distribution of load may be found in papers such as those by Harman and Davenport (1979) and Harman, Davenport and Wong (1984).

Work done by Nowak on the calibration of the 1994 AASHTO code was described in Section 5.3. As stated there, the analysis of loads was based on the 1975 Ontario truck survey. The effects of multiple presence were discussed also by Nowak and Hong (1991), using the same survey data, for successive vehicles on a single-lane bridge of moderate span, and for parallel vehicles in two adjacent lanes. References by Nowak to the survey size vary somewhat: here it is described as representing about 10,000 trucks in a 2-week period of heavy traffic on an interstate highway, with an equivalent number of trucks, N, in a 75-year period of 15×10^6. The corresponding probability level is $1/15 \times 10^6$, leading to an inverse normal probability (Section 5.1) of 5.26. The corresponding truck moment for this recurrence interval was about twice that produced by a single HS20-44 AASHTO truck. With regard to multiple presence in a single lane, Nowak chose to consider the following three cases, 'based on observations and engineering judgement'.

1. Every 10th truck was assumed to be followed by an uncorrelated truck at a headway distance of less than 15.2 m.
2. Every 50th truck was assumed to be followed by a truck that was 'partially correlated' in terms of its weight.
3. Every 100th truck was assumed to be followed by a 'fully correlated truck'.

The term, correlation coefficient, has not been used previously in this book, but is a measure of the extent to which two variables are statistically related (Devore 1995: 215). For example, one could measure the ability of a class of students by a test at age 10, and then by another test at age 12. If scores given by the two tests are plotted – say the year 10 results plotted horizontally (x) against the year 12 results vertically (y) – then the results for each student plot as a point (x, y) in this domain. If the tests are uncorrelated, then the resulting points will be found to be completely scattered. If the test results are highly correlated, then the scatter is reduced and the points tend to plot close to a single line. The mathematical definition of correlation is by the *correlation coefficient*, where a value of 0 means no correlation, and 1 means perfect correlation. It is possible to generate, mathematically, a set of random values, and also two sets of values connected by a specific correlation coefficient. Nowak defined his correlation coefficients as 0 for case 1, 0.5 for case 2 (partial correlation), and 1 for case 3.

In case 1, one of the trucks was 'every 10th truck'. These trucks form a population of size $1/10 \times 15 \times 10^6$, corresponding to a recurrence interval of 7.5 years. The weight of this truck was taken as that with an inverse normal probability of 4.82. The corresponding values for 'every 50th truck' were 1.5 years and 4.53; and for 'every 100th truck', 0.75 years and 4.34. Having by this means determined the magnitude of the lead truck in a pair, the second trucks were allocated the following values: case 1, the average truck; case 2, the maximum daily truck; and case 3, the 0.75 year truck (where it appears that these were found by simulation studies to give the desired values of the correlation coefficient). An approximation to Nowak's curves for bending moment, calculated from the observed trucks and scaled against a single HS20-44 truck, is a straight line through the points (1.0, 1.0) and (2.0, 5.26) on normal probability paper. On this basis, the truck weights, found by simple linear interpolation, become: case 1: 1.90, 0.76; case 2: 1.83, 1.49; case 3: 1.79, 1.79; all multiplied by a single HS20 truck. The truck geometries were taken to be the same as for this truck. Two headway distances were assumed, of 4.6 and 9.1 m, and bending moments and shears were calculated and plotted for spans from 12.1 to 61 m. For comparison, the effects of the single 75-year vehicle ($2.0 \times$ HS20) were also plotted. The following conclusions may be drawn from these results.

(a) For a headway distance of 9.1 m, the effects of the single vehicle were largest for all spans.
(b) For a headway distance of 4.6 m, the single vehicle dominated for spans less than about 30–36 m (varying as between bending moment and shear).
(c) For the longer spans, the case 3 pair of vehicles gave the largest bending moments and shears.

This study gave some support, therefore, to the use of a single design vehicle for bridges of moderate span. However, the range of spans considered was not large enough to allow for significant conclusions for long span bridges.

Nowak and Hong (1991) also applied this technique to the study of multiple vehicles in two parallel lanes. They used simulation studies to show that the governing case was for two side-by-side trucks with full correlation, where this event occurred for every 500th truck. For each of these trucks, therefore, the population size was 1/500 × 15 × 10^6, corresponding to a return interval of 75 years/500, or 1.8 months, and an inverse normal probability of 3.94. The corresponding size of each vehicle was 1.69 × a single AASHTO HS20-44 truck. The total effect was therefore about 0.85 of that produced by a 75-year single lane vehicle (which was 2 × one HS20 truck).

In a separate paper, Nowak (1991) gives corresponding figures of 1.0, 0.85, 0.70 and 0.5 for 1, 2, 3 and 4 lanes. That is, if 1.0 represents the 75-year vehicle for one lane, then for 3 lanes, 0.7 times this vehicle should be applied in each of these three lanes. He goes further, and links these figures to average daily truck traffic (ADTT), which in the present context he defines as the total number of trucks travelling in the same direction, distributed between the 1, 2, 3 or 4 lanes. His survey size was 'about 10,000 heavy vehicles' in 'about two weeks', but these included only 'trucks that appeared to be heavily loaded'. Presumably for this reason, he interprets the sample as corresponding to an ADTT of 1000 (somewhat greater than 10,000/14 = 714). Take as an example the case with an ADTT of 100, carried by two lanes. The 500th truck corresponds then to a probability of 500 in 1.5 × 10^6 in 75 years, giving an inverse normal probability of 3.4 and a size of each truck equal to 0.78 times the previous 75-year value. Nowak gives the slightly different rounded figure of 0.8. His complete table of multiple lane factors is shown in Table 6.1.

These results come from limited data and are based partly on 'engineering judgment'. Nowak, Nassif and De Frain (1993) subsequently weighed and recorded truck traffic on two major US highways, each with 2 lanes in each direction. The results are important and include the following direct observations of multiple presence, with these terms defined as follows:

(a) in-lane: two trucks following each other with a headway distance (axle to axle) less than 15.2 m;
(b) side-by-side in tandem: two trucks in adjacent lanes, travelling in the same direction, with front axles on the same line;
(c) side-by-side and behind: as (b), but with the distance between front axles less than 15.2 m.

In-lane multiple presence involved about every 100th truck. Side-by-side multiple presence (b + c) totalled about 4 to 7 m every 100 trucks, divided about equally between (b) and (c). The weights of these vehicle-pairs were not recorded in the study.

Table 6.1

ADTT	*1*	*2*	*3*	*4 or more lanes*
100	0.95	0.80	0.55	0.45
1,100	1.0	0.85	0.70	0.50
10,000	1.05	0.90	0.75	0.55

Heywood (1990, 1992a, 1992b) carried out two important studies on multiple presence in two adjacent lanes, important not only in themselves, but because they demonstrated correlation between a simulation and a theoretical analysis. For this purpose he used data collected at a site on the Stuart Highway at Darwin in the Northern Territory of Australia. This data was for 37,553 trucks, northbound, over a total period of 221.2 days, from May to December, 1989. It included information for all vehicles with weights above a certain threshold level, including axle spacings and weights, and the time of arrival at the site. This arrival time was recorded in seconds, corresponding to a resolution of approximately 25 m at highway speeds. To avoid an incorrect juxtaposition of vehicles in the simulation, a random fraction of a second was added to the arrival time of certain vehicles, with safeguards to ensure a minimum headway. Procedures were also adopted to remove gaps in the record.

In his simulation study, he ran this sequence of vehicles over a series of influence lines, including those for mid-span bending moment in a simply supported beam, with another similar sequence in the adjacent lane, having a start that was delayed in time. To ensure that traffic peaks occurred in both lanes about the same time, it was desirable to choose a delay time that was within an hour of an integral number of days, but was not less than 10 minutes. For the combined loading in two lanes, he recorded histograms of mid-span bending moment and plotted them on normal probability paper. He repeated the study for various delay times and for the second lane moving in the opposite direction. The effects of these variations on the histograms were small.

An event can be represented by a graph of effect versus time. This graph can be replaced by a step-wise approximation formed by dividing the time scale into short steps, δt, and assuming that the event is constant over each step. The total event can be seen as a series of these 'point-in-time' events, arranged in sequential order. Heywood's analysis, based on the work of Turkstra and Madsen (1980), assumes that it is the magnitude and frequency of these point-in-time events (each of duration δt) that is important, rather than their order. The actual curve of effect versus time will consist of graphs for separate vehicles, separated by periods when there is no vehicle on the bridge. The effect curve for a particular vehicle can be broken down into a number of point-in-time events, and these used to determine a probability density distribution. Similarly, the effects of a total sequence of vehicles can be combined by adding the effects of the separate vehicles to give a total probability density distribution and from that, a distribution of cumulative probabilities. Let this be $F^*(z)$, when z is the magnitude (say of bending moment) and F^* its cumulative probability, for all times when there is a vehicle on the bridge. Suppose that there are two lanes, numbered 1 and 2. Then, if the traffic sequence in each lane is defined, the separate distributions $F_1^*(z)$ and $F_2^*(z)$ can be derived.

Heywood computed distributions of this kind for the Stuart Highway truck survey, for vehicles in a single lane, with the result shown in Fig. 6.1. It may be taken here as $F_1^*(z)$. In his simulation study, he used a similar sequence of vehicles in the second lane. Furthermore, he approximated the cumulative distribution shown in Fig. 6.1 by three straight lines (upper curve in Fig. 6.2), where each represented a normal distribution valid over part of the range in z:

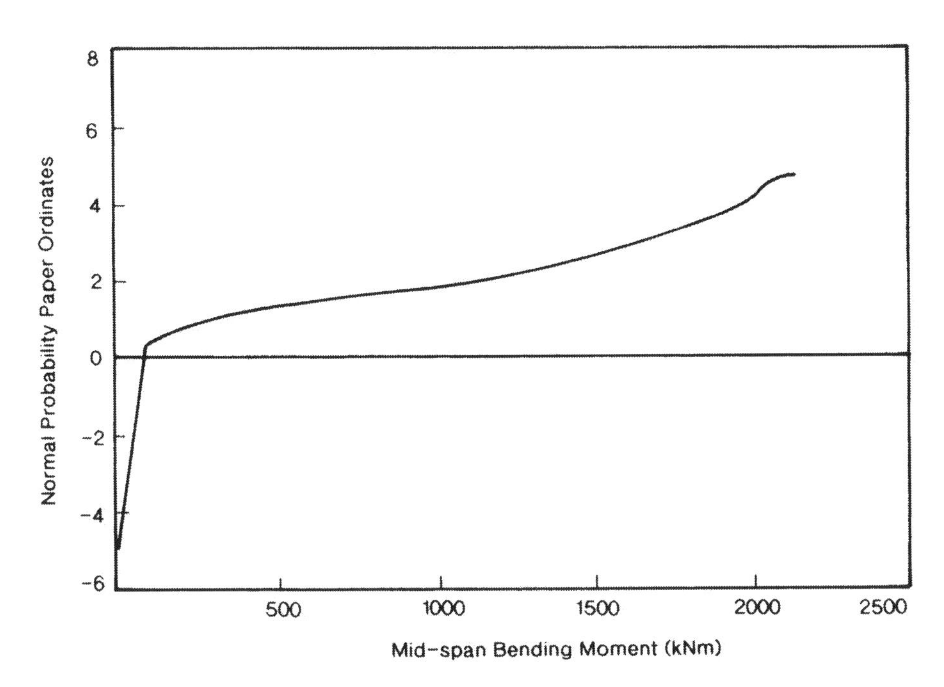

Fig. 6.1 Cumulative probability distribution for point-in-time events, calculated from truck survey on the Stuart Highway, Australia (Redrawn from Heywood 1992; used with permission)

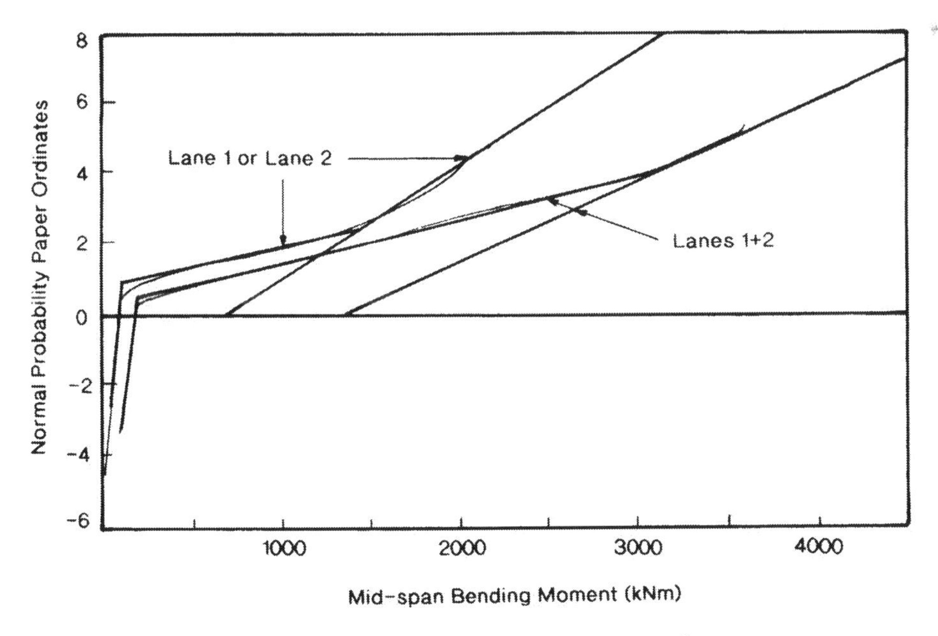

Fig. 6.2 Point-in-time distributions for one- and two-lane events, Stuart Highway, Australia (Redrawn from Heywood 1992; used with permission)

from $z = 0$ to 99.3; 99.3 to 1413; and greater than 1413. The properties of these distributions, in terms of mean and standard deviation were (85.0, 17.0); (−648, 872) and (680, 311). The intersection points of these lines can be calculated to give the proportions of the total population represented by the three lines. For example, the vertical ordinate, y, to the intersection point of the first two segments is given by

$$85.0 + 17.0y = -648 + 872y$$

Hence

$$y = 0.857$$

This value of the inverse normal probability function represents a cumulative probability of 0.804; that is, 80.4% of all points-in-time had values of mid-span moment less than 99.6 kNm and these varied with a mean of 85.0 and a standard deviation of 17.0. Similarly, the proportions of the other distributions may be evaluated.

Heywood then considered cases where the effects of a vehicle in both lanes was felt at the same time. To do this, he used Monte Carlo methods (Benjamin and Cornell 1970, 124ff) to generate two families, each satisfying the statistical description of Fig. 6.1. He added these to give a new family and again used normal probability paper to describe these statistically. For example, if the properties of the first values in the two samples were z_1 and z_2, then the sum, $z_{1+2} = z_1 + z_2$, was a member of the new family. The results of this addition are shown in Fig. 6.2, together with the curve obtained if the original distribution was used directly, rather than its approximation. There is reasonable agreement. This new curve represents the cumulative probability $F^*_{1+2}(z)$.

In practice, on a bridge with two lanes, there are large periods of time when the bending moment is zero; that is, when there is no vehicle in either lane. For a single lane carrying Q trucks per day, with a span, L, a truck length, ℓ, and a truck speed, v (kph), the probability of any vehicle being present is given by

$$p = \frac{Q(L + \ell)}{24{,}000\ V}$$

For example, for $L = 20$ m, $\ell = 10$ m, $V = 100$ kph, and $Q = 500$ trucks per day per lane,

$$p = 0.00625$$

Let p_1 be the probability of a truck being present on lane 1, and p_2 the probability for lane 2. Then the actual cumulative frequency, $F(z)$, for multiple presence is given by

$$F(z) = p_1(1 - p_1)F^*_1(z) + p_2(1 - p_1)F^*_2(z) + p_1p_2F^*_{1+2}(z) + (1 - p_1)(1 - p_2)$$

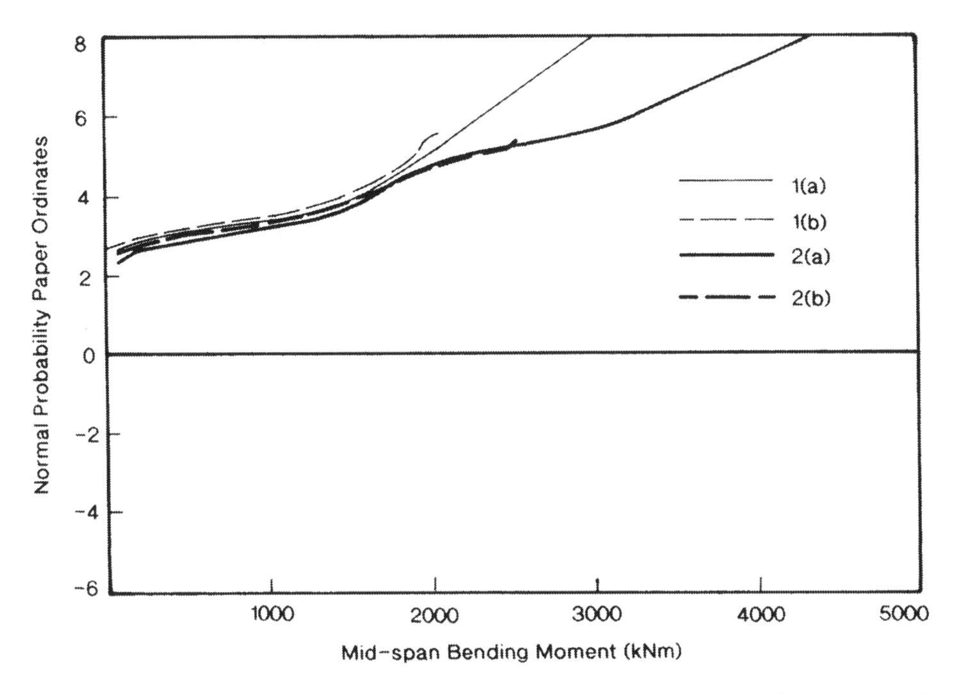

Fig. 6.3 Prediction of two-lane events (Redrawn from Heywood 1992; used with permission)

The last term is a correction term for the case where there is no vehicle on the bridge. All such points-in-time have the load effect, z, equal to zero. The number of such points-in-time must be added to the cumulative probability for any value of z. The simulation will also generate points-in-time with false, negative values of z and these may be discarded. $F(z)$ may therefore be evaluated for chosen values of z.

Figure 6.3 shows the results of this analysis and compares these with those obtained from the simulation study. The four curves are:

1. effect of traffic in one lane: (a) simulation (b) analysis;
2. combined effect of traffic in two lanes: (a) simulation (b) analysis.

It can be seen that a useful agreement has been obtained.

Levels of midspan moment for various return periods can be evaluated for particular values of the average daily truck traffic. Heywood did this and recorded the ratios of the resulting two-lane event with twice the value of the corresponding single-lane event. Some of these ratios are listed in Table 6.2.

It can be seen that these ratios reach a maximum of 0.75 for 8000 trucks, but this is only a minor increase on the value of 0.72 for 1000 trucks. These ratios are lower than those used in many codes; Nowak (1991), for example, gives a ratio of about 0.85. However, the traffic at this site is a rather special case, with very large single trucks spaced out between a large number of much

Table 6.2

ADTT	*2-lane/(2 × 1-lane event)*
1000	0.72
2000	0.73
4000	0.74
8000	0.75

smaller trucks. The value of Heywood's work lies not so much in the values, but in his contribution to the selection of a suitable method of analysis.

Other important studies of multiple presence may be found in the papers by Bakht and Moses (1988) and Bakht and Jaeger (1990).

6.2 Progressive failure

Duczmal (1988 and 1989; see also Duczmal and Swannell 1985, and Swannell, Miller and Duczmal 1985) developed a program for the analysis of the dynamic interaction between a vehicle and a composite steel and concrete girder bridge. His analysis was extensive and included many parameters, such as the use of non-linear stress–strain relationships for steel and concrete, and the effects of residual stress, but not the effects of strain rate on strength mentioned in Section 5.7. It allowed for the true moment–curvature relationship and provided for a bumpy road. It also allowed for repeated passages of the vehicle, including both permanent and elastic deformations of the bridge in the roadway profile. The bridge was assumed to be stationary at the beginning of each vehicle passage. The basic accuracy of his analysis was confirmed by comparing some of his predictions with those from independent but related studies (Chan and O'Connor 1990a and 1990b; O'Connor and Chan 1988a and 1998b; O'Connor and Pritchard 1982 and 1985; Pritchard and O'Connor 1984).

This Section describes results obtained for successive vehicle passages of a vehicle similar to the T44 truck (1.33 × the AASHTO HS-20 truck), but scaled up to the levels required for failure. Realistic, non-linear models of vehicle springs and tyres were included (see O'Connor, Kunjamboo and Nilsson 1980). For this study, the roughness of the road, both on the approaches and on the bridge, was modelled as a sinusoidal profile, with a wavelength of 1.2 m (equal to the minimum axle spacing) and a bump height or double-amplitude, from 4–20 mm. For one set of analyses, a vehicle velocity of 12.93 m/s was chosen so that the time to travel a distance of 1.2 m equalled the first natural frequency of the bridge (10.8 Hz). This combination of vehicle speed, sinusoidal roughness, wavelength and axle spacing was selected to achieve dynamic excitation of the bridge. Other analyses were carried out for a different vehicle speed. Figure 6.4 gives typical values for the second and third axle groups of a T44 truck travelling along a rigid, bumpy road at speeds of 12.93 and 19.4 m/s. The speed of 12.93 m/s, with a bump height of 10 mm, gave an impact factor of 0.22, com-

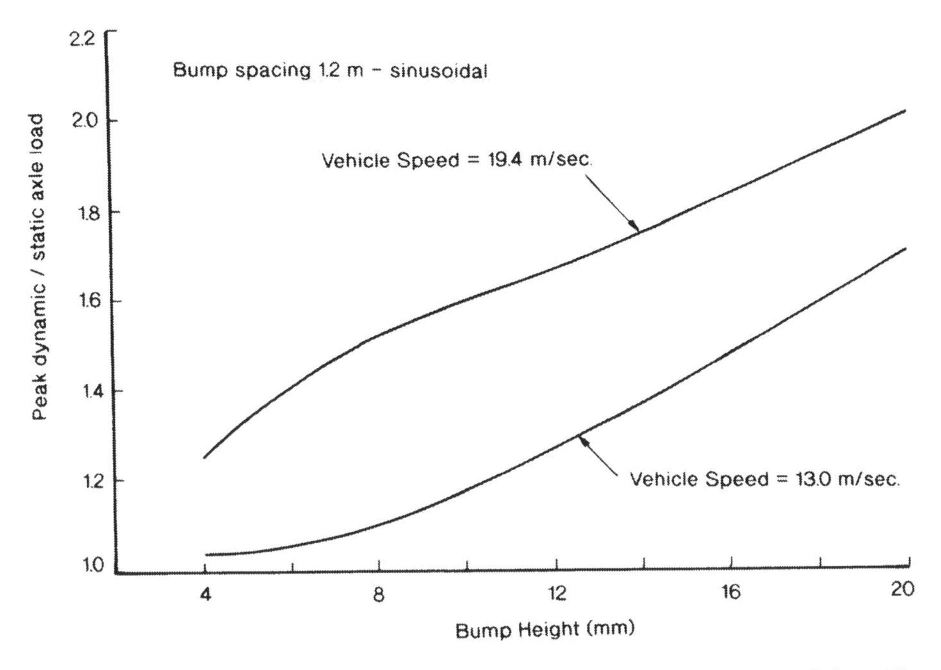

Fig. 6.4 Theoretical impact values for various bump height–sinusoidal profiles (Redrawn from Duczmal 1989; used with permission)

pared with the code requirement at the time of either 0.3 or 0.25. Much larger impact values were obtained for larger bumps and at the higher velocity. The bridge span was 11.28 m, chosen to correspond with that of an existing bridge used in related studies. The member dimensions were also the same as for this bridge. The analysis was not directed at the problem of transverse distribution of the load and considered effectively a single girder. However the load applied to this girder was scaled so as to give a direct comparison with the total bridge load. Here the magnitude of the total truck load will be expressed as αN, where N is the number of T44 trucks which, if applied as a static load, with the self-weight of the bridge, would cause failure.

Figure 6.5 is a typical plot of axle load against time, for the middle axle group of a single T44 truck, travelling over the bridge at 12.93 m/s, with a bump height of 10 mm. The plot is for the fifth passage of the vehicle; the maximum impact factor is about 0.32.

Figure 6.6 shows three examples of the permanent set left in the bridge after passages of the vehicle. The first, (a), is for $\alpha = 0.55$, a velocity of 12.93 m/s, and the bump heights shown. For heights of 8, 10 and 12 mm, the permanent set became stable after a few passages. For the remaining cases, the bridge failed during the passage following the point shown; that is, during the 24th passage for a 14 mm bump height. For 16, 18 and 20 mm bump heights, failure was during the 14th, 7th and 4th passages. Figure 6.6(b) is for $\alpha = 0.66$ and the same velocity. Not surprisingly, failure occurred at an earlier stage.

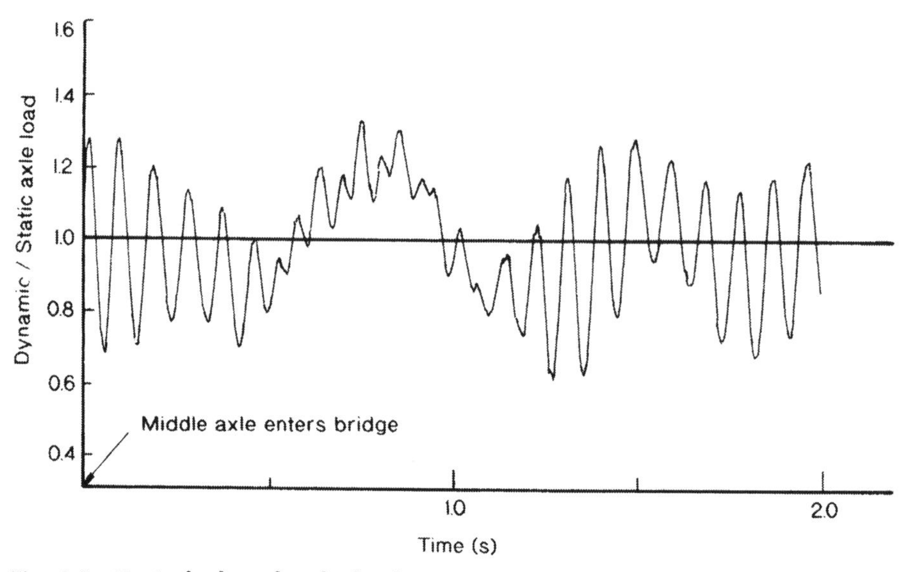

Fig. 6.5 Typical plot of axle load against time – theoretical (Redrawn from Duczmal 1989; used with permission)

Figure 6.6(c) is for the smaller vehicle, $\alpha = 0.55$, but at the increased velocity of 19.40 m/s. Failure occurred at an even earlier stage.

It is clear from these examples that dynamic wheel loads, when coupled with the deflected shape resulting from earlier overloads, may cause the collapse of a bridge, even when the applied load is well below that which, if applied statically, would cause failure. However, as can be seen from the figure, the risk to the bridge varies with the bump height. It may also be expected to vary with the detailed shape of the road profile.

6.3 Fatigue loading

Repeated loads may cause cracking and the eventual failure of steel members and components by fatigue, as discussed briefly in Section 2.6. Fig. 2.24, taken from Fisher (1977: 19), plots lower bound curves for the number of cycles to failure against stress range for three types of steel members. It is important to note here that the vertical ordinate represents stress range, that is, from the minimum to maximum stress. For a fully reversed stress, from $-\sigma$ to $+\sigma$, the range is, therefore, 2σ. Both ordinates are plotted on a log-scale. With these scales, each curve consists of two straight lines with the second horizontal; that is, there is for these cases a stress level, called the endurance limit, below which fatigue failure does not occur. For the cases shown, these fatigue limits are:

rolled beams	165 MPa
welded plate girders	110 MPa
coverplated beams (at coverplate end)	34 MPa

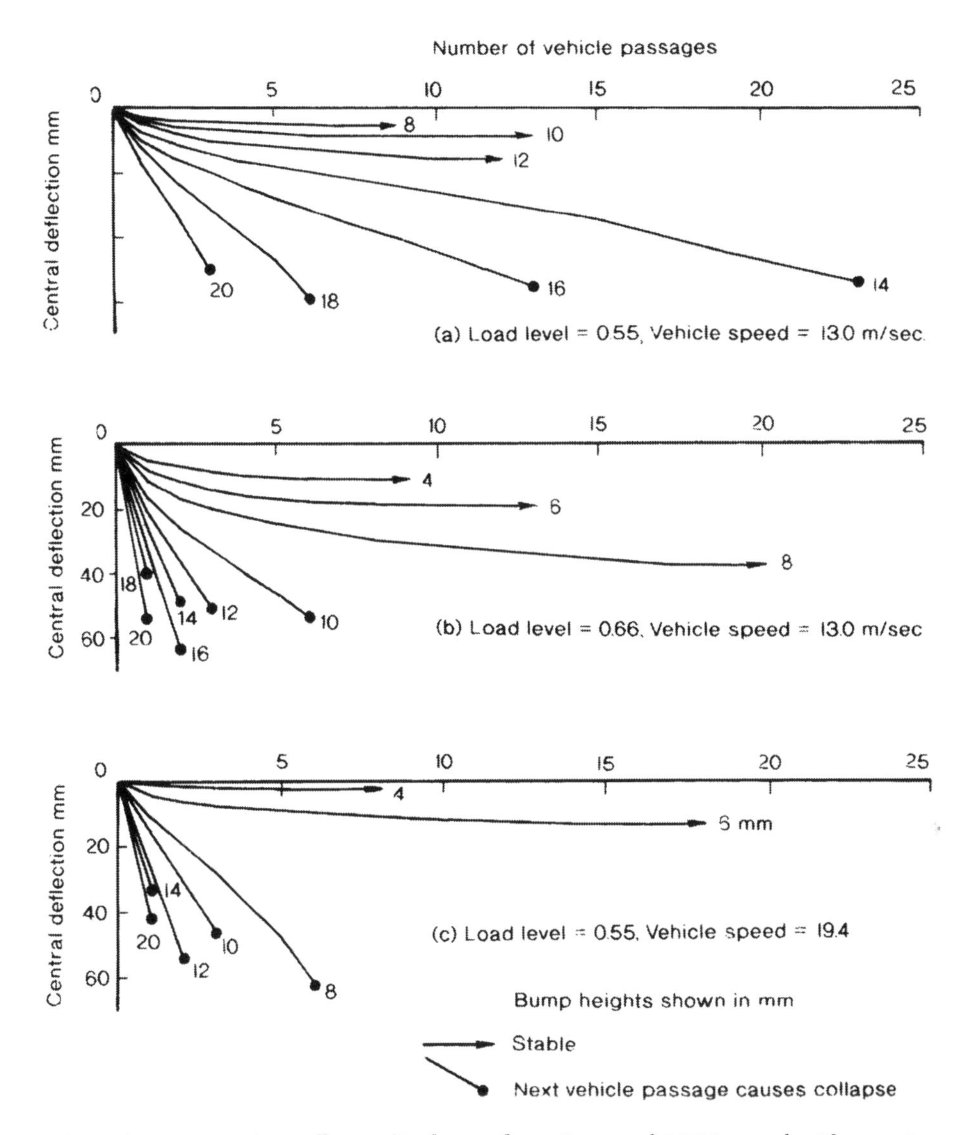

Fig. 6.6 Progressive collapse (Redrawn from Duczmal 1989; used with permission)

The third of these limits lies well within the range of stress likely to occur in practice. It is necessary, therefore, to consider the possibility of fatigue failure. For these cases, the endurance limit is defined by 2×10^6, 3×10^6 and 8.5×10^6 cycles on the horizontal scale. The previous section refers to various values of the Average Daily Truck Traffic (ADTT); Nowak's study of the magnitude of truck loads, and the effect of successive vehicles, used a value of 1000, with a total number of vehicles in a 75-year bridge life of $15–20 \times 10^6$. If a maximum load effect is caused by a single vehicle, then this figure suggests 15–20 × 10^6 stress cycles. On the other hand, the maximum effect at a local level, such as in a bridge deck, may be caused by the passage of a single axle. The HS20-44 vehicle had 4 major axles, plus a steer-axle. If the number of 4 were taken as the number of loads per truck, then 15–20 × 10^6 trucks would correspond to more than 60 × 10^6 load cycles, or, perhaps, 40 × 10^6 if the trucks are spread between a number of lanes. These counts are also sufficient to suggest that the possibility of fatigue failure deserves further consideration.

The first points to notice are that not all trucks will produce the same range of stress, nor do the combined total stresses vary over a range $-\sigma$ to $+\sigma$. Concerning the second point, the general assumption, based on experimental studies, is that the effect of mean stress, i.e. $\frac{1}{2}(\sigma_{max} + \sigma_{min})$, is unimportant, although this would be surprising if both σ_{max} and σ_{min} were compressive. Fisher (1977: 19) argues that high tensile residual stresses always occur at welds and, for this reason, the stresses applied by service live loads always correspond to tension at the critical locations for crack formation and growth: 'most of the fatigue life is exhausted by the time the fatigue crack propagates out of this high tensile residual zone'. Although this argument may not be completely persuasive in all cases, it is accepted as sufficient in the present context.

The second factor is the presence of varying ranges of stress. Here the normal hypothesis (Miner's Hypothesis) is that proposed by Miner (1945), based on earlier work by Langer (1937) and others. He introduced the idea of 'fatigue damage'. Figure 6.7 shows a typical plot of stress range S against the number of cycles to failure (here called N). Suppose that n_1 cycles of stress range

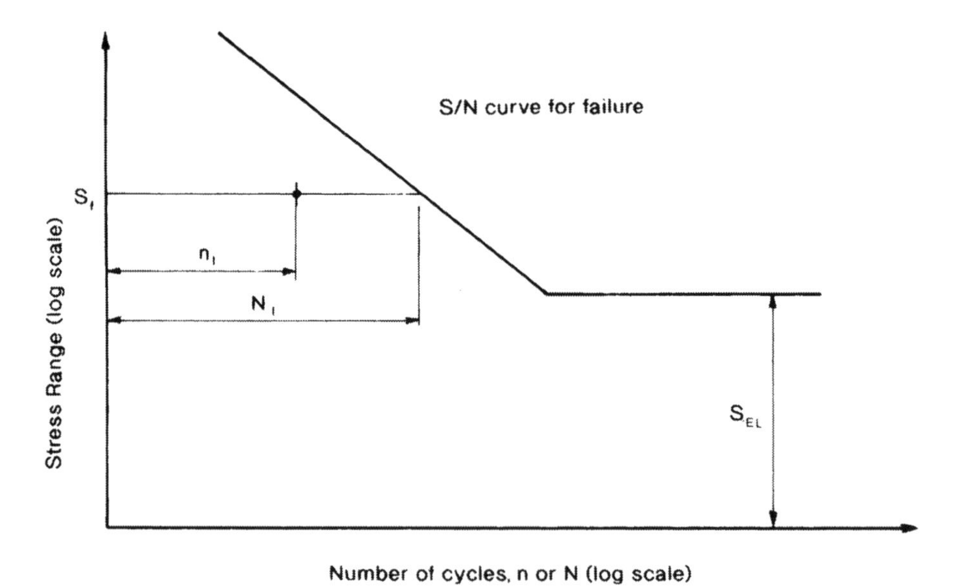

Fig. 6.7 Basis of Miner's Hypothesis for fatigue damage (Miner 1945)

S_1 are applied. Let the number of cycles to cause failure under the stress range S_1 be N_1. Then the ratio n_1/N_1 is taken by Miner to represent the damage caused by the cycles of S_1. Such terms can be evaluated for all levels of stress. The corresponding ratios are added to represent total damage. Then failure will occur when the total damage reaches 1.0; that is,

$$\Sigma(n_1/N_1) = 1.0 \text{ for fatigue failure}$$

Fisher observed that experimental S/N curves typically were parallel, as in Fig. 2.24, and for the inclined straight lines suggested the relationship,

$$N = AS^{-3}$$

or

$$\log N = \log A - 3 \log S$$

It may be shown that this slope (−3) is represented in all the sloping lines of this figure. Then Miner's Hypothesis suggests that, for safety,

$$\Sigma(n_1/AS^{-3}) \leq 1.0$$

Fisher also discusses an alternative Root Mean Square law, where the combination of a series of different ranges of stress may be replaced by the single range of stress S_{RMS}, evaluated as

$$S_{RMS} = \sqrt{\Sigma n_1(S_1 - S_{\min})} + S_{\min}$$

where

n_1 is the number of cycles at stress range S_1, and
$S_{\min}$ is the minimum stress range.

Previous sections have shown how such parameters as the maximum bending moments in a simply supported beam due to service vehicles may be plotted on normal probability paper and represented by a normal distribution of given mean and coefficient of variation. Such data may be used with one or other of these rules to predict the likelihood of fatigue failure. There are, however, some further matters that need to be discussed.

Miner originally proposed that only stress ranges in excess of the endurance limit (S_{EL} in Fig. 6.7) should be included as contributing towards damage. Fisher (1977: 20) carried out tests that included some stress ranges above, and some below S_{EL}. He concluded that the presence of some cycles '(in excess of) the constant cycle fatigue limit ... caused all stress cycles to contribute to fatigue damage'. As a result, all stress cycles were included in his calculations.

This raised a further point. Actual records of stress produced when a

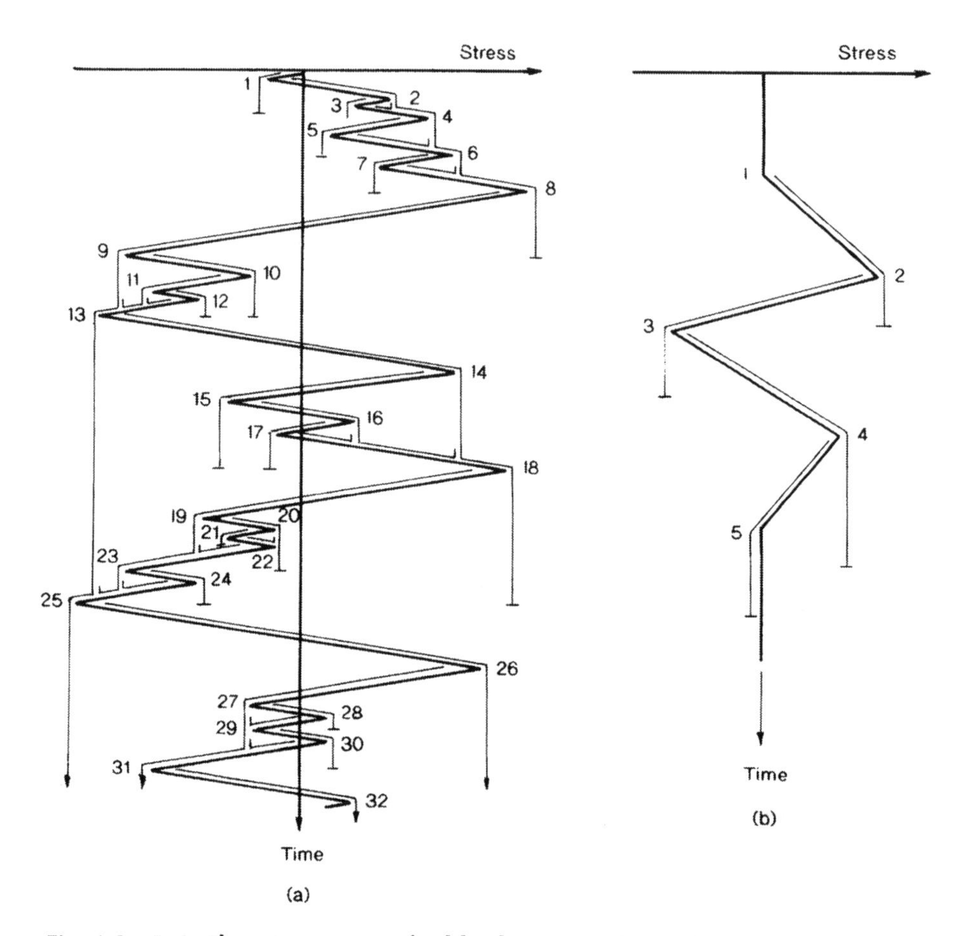

Fig. 6.8 Rain Flow Counting Method for fatigue (Redrawn from Dowling 1972)

vehicle passes over a bridge (see for example Chan and O'Connor 1990b: 1757; O'Connor and Pritchard 1983: 193; 1985: 648) include dynamic effects, superimposed on equivalent static effects. These cause not only higher stresses, but also increased cycles of stresses at lower stress ranges. One method of counting these cycles is called the Rain Flow Counting Method (Dowling 1972: Appendix).

Figure 6.8 is a hypothetical plot of stress (or strain) against time, with the time plotted vertically downwards. Maxima on the curve have the even numbers, 2, 4, 6 etc; with the odd numbers allocated to successive minima. The graph shows a hypothetical drop of water running from 1 to 8. As it drops from point 2, the part of the curve to the left, with a minimum point numbered 3, is not traversed. A second drop is allowed to run from 2 to 3, and so on. The count of cycles includes one half cycle for the stress range 1–8, plus full cycles corresponding to the stress ranges 2–3, 4–5, 6–7 etc. The nature of the model gives the source of its name.

The procedure may be programmed as follows.

1. Scan the record, and allocate the even numbers 2, 4 etc to the maxima, and the odd numbers, 3 etc, to the minima, with 1 as the starting point. Store the corresponding stresses.
2. Consider again the maxima in ascending order, 2, 4 etc. If a maximum such as 2 has its stress, σ_2, less than the succeeding maximum, σ_4 (as shown in Fig. 6.4(a)), regard the stress range (σ_2–σ_3) as a complete cycle, add 2 to the count for this stress range, remove points 2 and 3 from the record, and renumber all succeeding points; with point 4 renumbered as 2. Proceed through the entire record.
3. Rescan the record for the minima, starting with point 3; for example, consider point 21 in Fig. 6.4. If σ_{21} exceeds the subsequent minimum stress, σ_{23}, then count the stress range (σ_{22}–σ_{21}) as a complete cycle, delete points 21 and 22 and renumber all succeeding points.
4. Repeat steps 2 and 3 until a stage such as that shown in Fig. 6.4(b) is reached. The only interior points are 2, 3 and 4, where neither maximum is followed by a larger maximum, nor is the minimum followed by a lower minimum. Finalise the count by adding 1 to the count for each of the half cycles $|\sigma_2-\sigma_1|$, $|\sigma_3-\sigma_2|$ etc.

As presented here, the method counts half cycles; if all numbers are divided by two, the result is a count for full stress cycles. An example for the count of cycles is presented in Table 6.3 (O'Connor and Pritchard 1983). These represent stresses calculated from strains measured in the lower flange of a central girder in a small composite girder bridge during the passage of more than 42,000 vehicles, for a period of 117 days (October 1981 to April 1982). The results have been grouped in stress ranges, each of 4.6 MPa, with a maximum of 71.5 MPa. The term α is the ratio, stress ÷ 71.5.

The total number of events was 42,222. If the maximum effect of each event is counted, the figure α^3N becomes 2,101. The rainflow count has, therefore, caused an increase in the equivalent number of cycles. Nevertheless, the total number of equivalent cycles (3036) is much less than the total number of events, with a significant contribution from events a little below the middle of the range. Grundy (1980) has also presented rainflow counts for some Australian railway bridges.

Table 6.3

Centre of stress range (MPa)	*N*	α^3N	*Centre of stress range (MPa)*	*N*	α^3N
71.5	1	1	39.2	603	99
66.9	1	1	34.6	2,391	271
62.3	3	2	30.0	9,367	691
57.7	7	4	25.4	22,866	1,021
53.1	9	4	20.8	23,448	574
48.5	39	12	16.2	28,383	327
43.8	125	29	Sum =	87,243	3,036

Chapter 7

Dynamic vehicle loads

7.1 Introduction

Dynamic load effects to be considered in bridge design include the following.

1. The maximum vertical loads exerted by a moving vehicle will often exceed those produced by an equivalent static or slow moving vehicle. The effect has commonly been called *impact*, with the Impact Factor, I, defined as the ratio of the additional load (dynamic minus static) divided by the equivalent static load. Another term with the same definition is Dynamic Load Allowance. Alternatively, the Dynamic Amplification Factor has been defined as the ratio of maximum dynamic effect to the maximum effect produced by an equivalent static load. Here, the older term, impact, will be retained (but see 4. below).
2. Longitudinal loads may be applied by braking or accelerating vehicles.
3. There is also the possibility of transverse horizontal centrifugal forces, of the form mV^2/R, where m is the mass of a body moving with tangential velocity, V, around a circle of radius, R. These may arise in the case of a bridge curved in plan, or when a vehicle changes its direction of movement.
4. True impact occurs when two bodies that are originally separate strike against each other. Cases that arise in bridge design are when a vehicle strikes a guard wall, or when a log or floating ice strikes against a bridge pier. Indeed, there is a sense in which all stream or wind loads may be classed as dynamic loads.
5. Earthquake effects also produce dynamic bridge loads.

Not all of these will be considered here. Loads on crash barriers are mentioned in Chapter 10, the effects of moving water and floating objects are discussed in

Chapter 11, with wind and earthquake effects in Chapter 12. The primary focus of this chapter will be on items 2 and 3.

Dynamic effects may influence the collapse of a structure; it is common, for example, to use in bridge design not the equivalent static vertical load, but this load increased for dynamic effects. This is then multiplied by the appropriate load factor to check against collapse. Moreover, there is the possibility that dynamic effects may cause discomfort to the users of a bridge. This possibility arises chiefly when pedestrians share use of the bridge with moving vehicles. Pedestrian loads are discussed in Chapter 9, which includes Section 9.3, on the 'Perception of vibration'. That chapter considers the dynamic excitation of a pedestrian bridge by pedestrians, which may also be a problem. Another case is a bridge carrying a mixture of stationary and moving vehicles, where persons in stationary vehicles may be disturbed by dynamic movements caused by those that are moving. This problem is not generally addressed here, but reference may be made to ISO 2631/1-1985, *Evaluation of human exposure to whole-body vibration* (see Standards Association of Australia (1990) AS2670.1). Some highway bridge codes state specifically that this problem need not be considered for many bridges (see for example BS 5400: Part 2: 1978: Cl.6.11; BD37/88: Cl.6.13). Comfort levels for railway bridges are mentioned briefly in Section 7.5 (see Eurocode 1).

Another problem arising from the dynamic nature of primary loads is that the number of occurrences of a particular stress level may be increased, leading in some cases to an increased possibility of fatigue failure; an example of this effect was given in Section 6.3. Section 6.2 of that chapter also presented some results from a theoretical study of progressive failure due to repeated passages of a vehicle that had been dynamically disturbed.

Code provisions for the dynamic vertical effects of primary vehicle loads were included in the discussion of codes of practice in Chapter 4. This chapter will consider in some detail the basic nature of these vertical effects, with a review of observed impact values in Section 7.2, a treatment of road and railway track roughness in 7.3, followed by an account of the dynamic modelling of vehicle behaviour in Section 7.4. Horizontal dynamic effects are considered in Section 7.5, with a review of code provisions and some analysis of their significance and underlying assumptions.

7.2 Measured impact values

Field measurements of the vertical impact effect have been reported by many authorities for both road and railway bridges, but only a selection of these will be described here.

Page (1976) updated measurements of the dynamic wheel loads that occurred when a particular vehicle was driven across 30 short-span bridges and 6 long bridges in the United Kingdom. The vehicle is described as an AEC Mandator, two-axled lorry, with static axle loads of 60 kN on the front and 100 kN on the rear axle. Of the short-span bridges, 10 carried a motorway, with the other 20 carrying either Class A or Class B minor roads over a motorway. The maximum length was 84.0 m, where this bridge had three spans across a motorway. In one case, the vehicle crossed the bridge in both directions, giving a total of 31 cases. In addition, measurements were taken for four long viaducts, including the A4 Hammersmith Flyover, and the Severn (with a suspension span of 988 m) and Wye (with a cable-stayed span of 235 m) bridges (see O'Connor 1971a: 130, 380). In reviewing this work, it is essential to remember that the impact values all represent wheel loads (dynamic component ÷ mean); they do not record the response of the bridge. Viewed another way, they are measures of the effect on tyre loads that can be caused by the road profiles on a bridge, where these may be modified also to some small extent by movements of the bridge itself. The vehicle speeds ranged from 42.3 to 64.0 kph.

The maximum observed impact value for the short bridges was $I + 1 = 1.75$; the minimum peak value for a particular bridge was 1.09; the mean of the 31 reported peak values was 1.37 (all based on wheel loads). The paper records the position of the peak impact value on the bridge: in 6 cases, this was within the first 10% of the length of the bridge, and in 23, within the last 10%; that is, in only 8 cases out of 31, or 26%, was the peak impact value in the central 80% of the length.

In another relatively new bridge carrying an unclassified road over a motorway, a large irregularity was observed at its fixed end. Here $I + 1$ was measured as 2.77. Three other bridges, additional to the basic group of 30, were identified with surfaces that were classed as: A, good; B, average; and C poor surfaces. The vehicle was run over these in both directions at speeds from 18 to 60 kph. For bridge A, the peak impact value ($I + 1$; wheel load) for travel in the worst direction, varied from 1.54 to 1.67; for bridge B, 1.58 to 1.96; for bridge C, from 1.67 to 2.29.

For the four viaducts, peak values of $I + 1$ varied from 1.36 to 1.53. Tests were carried out at a nominal vehicle speed of 64 kph, with the locations as well as the magnitudes of the larger impact values recorded. For the Severn bridge, peak values of about 1.66 were observed at the south-eastern end of the side span, and at the two piers. Elsewhere, typical values were of the order of 1.28–1.33, with ten peaks in excess of 1.25 in the main span. The maximum value observed on the Wye bridge was only a little over 1.25, at the western tower.

Page also carried out tests on the Transport and Research Laboratories test track for this vehicle with two types of tyres, at different tyre pressures, travelling over planks with heights of 20, 30 and 40 mm, at various speeds. At 60 kph, $I + 1$ (for the first peak) varied from about 1.70 to 2.23. On this evidence, peak wheel impact values ($I + 1$) in the range 1.09 to 2.77 could occur, with some dependence on speed – higher values occurring at the higher speeds. Page (1973a and 1973b) also carried out associated work on the dynamic analysis of vehicle suspensions.

At about the same time, Leonard, Grainger and Eyre (1974) observed dynamic axle loads when eight vehicles, of gross weight 32.5 to 44.7 tonnes were driven over artificial bumps of 40 mm height, not on a bridge, with varying profiles and lengths from 250 to 750 mm. To quote from their conclusions:

(a) No clearly defined relationship was observed between gross vehicle weight and dynamic loading or vibration. . . .
(b) The work suggests that the most important influence on loading and vibration comes from the dynamic properties of the vehicle suspension systems.

Observed values of $I + 1$ (their Figure 8) varied from 1.56 to 3.04, or from 1.56 to 2.16 if the more lightly loaded axles are ignored, for vehicles travelling at 64 kph over a 40 × 250 mm plank. Four of the vehicles were run over a smooth surface, with a profile said to be better than that achieved on most trunk roads and motorways. Results are given for two of these vehicles, moving at speeds from 62 to 71 kph, with $I + 1$ for the peak axle load varying from 1.19 to 1.27.

A major programme for the dynamic testing of highway bridges was initiated in 1922 by M. Roš of the Swiss Federal Laboratories for Materials Testing and Research (EMPA). In 1958, a decision was made to standardise test procedures, and from then until 1981 static and dynamic load tests were carried out on 226 slab and girder highway bridges. Much of this (and later) work has been described by R. Cantieni and his associates (Bez, Cantieni and Jacquemoud 1987; Cantieni 1984a, 1984b, 1987, 1992; Krebs and Cantieni 1997). Cantieni (1984a and 1984b) reported first the measurement of natural frequencies (f) for 224 bridges, with spans from 13 to 86 m. He plotted f (Hz) against the maximum span L and derived the best-fit curve:

$$f = 90.6 \times L^{-0.923}$$

The standard deviation of departures from this curve was $\pm 0.8f$ (Hz). He also plotted the expression,

$$f = 100/L$$

or

$$fL = 100$$

but this tends to fall a little below his test results (see Section 7.4). Damping in 198 concrete bridges was indicated by the following values of the logarithmic decrement: minimum, 0.019; mean, 0.082; maximum, 0.360.

Tests for the dynamic increment (called here, the impact, I) were carried out using two-axle trucks with gross weights from 110 to 193 kN (mean 158 kN), with axle spacings in the range 3.4 to 6.0 m, but of these spacings, more than 70% lay in the range 4.5 ± 0.12 m. These vehicles were driven at constant speed, in steps of 5 or 10 kph up to the maximum achievable speed, along the longitudinal axis of the bridge. Cantieni (1984a: Figure 9) presents a histogram of the maximum speed for each of 224 tests, with the following values: minimum, 10 kph; mean 47 kph; maximum, 80 kph. Of 226 bridges, 205 were prestressed concrete; 5, reinforced concrete; 14, composite steel and concrete; and 2, prestressed lightweight concrete. The most common structural arrangement was a bridge continuous over more than one span, and the maximum bridge spans varied from 11.0 to 118.8 m with a mean of 39.5 m. The dynamic increment (I) was calculated from measurements of deflection (usually at mid-span) as the ratio of the maximum additional dynamic component to the measured static deflection. The final recorded programme of dynamic increments was limited to tests on concrete bridges; in 73 tests, the vehicle moved along the undisturbed pavement; in another 69, the vehicle crossed a plank, 50 mm thick, 300 mm wide, placed at the point where the deflection was recorded.

Cantieni (1984a: Figures 16, 17) plotted the measured values of I first against the span and then against the bridge frequency. Figure 7.1(a) plots two limiting bounds for the observed values of I: (i) without a plank, and (ii) with the plank. The first has an upper bound of 0.70 for frequencies between 2.5 and 4 Hz; the second has two upper boundaries: at $I = 2.5$ between 2 and 3 Hz, and for frequencies of 8 Hz and above. The measured values of I are not shown here; none lie above the upper bounds and these form reasonable limits to the experimental values. For example, the maximum recorded values of I in the first set are about 0.65 at 2.7 Hz, 0.66 at 3.3 Hz, and 0.34 at 4.9 Hz. Similarly, for the cases with a plank, typical upper bound experimental values are 2.27 at 2.7 Hz, 1.22 at 4.7 Hz, and 2.26 at 8 Hz. However, the experimental values are not evenly distributed over the full range of bridge frequencies, with only about 15% of the test values above 5 Hz, and about 10% above 7 Hz. The maximum frequency was 14 Hz.

In assessing these results, one must observe also the truck characteristics. Cantieni describes the trucks as 'fully loaded two-axle tip trucks with leaf springs. Under normal conditions of pavement roughness, the dynamic wheel loads of such vehicles predominantly occur in two frequency ranges' – 'body bounce ... between 2 and 5 Hz, and wheel hop ... between 10 and 15 Hz'. He adds the comment that 'as a leaf spring truck is a nonlinear mechanical system, its natural frequencies are dependent on the excitation amplitude', being less with increasing amplitude. It can be seen that the first of the peaks in Fig. 7.1, from 2 to 4.5 Hz, corresponds to the body bounce frequencies, while the second, about 8 Hz and above (in the experimental values) is a little less than the range of 10–15 Hz quoted above for wheel hop (see also Section 7.4). These results, as with those by Leonard et al. (1974), suggest strongly that bridge impact varies with the truck characteristics.

Bez, Cantieni and Jacquemoud (1987) describe associated work used to suggest highway traffic loads and impact values for the 1989 edition of the Swiss loading code SIA160 (see Section 4.5); as stated in the paper, the final code provisions differ in some respects. Load model 1, which in the final code consists of a two-axle group, together with a distributed load, is described in the paper as corresponding to an isolated truck moving as part of the normal bridge traffic. For this case, recommended impact values are as shown in Fig. 7.1(b). They correspond generally to Cantieni's (1984a) upper bound for trucks moving

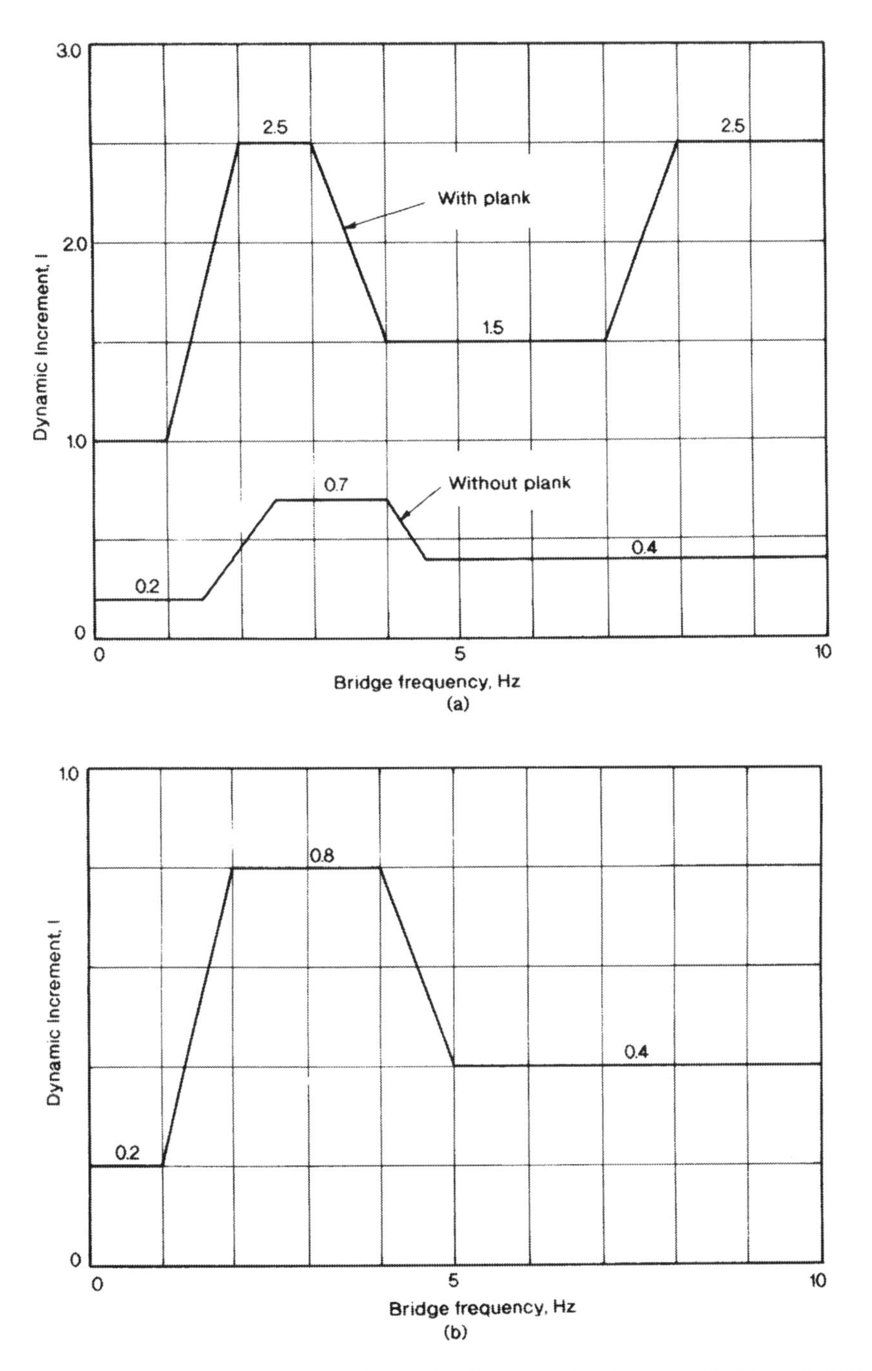

Fig. 7.1 (a) Upper bounds to values of *I* observed by Cantieni (1984a, 1984b) for truck passages with and without a 50 mm plank; (b) Values recommended by Bez, Cantieni and Jacquemoud (1987) for dynamic load coefficients to be used with the Swiss Load Model 1

across the normal bridge pavement, but with an upper bound of 0.8 specified for bridge frequencies from 2 to 4 Hz.

Cantieni (1987, 1992) and Krebs and Cantieni (1997) have also described work of this type carried out on specific Swiss bridges, notably the Deibüel Bridge, a single-cell prestressed concrete bridge with spans of 37, 41 and 32.4 m, and the Sort Bridge, continuous over five spans, with a maximum of 70.0 m. In his 1992 paper, Cantieni also summarises work carried out in thirteen countries, including Canada and the United States. It is not intended to review this work here, although attention should be drawn to the AASHO Road Test carried out in the United States between 1956 and 1960, and reported between 1961 and 1962. It is fair to describe it as probably the greatest civil engineering research project ever carried out; one of its major objectives was to observe the dynamic behaviour of bridges. Six special loops of roadway were constructed and these included 16 bridges, all of 15.2 m span, with four of prestressed concrete, four of reinforced concrete and eight of composite steel and concrete construction. Fifteen of these were studied under approximately 1900 passages of 70 to 80 vehicles of ten different types, with axle hop frequencies from 10 to 13.2 Hz, and frequencies for body bounce, under various conditions, from 1.7–5.6 Hz. The maximum recorded values of the dynamic increment, I, were 0.63 from deflection measurements (with 88% between 0.1 and 0.4), and 0.41 from strains (90% between 0.05 and 0.30). These values are relatively low, and undoubtedly contributed to the relatively low values of I (0.30 or less) specified in the AASHTO design code (see Section 4.3). Although the number of passages and vehicles is large, the number of bridges still constitutes a relatively small sample.

An early American study, *Impact in Highway Bridges. Report of a Special Committee* was published in 1931. Another review is by Varney and Galambos (1965). Buckland (1982) listed needs for research in this area. Hwang and Nowak (1991) have presented a relatively recent theoretical study. There is a close link between work carried out in the United States and Canada, particularly in connection with the Ontario Highway Bridge Design Code (OHBDC), first published in 1979, and the later editions of the AASHTO code. Cantieni (1992: 32) refers to work done by Csagoly, Campbell and Agarwal (1972) in connection with the OHBDC and says that, 'probably for the first time, this report contains a presentation of the dynamic increment as a function of the fundamental frequency of the bridge'. The 1979 Ontario code was, perhaps, the first to specify design values for I in terms of the bridge frequency. It is appropriate, therefore, to review some Canadian work here.

Billing (1984) and Green (Billing and Green 1984) carried out dynamic tests on bridges at 22 locations in Canada, including five twin bridges, with 14 of steel with spans 22–122 m, 10 of concrete with spans 16–41 m, and 3 of timber, all with spans of about 5 m. The road profiles on these bridges are described as 'good to excellent'. Four test vehicles owned by the Ministry of Transport and Communications were used, two of five axles with gross weights of 391 and 414 kN; one of eight axles with a gross weight of about 580 kN; and a three-axle service vehicle of gross weight 241 kN. Values of strain and deflection were first recorded in slow speed runs (16 kph) to enable the calibration of

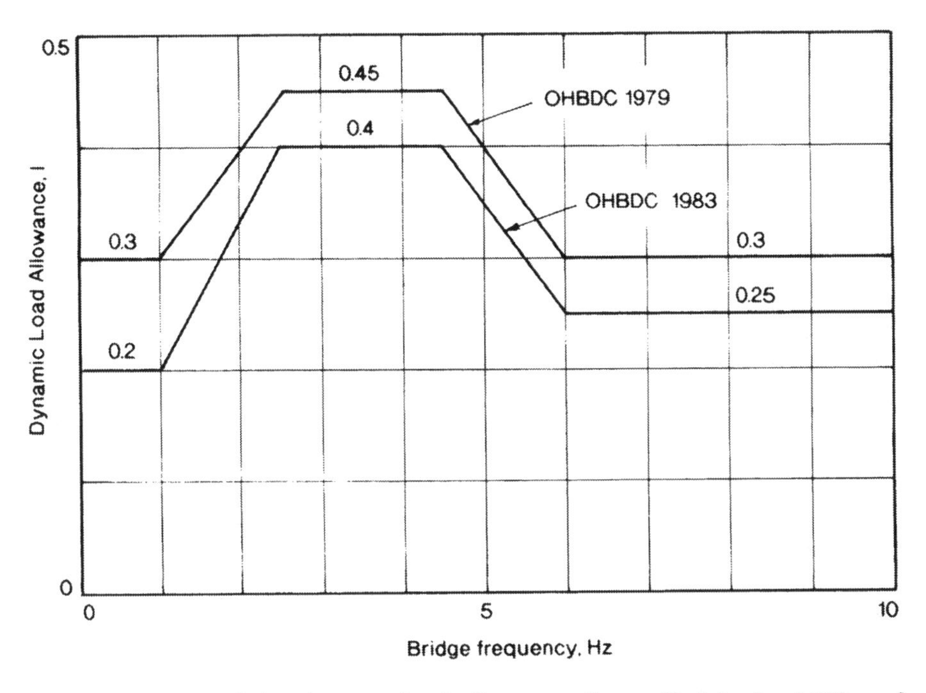

Fig. 7.2 Values of the dynamic load allowance (*I*) specified in the 1979 and 1983 editions of the Ontario Highway Bridge Design Code (Billing 1984)

the bridge for estimation of the weight of passing trucks. Dynamic records were then taken for 100 or more individual runs by the test vehicles, and also from trucks that passed during the test. The dynamic amplification factor was calculated as the peak value divided by 'an equivalent static response' obtained by smoothing the records. For each bridge, a peak value, taken as a maximum from the various channels, was recorded. The resulting impact values (I) were relatively small, with a maximum of about $I = 0.52$ for a two-span prestressed concrete bridge (spans 36.6 and 38.1 m), with a natural frequency of 2.6 Hz. Figure 7.2 shows specified values of I from the 1979 and 1983 Ontario codes, with the upper curve for the 1979 code. The reduction in the specified values resulted largely from this work by Billing. If the lower curve is taken for reference, then only five bridges gave higher values of I: 0.52 at 2.6 Hz (as already quoted); 0.42 at 2.9 Hz; 0.29 at 7.2 Hz; 0.33 at 10.3 Hz; and 0.32 at 11.9 Hz. The values of 0.42 and 0.33 were for steel bridges. Apart from these five exceptional values, the lower curve forms a reasonable upper bound to Billing's results. The dynamic amplification factor was found to reduce as the truck weight increased. Billing also observed coefficients of variation for the values of I recorded for particular bridges. They were typically large, varying from 0.56 to 1.11, with a mean of about 0.82.

The fourth group of impact studies to be described here was carried out on Six Mile Creek bridge on the Cunningham Highway, near Brisbane (O'Connor and Pritchard 1982, 1985; Pritchard 1982; Pritchard and O'Connor 1984). As mentioned in Section 4.7, this bridge was instrumented for the Australian Road Research Board in 1981 for the purpose of recording truck weights and dimensions. However, the siting of the bridge is more or less unique, for there is a compulsory weigh-bridge 7 km from the bridge and, at the time of this test, all vehicles with a large gross weight were required by law to be weighed. Although some trucks leave the highway between the weigh-bridge and the bridge itself, the majority do not. The site, therefore, was admirably suited for an impact study, with the bridge crossed by a random sample of service vehicles, of known weight and dimensions. Accordingly, in 1983, an impact study was added to the original programme.

The instrumented bridge carries the two east-bound lanes of a divided highway. It is of composite construction, with five rolled steel joists, 610 × 190 mm × 140 kg/m, supporting a 178 mm slab; and five simply-supported spans from 11.28 to about 13.5 m. Of these, the 11.28 m span, the first span reached by traffic, was instrumented with strain gauges on the bottom flange, 0.45 m on the approach side of the centreline. The bridge was calibrated by loading it with a static, known vehicle and the central bending moment was expressed as a constant multiplied by the weighted sum of the strains (with weighting factors, 1.28, 1.13, 1.0, 1.13 and 1.38). The bridge had been selected after trials with another bridge, and stainless steel/teflon inserts were added to the moving bearings to reduce friction.

The first impact study was carried out on 8 December 1981. Observers were stationed at the bridge and weigh-bridge and of 356 vehicles crossing the bridge, 137 effective correlations were achieved. Data observed at the weigh-bridge included the vehicle description, gross weight, weight of each axle group and the distances between the centres of the axle groups. Only single and dual-axle groups occurred; in the second case, the distance between axles was assumed to be 1.5 m, based on other Australian studies. The results are shown in Fig. 7.3, with $I + 1$ plotted against the gross vehicle mass. The graph also shows a mean line produced by a linear regression analysis, and estimates of the 95% and 5% confidence lines. The recorded values of I vary between -0.03 and $+1.32$ with a mean of $+0.38$. The upper bound values are high. It was decided to repeat the test, on 7 September 1983, with another 33 correlations, as shown in Fig. 7.4. In this test, I varied from $+0.21$ to 1.31, with a mean of $+0.72$.

The magnitude of these values, compared with those from other field studies, prompted further work, both experimental and analytical: some of which is described in Section 7.4. Chan (see Chan 1988; Chan and O'Connor 1986, 1990a, 1990b; O'Connor and Chan 1988a, 1988b) carried out further tests on Six Mile Creek bridge, with additional strain gauges provided to allow the estimation of curves of axle load against time for moving vehicles (see Section 7.4). He re-calibrated the bridge and then repeated Pritchard's work for 130 vehicles crossing the bridge during a period in June 1987, of which 27 were weighed at the weigh-bridge; the static weights of the others were estimated from the dynamic records using his program for the estimation of axle loads. As shown in Figure 7.5, values of I ranged from -0.20 to $+1.25$ with a mean of 0.51.

If these tests are compared, it is apparent that all gave high values of I:

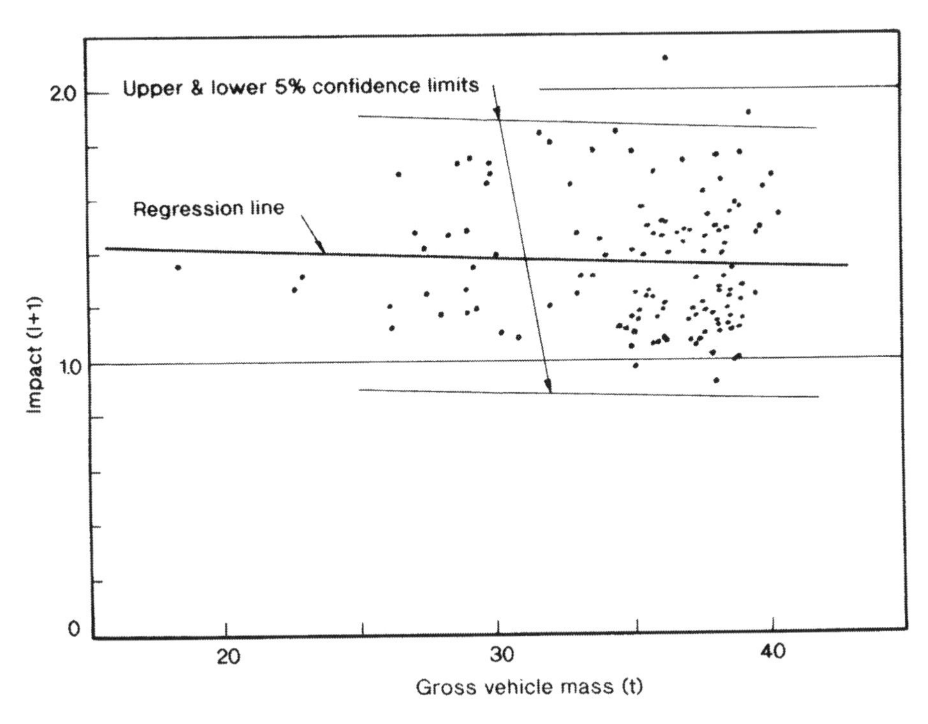

Fig. 7.3 Values of $I + 1$ observed by Pritchard in his first impact study of Six Mile Creek Bridge (Redrawn from Pritchard 1982; O'Connor and Pritchard

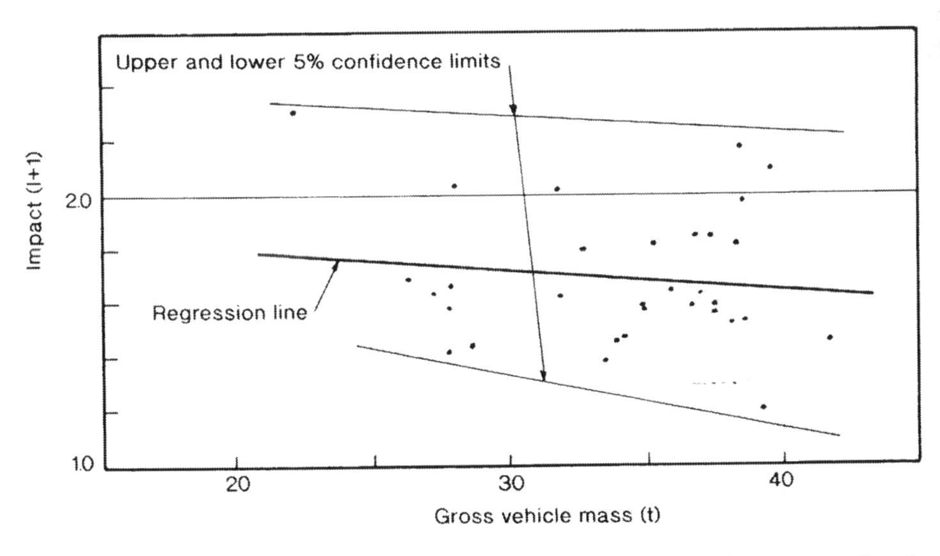

Fig. 7.4 Values of $I + 1$ observed by Pritchard in his second impact study of Six Mile Creek Bridge (Redrawn from Pritchard 1982; O'Connor and Pritchard 1985)

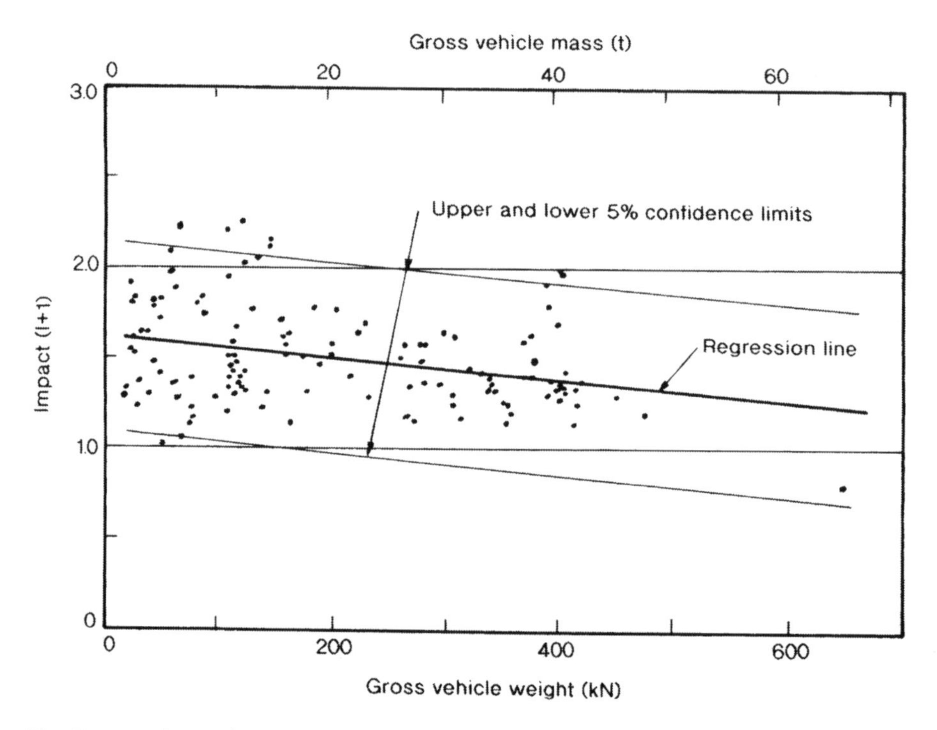

Fig. 7.5 Values of $I + 1$ observed by Chan on Six Mile Creek Bridge (Redrawn from Chan 1988; Chan and O'Connor 1990a)

1.32, 1.31 and 1.25 ($I + 1 = 2.32$, 2.31 and 2.25). The means varied in the following chronological order: 0.38, 0.72 and 0.51. On inquiry of the Main Roads Department, it was found that between the first and second impact studies, the road on either side of the bridge had been resurfaced, with no resurfacing applied to the bridge itself; as a result there was an increased drop from the approach road to the bridge surface. Then, in April 1985, an overlay of 20 mm of bitumen was applied to the highway without interruption over the bridge, leading to a reduced drop on entry. This history is consistent with the variation in the observed mean values. Chan estimated the bridge frequency as 10.8 Hz. It is useful to observe that this lies within the second peak of impact values found by Cantieni (see Fig. 7.1(a)) in his work with a plank on a bridge. Wheel hop was recorded by Cantieni as occurring commonly between 10 and 15 Hz. Shaw (1993) estimated values in the range 10–12 Hz from dynamic analyses of two suspensions (shown in Figs. 7.15 and 7.16). It also should be repeated that this study is unusual in that the measured values of I are for a wide range of service vehicles.

There have also been studies on the effects of braking on the vertical load applied by vehicle to a bridge or pavement.

Section 4.8 showed that there was an early interest in the dynamic loads applied to railway bridges. Some of this was due to the nature of early steam locomotives, but the problem still remains. It differs from the case with highway bridges in two ways: (a) the suspension characteristics and geometry of railway bridges are different; and (b) the nature of rail and wheel roughness also is not the same. A few Australian field studies will be recorded here.

Muller and Dux (1988; 1992) have reported field studies carried out during December 1987 and March 1988 on a prestressed concrete girder railway bridge in Central Queensland, built in the early 1980s. It has three spans of 15 m and three of 25 m, each with three girders per span. The deck is ballasted. The bridge carries coal trains, each with typically 111 to 148 wagons of gross mass 80 or 72 tonnes, equally distributed between four axles. An industrial train may be up to 2 kilometres in length, with up to five electric or six diesel electric locomotives distributed along its length, and these have nominal axle masses of 18.3 (electric) or 16.3 tonnes. It is apparent that the wagon axle loads (20 or 18 t) may exceed those from the locomotives. The normal running speed of the trains is 40 to 60 kph, with a legal restriction at the time of 60 kph. The current codes of practice (AREA or Australian and New Zealand Railway Bridge Design Manual; see Section 4.11) specified impact values of 30% for the 15 m span and 22% for the 25 m span.

These studies indicated that measured impact values were small for smooth track. However the chief cause of concern related to two common defects. If a train brakes sharply so as to cause the wheels to slide along the rails, wheel flats are formed. The general practice on this line was that wheels were not removed for remachining until flats showed a deviation of around 2 mm from the original circular shape. Muller and Dux (1988: 23) note that 'records to date indicate that extreme cases of wheel defects with deviations of up to 5 mm and flat lengths of up to 125 mm exist for some short time'. The second related defect is 'wheel burn', caused by a locomotive spinning its wheels to obtain traction. Here the defect is left in the rail; not only is part of the rail head ground out to form a depression, but a molten spray of droplets of metal is deposited on the rail behind the wheel and hardens to form a lump. Muller and Dux (1988: 23; 1992: 362) have recorded evidence of a depth of wheel burn of around 3 mm, but extreme wheel burns of greater depth have been observed – 'for example, burns completely through the head of the rail and into the web'. Other track irregularities include battered or mismatched joints, weld irregularities, cracks in the rail, and surface corrugations; the last will be referred to briefly in Section 7.3.

It was decided to introduce artificial irregularities ground into the top surface of the rail: either one defect 3 mm deep and 250 mm long, or 6 mm deep and 300 mm long in one rail; or two defects 6 mm deep and 300 mm long spaced 1680 mm apart in one rail; where in both cases one defect was located at a strain-gauged cross-section near mid-span. For practical reasons, it was not possible to stop trains on the bridge. For trains with electric locomotives a record was available for a traverse at 10 kph on a smooth track, and this is used as a datum for I. In other cases, the recorded traces indicated that impact caused by vehicles on this bridge with its initial smooth track were small, and these measured strains have been used as a datum (where shown in brackets in Table 7.1) for the peak values of I listed on the next page.

It can be seen that values in excess of the code values ($I = 0.30$ for 15 m; 0.22 for 25 m) were recorded in these tests. However, the observed base stress

Table 7.1 Peak values of *I*

	15 m span			25 m span		
Defects:	*1 × 3 mm*	*1 × 6 mm*	*2 × 6 mm*	*1 × 3 mm*	*1 × 6 mm*	*2 × 6 mm*
Electric locos	0.23	0.45	1.0	–	0.33	0.56
wagons	0.38	0.50	0.75	0.22	0.28	0.47
Diesel locos	(0.67)	–	(0.86)	(0.22)	(0.44)	(0.47)
wagons	(0.22)	–	(0.67)	–	(0.26)	(0.25)

levels were found to be less than those calculated by normal theory, of the order of 67–80% for the 15 m span and 82–89% of the theoretical values for a 25 m span, possibly because of the effectiveness of the track and ballast in distributing load.

Muller and Dux also draw attention to the large number of peak stresses, a train with 148 wagons and five electric locomotives will apply 622 axle loads in a single passage, and superimposed on these will be additional dynamic stress fluctuations. Chitty, Grundy and McTier (1990) have carried out rainflow counts (see also Section 6.3; Grundy 1980, 1982; Grundy and Teh 1986) on stresses measured in seven short span bridges on the major line from Melbourne to Albury (and Sydney). Grundy (1996) comments on this work, 'Extensive studies of a standard stringer-girder underbridge on the standard gauge line between Melbourne and Albury failed to find impact factors exceeding 20% of live load at any time'. However, he goes on to compare observed and calculated stresses at a cover plate end and again finds that the observed stresses are low. He also attributes this 'to the load spreading effect of the rail and sleepers'.

7.3 Road and rail roughness

It should be mentioned, initially, that most measurements of road and rail roughness are not primarily concerned with bridges but rather with such matters as vehicle riding quality and safety, and with the maintenance of roads.

The General Motors (GMR) road profilometer was developed in the 1960s (Spangler and Kelley 1965). This used a 152 mm road-following wheel mounted on a trailing arm beneath the measuring vehicle. The principle of the system is shown in Fig. 7.6: the sensor wheel is connected by a spring (k) and dashpot system (c) to a mass, m. Two measurements were made: a potentiometer was used to measure the relative displacement between the wheel and the mass, and an accelerometer mounted on the mass gave readings of acceleration that were integrated twice with respect to time to give displacement. The sum of the two displacements gave the road profile.

Figure 7.7 shows two photographs of a profilometer that was developed by O'Connor and Yoe (Yoe 1986) to operate on the same principle. It had essentially three components: a bicycle wheel mounted on a central frame pivoted on an axle, mounted at the rear of a vehicle; two side arms pivoted on

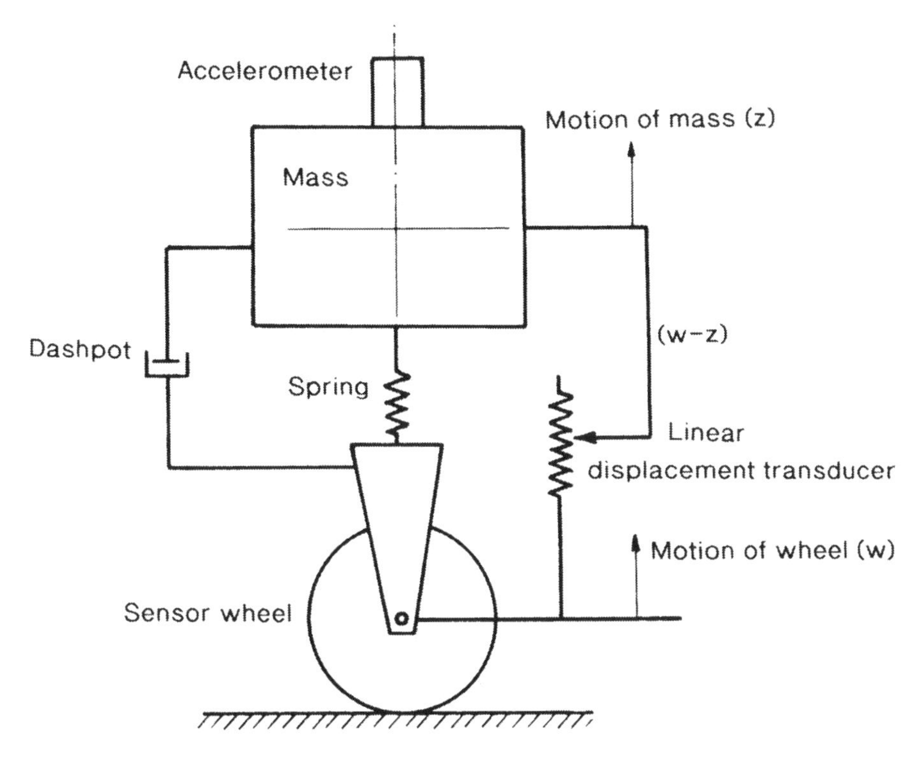

Fig. 7.6 Diagrammatic view of GMR Profilometer

Fig. 7.7 Profilometer developed by O'Connor and Yoe (Yoe 1986)

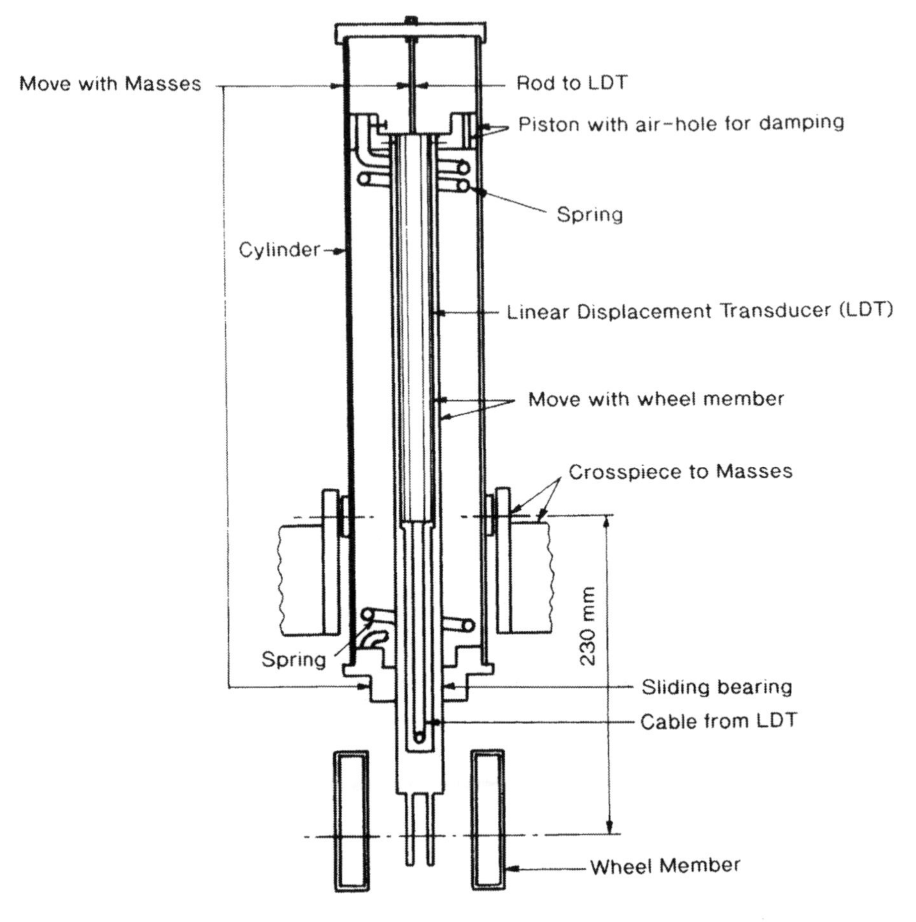

Fig. 7.8 Later suspension detail for the Yoe profilometer

the same axle and supporting two masses whose centres of gravity were concentric with the axes of the bicycle wheel; and a connection detail serving the same functions as the connection in Fig. 7.6. Figure 7.8 shows a revised connection detail for the Yoe profilometer, with a centrally mounted linear displacement transducer, a concentric coil spring mounted between an upper piston and the surrounding cylinder, and provision for air damping. Later profilometers, such as an early example developed by the British Transportation and Road Research Laboratory (Still and Winnett 1975–6; Still and Jordan 1980), rely on contactless sensors of the road profile, but still use for part of the measured displacement an integration of vertical acceleration measurements; this can cause difficulty in the selection of the initial constants of integration.

Given that longitudinal (and also transverse) road profiles can be measured, there is still a problem in the assessment and use of the resulting records, particularly in their application to bridge design. One procedure is to apply the Fast Fourier Transform and replace the record by its frequency components.

Another is the use of a so-called quarter-car model that is run either physically or theoretically along the bumpy road. This consists of a sprung mass, M_s, supported by a spring of stiffness, k_s, and damper with a linear viscous damping coefficient of C_s, on an axle of unsprung mass, M_u, resting on a tyre of stiffness, k_t (as proposed initially by Gillespie, Sayers and Segel (1980)). In more recent papers, Sayers (1989, 1995) has proposed such a model – the IRI or Golden Car – with the following parameters:

$C_s/M_s = 6.0 \text{ sec}^{-1}$
$k_t/M_s = 653 \text{ sec}^{-2}$
$k_s/M_s = 63.3 \text{ sec}^{-2}$
$M_u/M_s = 0.15$

This model is run along the profile at 80 kph; the total absolute value of the vertical movement of M_s is accumulated, and then divided by the length, L(m), of the record. The resulting dimensionless parameter, of the form of a slope, was accepted in 1986 as an international roughness index (IRI).

It is not immediately apparent how this measure may be applied to bridge design. One suggestion (made by O'Connor in 1990, but unpublished) is to observe that the quarter-car model has two degrees of freedom, with two corresponding first natural frequencies of vibration. It is possible to select the quarter-car parameters so that these natural frequencies correspond to those of interest in bridge design. The set in Table 7.2 envisage M_s = 4000 kg and have their frequencies (in Hz) embodied in their designations.

The values of C_s/M_s are such that, based on a simplified analysis, an initial axle disturbance reduces to about 0.2 × the initial value in about 2.0 cycles. The effective chassis damping is less: to 0.2 in about 33.6, 6.2, 10.5 and 20.6 for the quarter trucks in the order quoted; QR 3.5/– is undamped.

The suggested strategy is that profiles be rated by running each truck theoretically over the profile at chosen speeds, recording reactions at the pavement to tyre interface. Peak values of I would be calculated and histograms of I would be stored, together with the maximum value. These would then be examined, and an appropriate portion of the profile around the point of maximum I would be kept, for each quarter truck and speed. In this way, specific road profiles and speeds, known to cause high I for a particular pair of frequencies, would become available for use in analyses for bridge-vehicle interaction. The IRI Quarter Car would also be run along the profiles, at 80 kph, and the

Table 7.2

	C_s/M_s	k_t/M_s	k_s/M_s	M_u/M_s
QT 1.5/10	2.2	425	113	0.137
QT 2.0/12	2.4	475	242	0.128
QT 2.5/14	3.0	525	495	0.136
QT 3.0/16	4.6	575	1185	0.190
QT 3.5/–	4.6	575	∞	0.190

International Road Roughness Index recorded. It is possible that a step of chosen size could be added to the profile at the point of entry to the bridge.

The discussion of rail and wheel roughness in Section 7.2 is sufficient to provide a general introduction to much of the problem for railway vehicles. There is, however, another form of rail roughness described generally as rail corrugation. Grassie and Kalousek (1993) classify six types of corrugation:

1. heavy haul, with a wavelength 200–300 mm;
2. light rail bending, 500–1500 mm;
3. due to sleeper resonance, 45–60 mm;
4. due to contact fatigue, 150–450 mm;
5. rutting, 50 mm (trams), 150–450 mm (trains); and
6. roaring rails, 25–80 mm, from an unknown cause.

The corresponding frequencies for a vehicle travelling at, say 60 kph, are high, from the order of 600–200 Hz for roaring rails, to about 10 Hz for light rail bending. One would expect that only the lower frequencies would be significant in bridge design.

7.4 Analyses for bridge–vehicle interaction

This section describes two classes of analysis for highway bridge–vehicle interaction carried out at the University of Queensland under the general supervision of O'Connor and P. Swannell. It also refers briefly to some work on railway bridge impact.

Field impact studies by Pritchard (1982; O'Connor and Pritchard 1982, 1984, 1985) suggest that impact in a highway bridge is vehicle dependent, for it is observed to vary widely from vehicle to vehicle. There is a great range in the geometries of modern trucks and their suspensions, but, in addition, it is possible that suspension characteristics may vary between units made to the same design. The use of a single vehicle to explore impact effects is hardly likely to give complete results, and there are logistical difficulties in using a sufficient number of vehicles for field studies. For this reason, Shaw (1993, Shaw and O'Connor 1986) explored the following strategy.

1. Establish a laboratory facility for the rapid observation and recording of a truck suspension unit.
2. Record this essential data for a wide range of commercial vehicles by hiring them for the short time necessary for this task.
3. Develop a general purpose analysis program which can take this data and simulate the dynamic behaviour of the vehicles as they pass along bumpy roads of specified profile, and across a bridge.

Suspension properties which are required include:

(a) the geometry of the suspension;

(b) the mass and mass moment of inertia of each member;
(c) the force-deflection relationships for the springs;
(d) the force-deflection relationships for the tyres at various tyre pressures; and
(e) the force-velocity relationships for the shock absorbers.

Some of these properties may require calculation, but many may be measured; for example, load cells built into the floor may be used to measure axle weights for the complete vehicle, or with the chassis supported by an overhead crane; and a rig could be used to apply static loads to the spring-tyre assembly without dismantling the vehicle. As part of this process, identification numbers would be allocated to all nodes and members. The program developed by Shaw allowed for six member types: the chassis, lumped masses, rigid members, shock absorbers, springs and wheel members; it could be extended to include air bags or torsion bars. The experimental, laboratory phase of this strategy has not been carried out and is available to others to use. The computer program has, however, been fully developed. It is useful to describe here its main features and some typical results.

Earlier work was carried out by Kunjamboo on the suspension characteristics of a rear-axle unit from an International ACCO 1810A truck (O'Connor, Kunjamboo and Nilsson 1980; Kunjamboo and O'Connor 1983). This included a main spring of fourteen leaves, with an auxiliary spring of four leaves coming into play at a deflection of 70 mm. The load-deflection relationship is shown in Fig. 7.9(a). The two phases of behaviour, with and without the auxiliary spring, are clearly visible. There is also a marked separation of the loading and unloading curves. Both are non-linear, but linear approximations are shown. The two non-linear curves are defined by the following relationships:

loading curves :

$$P = 0.1268d + 0.000893d^2 \qquad \text{for } d \leq 70$$
$$P = 2.554 - 0.090204d + 0.003472d^2 \quad \text{for } d > 70$$

unloading curves:

$$P = -1.90 + 0.115d + 0.0005d^2 \qquad \text{for } d \leq 70$$
$$P = 2.805 - 0.1544d + 0.003389d^2 \quad \text{for } d > 70$$

the linear approximations are:

$$P = 0.1575d \qquad \text{for } d \leq 70$$
$$P = -28.448 + 0.5639d \qquad \text{for } d > 70$$

In these expressions, d is the deflection, in mm, and P the force, in kN. The separation between the loading and unloading curves requires that, if the spring is loaded to a certain point and then unloaded, there must be an unloading

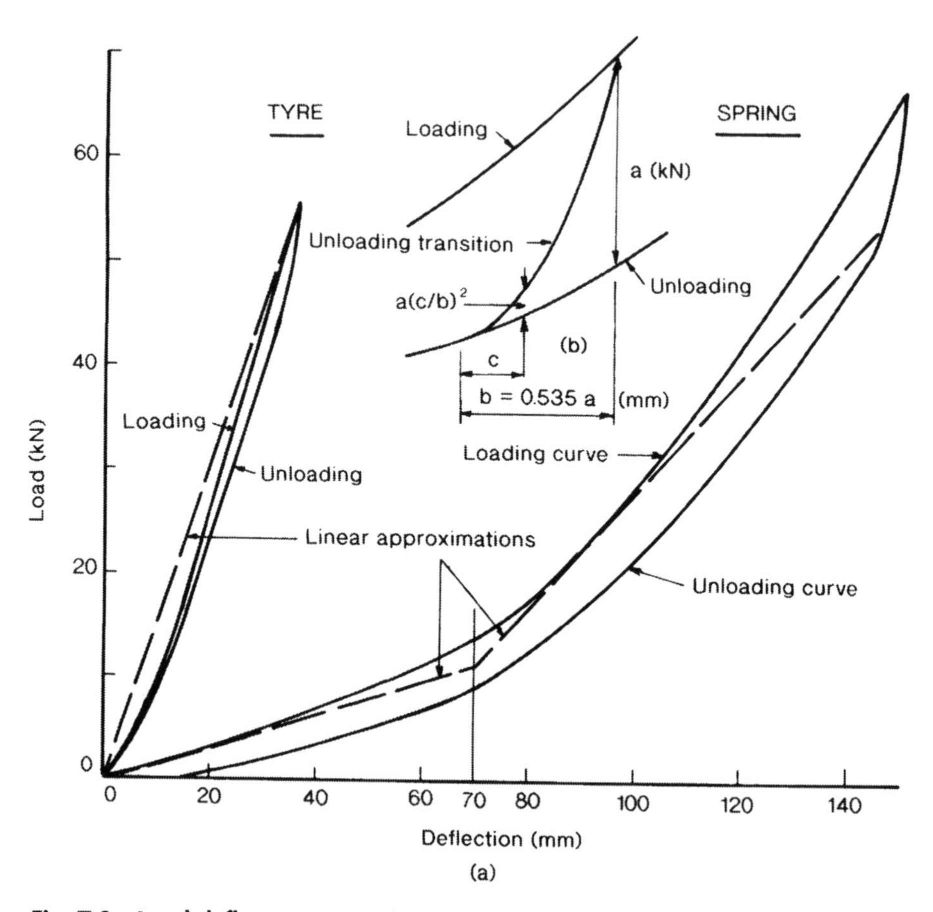

Fig. 7.9 Load-deflection curves for spring and tyre (Redrawn from Kunjamboo and O'Connor 1983)

transition to connect the two curves as shown in Fig. 7.9(b). If a (kN) is the vertical distance between the curves at the point where unloading commences, then Kunjamboo found that the horizontal length, b (mm) of the unloading transition was given by

$$b = 0.535a$$

He explored the use of a hyperbolic transition curve of the form shown in the figure, with relative coordinates, c, and $a(c/b)^2$. It should be emphasised that, in the present context, the shape of these transition curves is of paramount importance, for as a truck moves along a bumpy road, the load on the spring will vary, and the effective spring stiffness during these variations will be defined by the transitions rather than the basic loading and unloading curves. For a mass, m, on a spring of stiffness, k_s, the natural frequency, f, is given by

$$f = 1/T$$

where the period, T, is

$$T = 2\pi\sqrt{m/k_s}$$

The characteristic frequency of the suspension is, therefore, largely influenced by the effective slope of the transition curves. The area between an unloading and loading transition also corresponds to a loss in energy and provides a source of damping. Deflection of a leaf spring results largely from sliding between the leaves. The shape of these transitions will depend heavily on inter-leaf friction. It is conceivable that this will vary with the condition and age of the springs, so that the vibration frequency of a leaf-spring suspension system may also vary between units that are nominally identical.

Kunjamboo also obtained load-deflection curves for the pair of 9.0 × 20 Olympic cross-ply tyres fitted to the back axles of the International truck; one set of these is also shown in Fig. 7.9. He defined the loading curve as made up of an initial second order polynomial, from 0 to the point (H,V), followed by a straight line, with all the governing parameters expressed in terms of the tyre pressure, p (kPa). Kunjamboo's experimental work used tyre pressures from 483 to 690 kPa; his expressions are assumed to be valid for $p > 400$ kPa. As before, the vertical force scale (P) is in kN, and the displacement, d, and other horizontal parameters in mm. Then:

$$H = 9.327 + 0.01146p$$
$$V = -8.1555 + 0.048p$$
$$P = Dd + Cd^2 \qquad \text{for } d \le H$$
$$P = V + S(d - H) \qquad \text{for } d > H$$

where the coefficients S, C and D are given by

$$S = 0.3098 + 0.00277p$$
$$C = (SH - V)H^2$$
$$D = S - 2HC$$

These expressions define the loading curve; the vertical ordinate of the unloading curve is taken as 0.9 × that for the loading curve. The linear approximation shown is given by

$$P = 0.2855d + 0.0155pd$$

Kunjamboo found that the use of these linear approximations gave a significant loss in accuracy. Shaw's program allows for the more precise definitions, but with the coefficients replaced, if necessary, by values determined by other tests.

Shaw provided for four types of spring support. His basic leaf spring is shown in Fig. 7.10, with a pin at one end and a slipper at the other which allows frictionless movement along the line joining the slipper and the pin.

Figure 7.11 shows the slipper type spring, with a slipper at each end. The body of the spring is modelled by the outline shown. Nodes 1 and 2 are joined

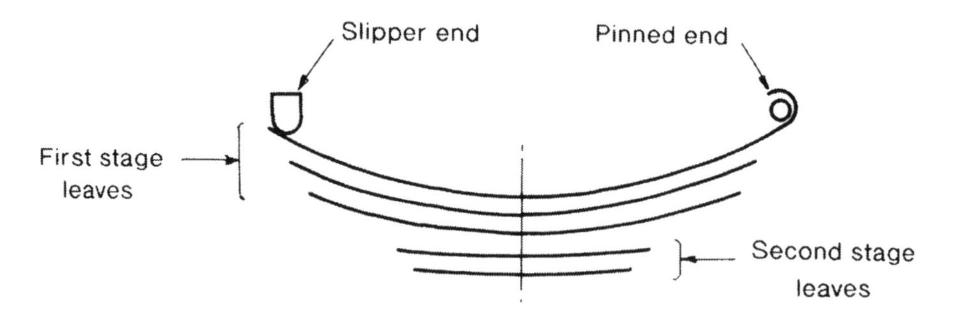

Fig. 7.10 Two-stage leaf spring

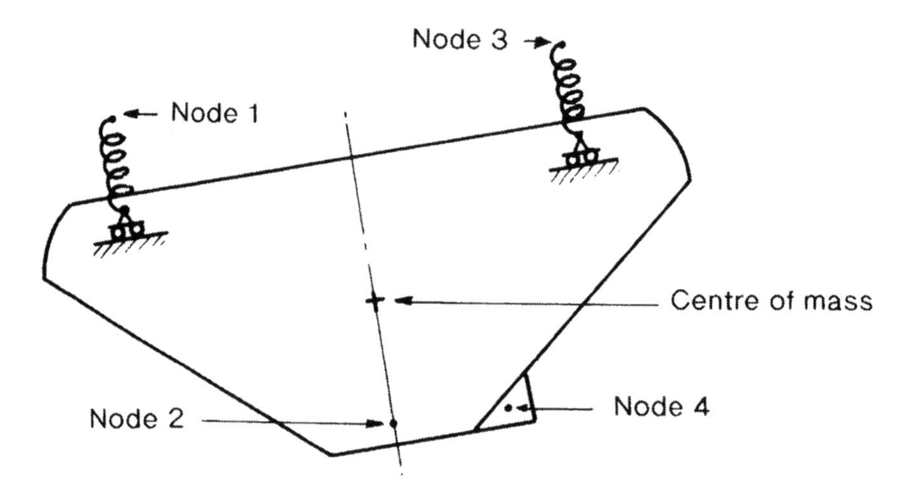

Fig. 7.11 Model for slipper-type leaf spring (Shaw 1993)

to the ends of the spring, with movements relative to the spring defined by the spring load-deflection curve. Node 2 is attached to the spring, and node 4 may be used for an additional link connection to the spring. This spring has both mass and a mass moment of inertia. Additional lumped masses may be added; at nodes 1 and 2 for example.

The pin-ended leaf spring is shown in Fig. 7.12. Here, both ends are pin-jointed to other members of the suspension system but are forced to slide in paths relative to the body of the spring by the slots shown in the model, so as to correctly model spring deflections. Shaw's combination type leaf spring has a pin at one end and a slipper at the other, as shown in simplified form in Fig. 7.13. He also allowed for the inclusion of a linear, massless spring, to enable comparisons to be made with this simple, classical case.

A suspension system may also include a shock absorber of the type shown in Fig. 7.14(a). Kunjamboo (1982) also carried out tests on a shock absorber, to measure the relationship between longitudinal force and velocity. His relationships are shown in Fig. 7.14(b). It can be seen, as would be expected, that the

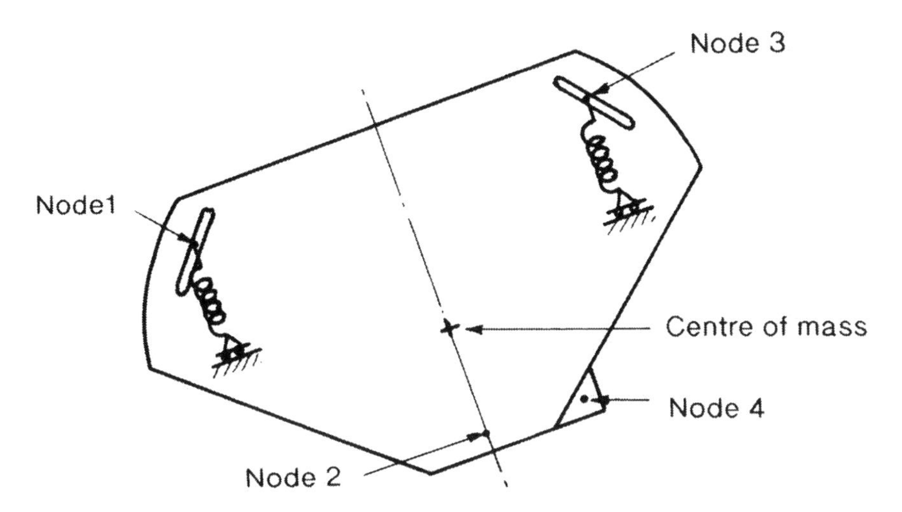

Fig. 7.12 Model for pin-ended leaf spring (Shaw 1993)

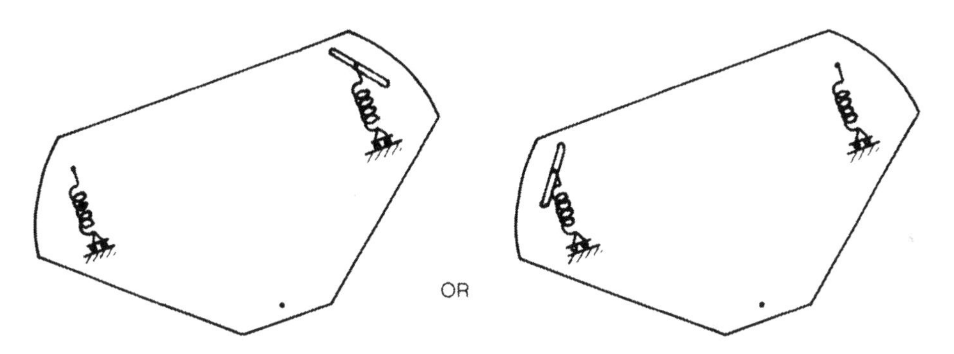

Fig. 7.13 Model for combination-type leaf spring (Shaw 1993)

loading (or bound) and unloading (rebound) curves differ. Both include a transition between upper and lower curves (F_{BU} and F_{BL}, F_{RU} and F_{RL}, are the bound and rebound upper and lower forces) at a transition velocity of 200 mm/sec. His expressions for these curves are as follows:

$$F_{BL} = 0.00657V - 0.0000127V^2$$
$$F_{BU} = 0.508 + 0.00149V$$
$$F_{RL} = 0.0201V - 0.0000307V^2$$
$$F_{RU} = 1.228 + 0.00782V$$

where the forces are in kN and the relative velocity, V, is in mm/sec. Both upper curves are linear.

Figure 7.15 shows Shaw's model for the McGrath MF2B tandem axle trailer suspension and illustrates three of his member types:

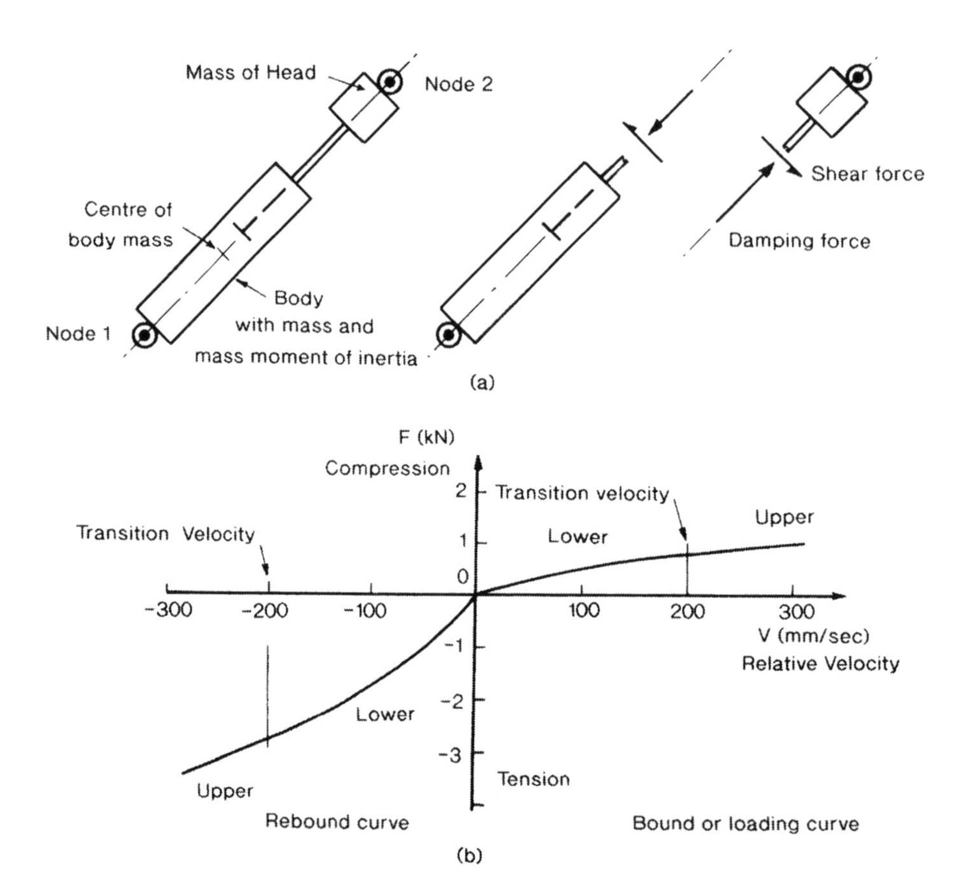

Fig. 7.14 (a) Model for shock absorber (Shaw 1993); (b) Force-velocity relationships for shock absorber (Redrawn from Kunjamboo 1982)

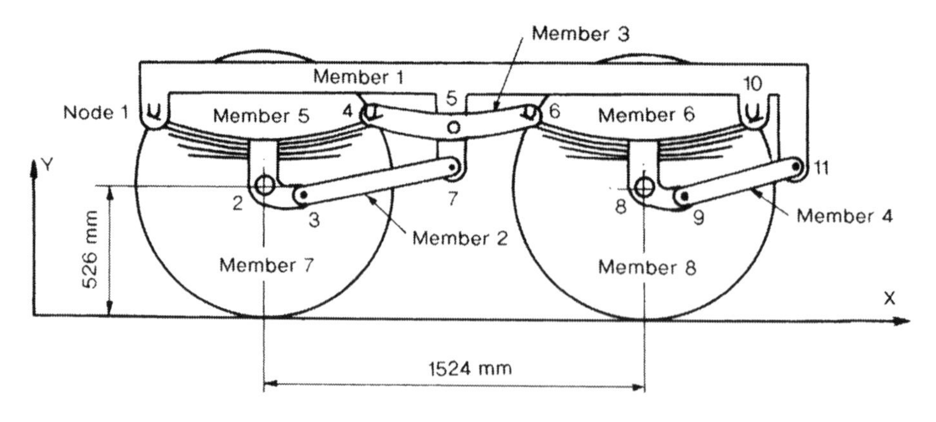

Fig. 7.15 Model of McGrath tandem-axle trailer suspension (Shaw 1993)

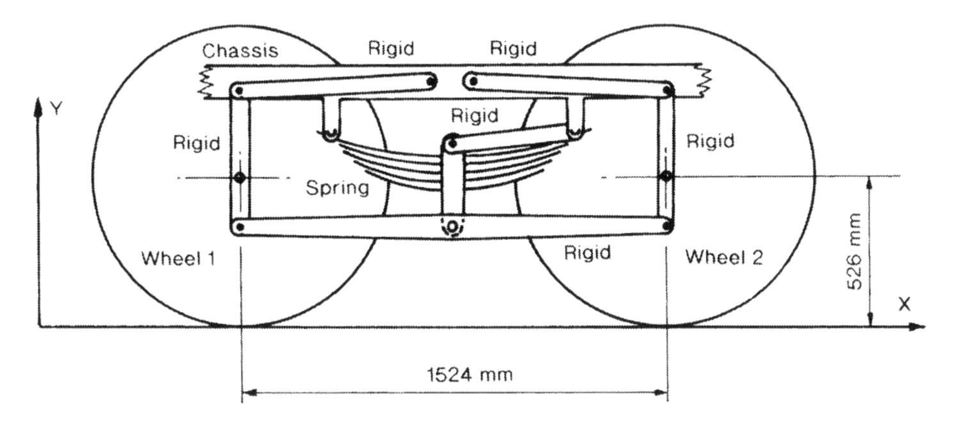

Fig. 7.16 Hypothetical rigid walking-beam tandem suspension (Shaw 1993)

(a) member 1 represents part of the chassis, and is assumed to be rigid;
(b) members 2, 3 and 4 are also rigid members;
(c) members 5 and 6 are both slipper-type leaf springs; and
(d) members 7 and 8 are tyres.

Another example is the hypothetical tandem walking beam unit shown in Fig. 7.16, based on the Hendrickson RT340 unit.

The program, however, permits the analysis of many other suspension geometries. As in a typical frame analysis program, the user may proceed by specifying nodes, members (or connectivity) and member types. As mentioned earlier, additional lumped masses may be included; payload may be added, and this may either be fixed to the chassis or allowed to bounce (free) with a specified coefficient of restitution. There is provision also for the application of out-of-balance wheel forces.

Shaw's dynamic analysis runs such a suspension unit at constant velocity along a road profile with one of three shapes:

(a) the sinusoidal profile has sinusoidal bumps, separated if desired by flat sections of road;
(b) a second type has a length of road of any profile, specified by a set of equally spaced ordinates, and repeated indefinitely; and
(c) the profile may be arbitrarily defined by a set of numerical ordinates.

The analysis divides into two stages:

1. An initial program calculates the displaced shape of the suspension under its own weight and the specified static loads.
2. The suspension is then moved at constant speed along a road profile of specified shape. This dynamic analysis allows for non-linear spring, shock-absorber and tyre characteristics, and also for changes in geometry.

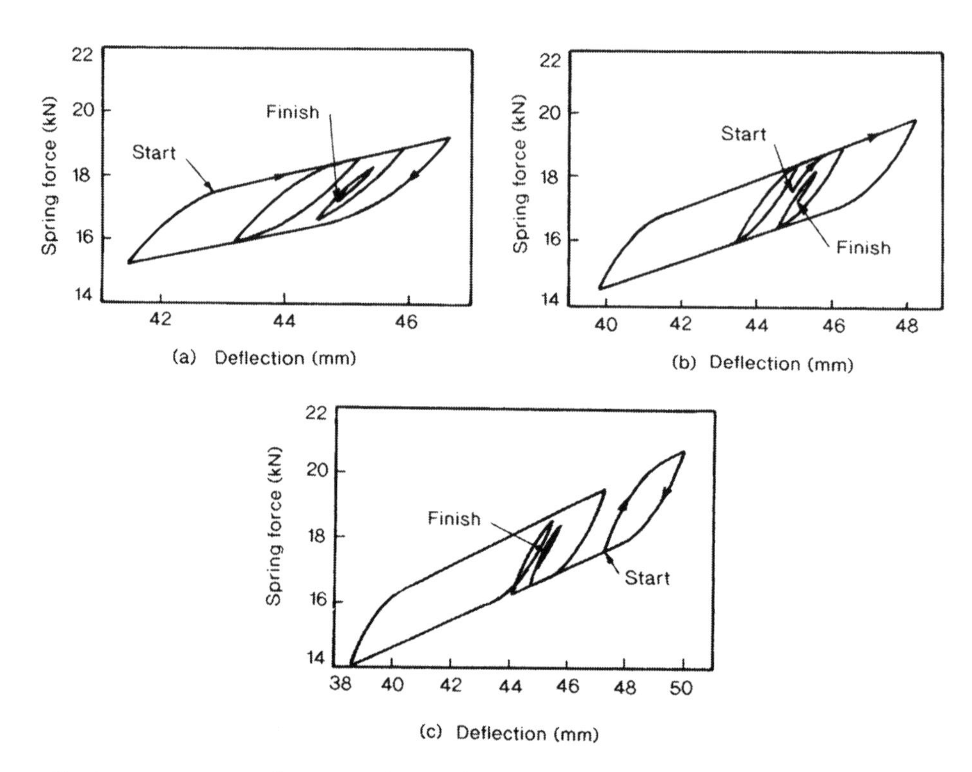

Fig. 7.17 Curves of spring force/displacement during the movement of a vehicle: (a) starting from loading curve; (b) starting mid-way between loading and unloading curves; (c) starting from unloading curve (Shaw 1993)

In the early stages of the work it was assumed that the spring and tyres started at points on their loading curves (Fig. 7.9), but it was later realised that this may be an unrealistic assumption. Figure 7.17 shows plots of spring force against displacement assuming starting positions: (a) on the loading curve, (b) mid-way between the loading and unloading curves, and (c) on the unloading curves. Not only are there differences in the resulting behaviour, but the final points are all about midway between the two curves. It was decided, therefore, to permit the specification of the initial starting position by the parameter, ζ, where

$$\zeta = (F_L - F_S)/(F_L - F_U)$$

where

F_S = the starting load
F_L = the vertical ordinate to the loading curve, and
F_U = the vertical ordinate to the unloading curve

The cases described below all have $\zeta = 0.5$.

Figures 7.18(a) and (b) show graphs of wheel impact (maximum dynamic/

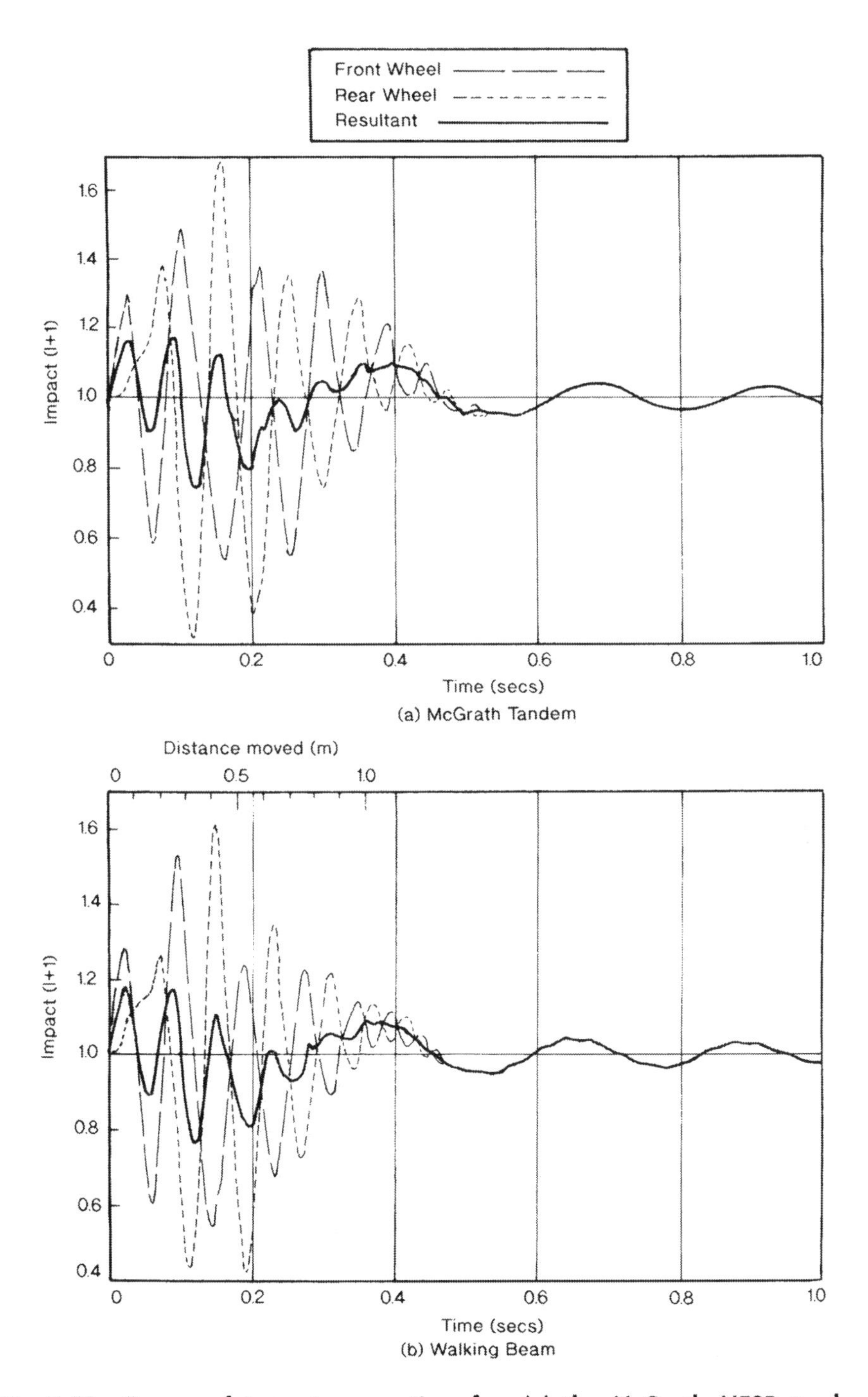

Fig. 7.18 Curves of impact versus time for: (a) the McGrath MF2B tandem suspension; and (b) the hypothetical walking-beam tandem suspension, both travelling over a 10 mm bump at 100 kph (Shaw 1993)

static axle load) for the McGrath MF2B and the assumed walking beam suspension units travelling over a 10 mm high bump at 100 kph. The bump shape was sinusoidal, with a total length of 2.22 m. At 100 kph, the time necessary to travel 2.22 m is 0.08 seconds. It can readily be seen that peak axle impact values are of the order of 1.68 and 1.61. However, if the two axle loads are added to give the resultant load, impact values are much less: Figs. 7.19(a) and (b) plot the peak wheel and resultant impact values for the two suspensions against speed and bump height. In both cases, the maximum resultant impact for a speed of 100 kph and a 20 mm bump height is of the order of 1.47. Of the two, it is likely that the resultant impact is more important in bridge design. The total effect can be seen as the resultant load, divided equally between the axles, together with a pair of equal and opposite loads from the axles, spaced at 1.524 m. These equal and opposite loads could hardly be expected to have a significant effect on the main structure of a bridge, although they may be relevant locally, such as in the design of the floor slab.

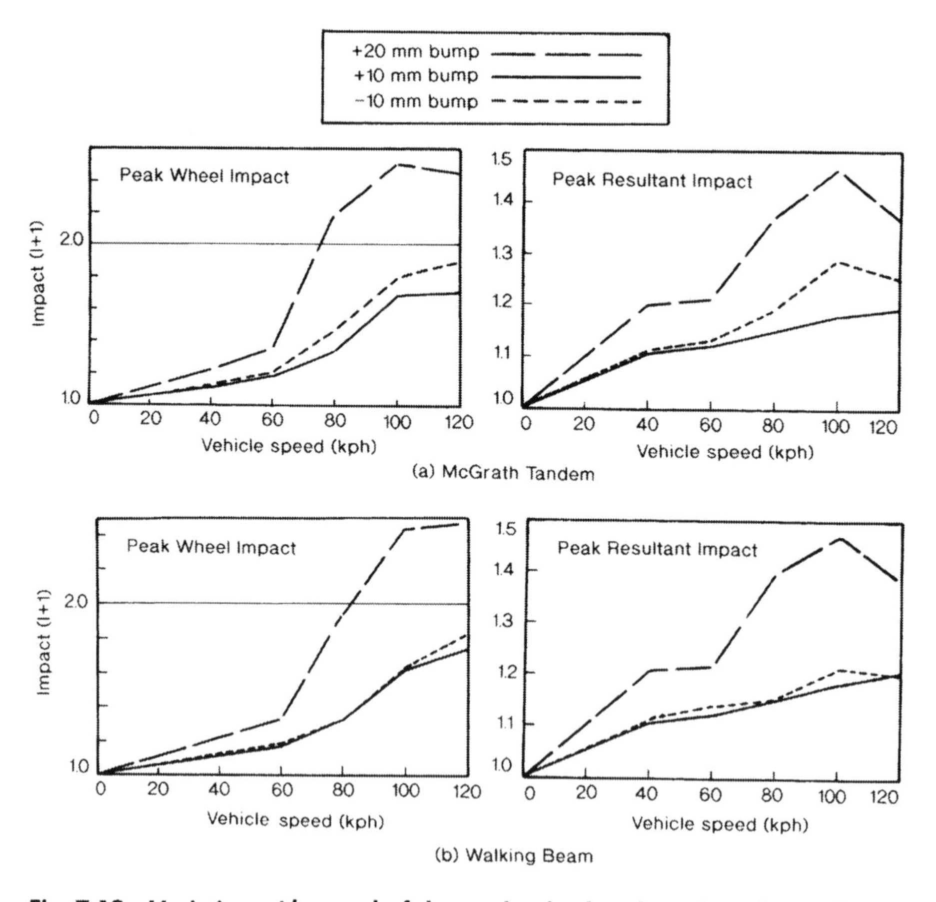

Fig. 7.19 Variation with speed of the peak wheel and resultant forces for: (a) the McGrath MF2B tandem suspension; and (b) the hypothetical walking-beam tandem suspension (Shaw 1993)

Two phases of behaviour may be identified in Fig. 7.18. For both suspensions the first 0.5 seconds is characterised by equal and opposite axle-forces superimposed on forces due to a general chassis movement. Beyond 0.5 seconds, these initial wheel oscillations are damped out; the axle forces are equal and their periodic variation is due solely to chassis movements. The axle-force variations in the first phase are due to what is called 'axle hop', and have frequencies in the range 10–12 Hz for both suspensions. In the second phase, the wheels do not oscillate; the behaviour is called 'body bounce', and has a frequency of about 4.4 Hz. Shaw observed that, during the first 0.3 seconds (at $V = 80$ kph), all tyres and springs used their loading and unloading curves (see Figs. 7.9 and 7.17), whereas after 0.3 seconds, the tyres and springs vibrated only in their transition zones. Figure 7.9(b) showed a typical transition, with a length $b = 0.535a$ (mm/kN). Shaw called 0.535 the BFAC (or 'b' factor). He carried out further studies (at $V = 80$ kph) with BFAC = 0.3 and 0.1, and showed that the frequencies for body bounce increased from 4.4 to about 5.7 and 9.5 Hz. In view of the significant relationship between the frequency of the truck loads and the natural frequency of the bridge, this variation is important. Values of $I + 1$ were not greatly altered. Another point to notice is that, at 100 kph, a vehicle will take 1 second to move 27.8 m, the length of a 27.8 m span bridge. Compared with this, the duration of 0.5 seconds for the axle-hop model is relatively short, but it is evident that repeated bumps could cause the axle-hop mode to extend over the full time taken for a vehicle to traverse the bridge.

These cases had a payload of 3000 kg. At the time, the maximum allowable load for each axle group was 16 tonnes, corresponding to a payload of 6093 kg. Figure 7.20 shows the effect of payload on the peak resultant impact for the McGrath MF2B tandem suspension, for speeds of 60 and 80 kph. For the higher speed, $I + 1$ reduces from 1.28 at 3000 kg to 1.15 at 6093 kg, for a 10 mm bump height.

Shaw also examined the effects of a change in tyre pressure, over the range 400 to 650 kPa, with a payload of 6093 kg, for speeds of 60 and 80 kph. Although impact tended to increase with tyre pressure, the effect was small, with $I + 1$ varying from about 1.13 to 1.15 for the higher speed.

At about the same time, Chan carried out related work at the University of Queensland (Chan 1988; Chan and O'Connor 1986, 1990a, 1990b; O'Connor and Chan 1988a, 1988b). As mentioned in Section 7.2, one phase of his work was to repeat the impact study carried out by Pritchard (O'Connor and Pritchard 1985) on Six Mile Creek Bridge. These studies recorded impact values calculated from measured bending moments; it should be realised that impact values at this level often exceed those within the axle loads that cause them. Chan's purpose was partly to clarify the reason why the values recorded in both studies at Six Mile Creek Bridge tended to be higher than those observed elsewhere. His first work (O'Connor and Chan 1988a, 1988b) investigated the use of bridge strains to measure actual dynamic loads. He then (Chan and O'Connor 1990a) measured field impact values for Six Mile Creek Bridge and used these to identify trucks that caused larger impact and to analyse the frequency spectra of the loads they applied to the bridge.

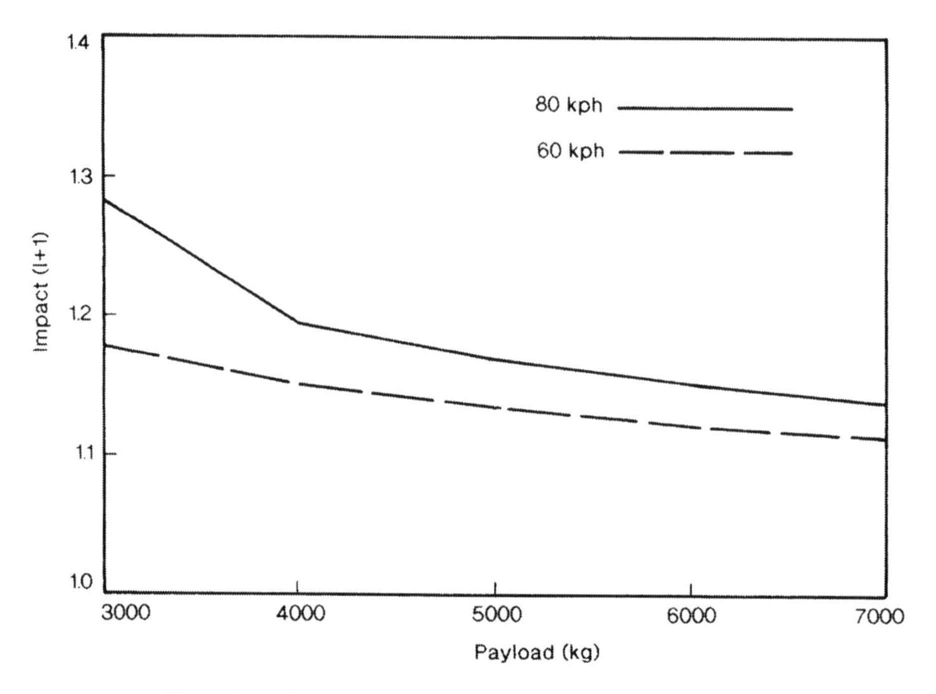

Fig. 7.20 Effect of payload magnitude on peak resultant impact for McGrath MF2B tandem suspension, travelling over a 10 mm bump at 60 and 80 kph (Shaw 1993)

He went on (see also Chan and O'Connor 1990b) to define a simple vehicle model for highway bridge impact,

> . . . in which each axle load is assumed to consist of a constant component plus a dynamic component varying sinusoidally about the mean. Factors such as the road roughness and prior excitation of the vehicle are taken to be transmitted into the amplitudes and the relative phase angles between the loads.

He defined a number of terms.

1. The dimensionless bridge factor (BF) is given by

   $$BF = Lf/V$$

   where

   f is the first natural frequency of the bridge and, for a uniform girder bridge, is given by

   $$f = (L^2/2\pi) \times \sqrt{(EI/m)}$$

where

L is the span
EI the bending stiffness and
m the mass per unit length

V is the velocity of the vehicle. Within the period $(1/f)$, the vehicle travels a distance, $s = V/f$. Then

$BF = L/s$

2. To allow extension of the work to bridges of various span, he defined the Dynamic Moment Ratio,

   DMR = maximum dynamic moment ÷ $(L \times W)$,

   where

   W is the sum of all axle loads for a particular vehicle.

3. He considered the case with multiple axle loads and defined the phase with respect to the bridge as

   ϕ = (the distance from mid-span to the nearest load peak before mid-span)/(V/f)

   It follows that, for $\phi = 0$, a peak dynamic load occurs with the load at mid-span.
4. He also studied the first natural frequencies of simply-supported girder bridges and found that they could be expressed with reasonable accuracy by the expressions,

   Lf = constant
   = 120 for prestressed concrete girder bridges and composite steel girder and concrete slab bridges
   = 60 for precast prestressed concrete box girder units

   The value, $Lf = 120$, was found to agree quite well with values given by Cantieni (1984a).

For Lf constant, the bridge factor, BF, is a function only of V. With $Lf = 120$ and $V = 20$ m/sec (72 kph), $BF = 6.0$.

Chan developed two programs for the analysis of a bridge under the action of dynamic loads, including damping in the bridge. Figure 7.21 shows the variation of central bending moment with span for a single unit mean load (kN), a dynamic component of 10%, $BF = 6.0$, a load frequency f_L equal to the bridge frequency f, a span of 20 m, and ϕ equal to 0. For an equivalent static load, the

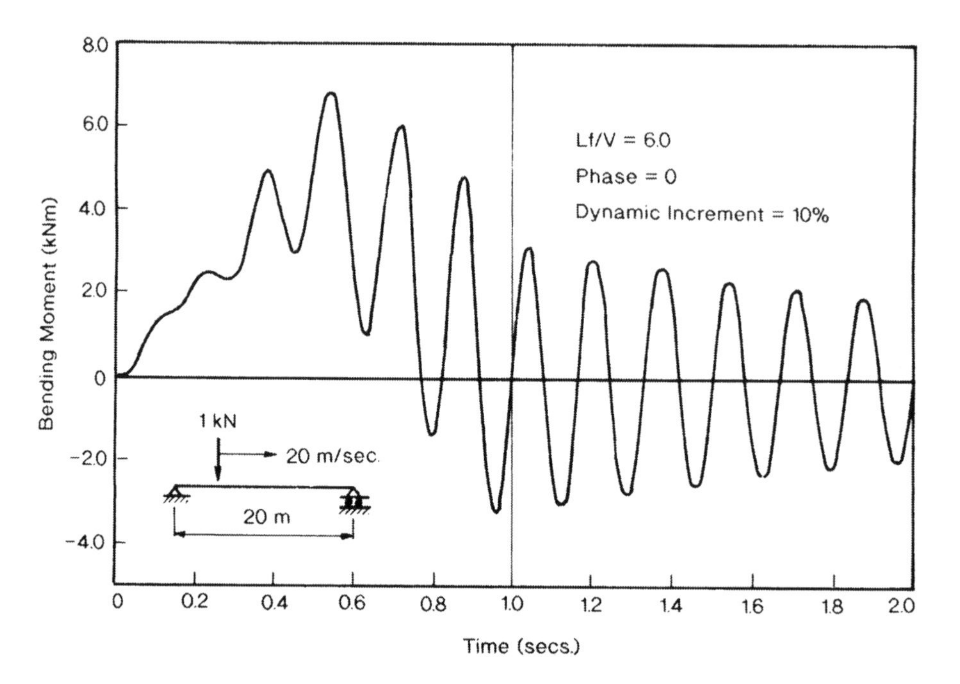

Fig. 7.21 Curve of mid-span bending moment against time for single moving load with dynamic load component of ±10% and phase of zero, for span of 20 m and bridge factor of 6.0 (Redrawn from Chan 1988)

maximum bending moment would be 5 kNm. The actual maximum is about 6.5 kNm, giving $I + 1$ equal to 1.3. Similar analyses were carried out at $BF = 3.0$, and phases of 0 and 0.4; the corresponding values of $I + 1$ were 1.29 and 1.12. Clearly, impact is affected by the phase of the dynamic load with respect to the bridge. It is also affected by the percentage dynamic load and its frequency, as shown in Fig. 7.22. The maximum effect is when $f_L = f$, and for this case I is linearly related to the percentage dynamic load.

Chan also investigated the effect of axle groups. For a double axle group, he found that, for all values of the phase, the maximum bending moment was always less than that produced by an equivalent single load. Accordingly, in later work, he replaced two- or three-axle groups by their resultant. For multiple axle groups, each replaced by a single load, the worst combined effect occurs when each has its phase with respect to the bridge equal to zero. A simple case occurs when the load spacing, S, is given by

$$S = K(V/f)$$

where

K is an integer

In this case the phase angle of the loads with respect to time are also equal.

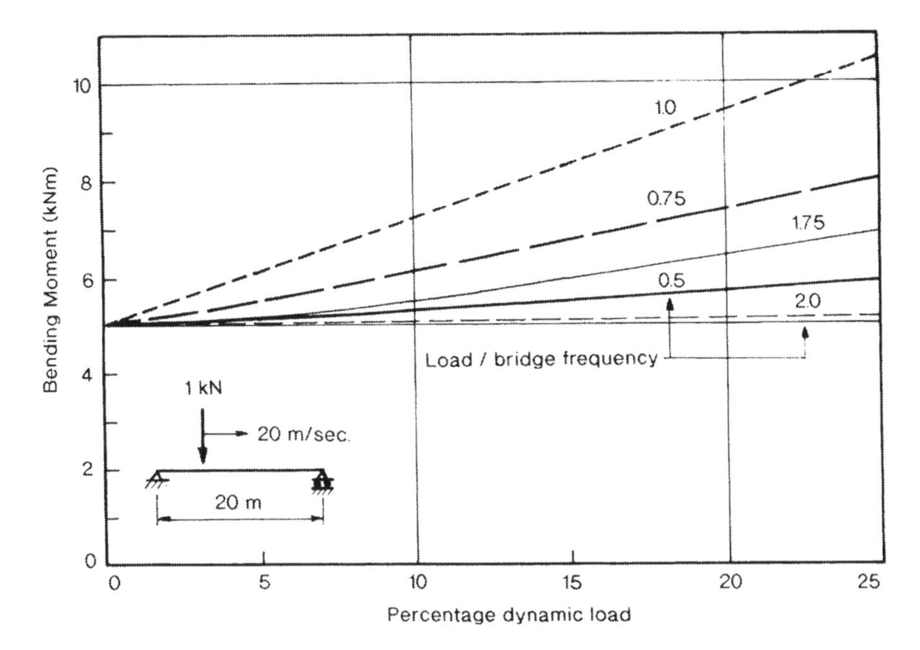

Fig. 7.22 Effects of magnitude of dynamic load component and ratio of load to bridge frequency on maximum bending moment (Redrawn from Chan 1988)

Chan, therefore, chose to consider vehicles:

(a) with three sets of axles, and equivalent static, axle-group loads of 48, 192 and 192 kN, as in the Australian T44 truck (see Section 4.7);
(b) with all axle groups having a dynamic load component equal to 10% of the mean load;
(c) with the spacings of axle groups all integral multiples of (V/f); and
(d) with all axle-group loads having the same phase, both with respect to time and the bridge.

He limited the axle group spacings to the range 3.7–8.5 m, and this imposed span limitations if the above conditions were to be satisfied. In his terminology, a (1,2) vehicle had spacings equal to $1 \times (V/f)$ and $2 \times (V/f)$. Figure 7.23 shows the maximum static and dynamic bending moments for simply-supported bridges with spans from 7.4 to 51 m and $BF = 6.0$, for a vehicle velocity, V, equal to 20 m/sec. The significance of the span restrictions may be seen from the following example:

For

$$L = 51 \text{ m and } K = 1$$
$$BF = 6.0 = Lf/V$$
$$S = K\,(V/f) = K\,(L/6.0)$$
$$= 8.5 \text{ m, for } K = 1.$$

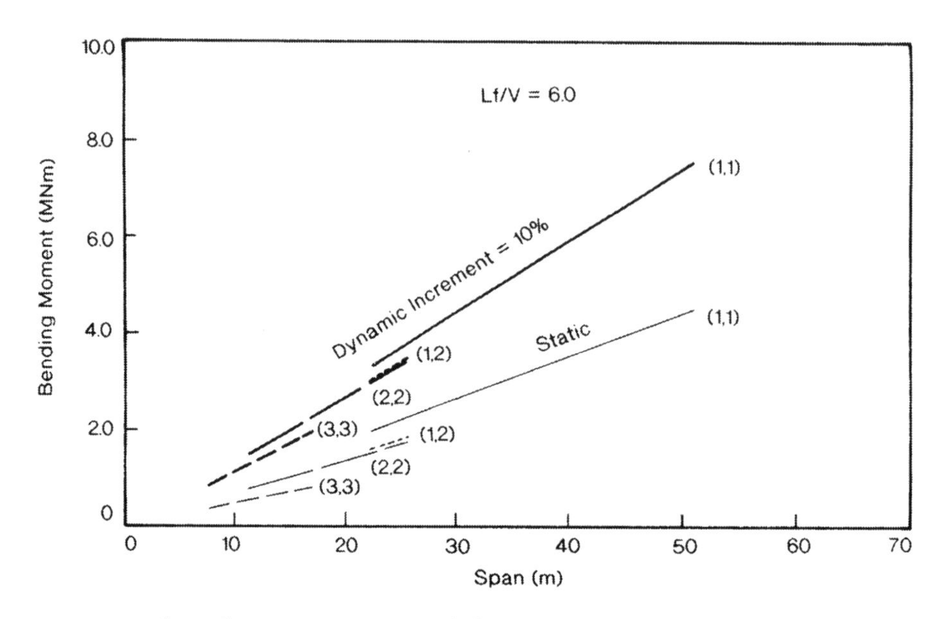

Fig. 7.23 Plots of maximum static and dynamic bending moments versus span, for vehicle with three axle groups, dynamic force component of 10% and speed of 20 m/sec (Redrawn from Chan 1988)

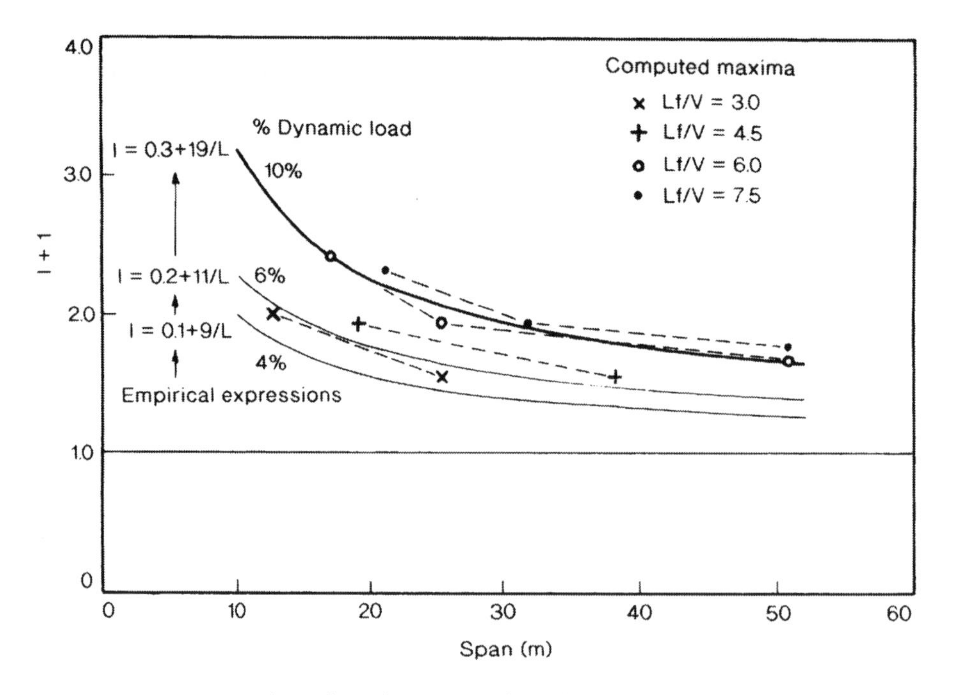

Fig. 7.24 Impact values for Chan's simple vehicle (BF = 6.0) (Redrawn from Chan 1988)

Table 7.3

% dynamic load	*I*
4	$0.1 + 9/L$
6	$0.2 + 11/L$
10	$0.3 + 19/L$

For longer spans, this expression would require $S > 8.5$ m. However, it must be remembered that this limitation is based on the requirement that the phase of the dynamic loads is the same, both with respect to the bridge, and with respect to time. There is no reason why the phase with respect to the bridge may not be kept constant for other axle spacings, even if the phase angle with respect to time varies.

For the cases analysed, maximum values of $I + 1$ are shown by the points plotted in Fig. 7.24. Relaxation of the constraint on axle-group spacings expressed in (a) above permits the calculation of $I + 1$ for other spans. Chan derived the empirical approximations to these values shown in Table 7.3.

For spans from 30 to 60 m, a value of 10% gives a reasonable approximation to some code provisions (see Chan and O'Connor 1990a: Figure 17). For shorter spans, the resulting values of I are relatively high. However, test results for Six Mile Creek Bridge (Section 7.2), with a span of 11.28 m, gave the maximum and mean values of $I + 1$ in Table 7.4.

For a span of 11.28 m, the previous expressions for I give the following values of $I + 1$:

4% dynamic load $I + 1 = 1.90$
6% dynamic load $I + 1 = 2.18$
10% dynamic load $I + 1 = 2.98$

Expressed alternatively, an 8.1% dynamic load is required to produce the maximum value of $I + 1 = 2.32$ at Six Mile Creek.

These studies do not resolve the matter of impact in highway bridges; their chief value is that they focus attention on the issues that are involved. In the case of Chan's work, for example, a dynamic load of 10% is not unreasonable; the chief questions are whether such a dynamic load will continue during the whole time a vehicle is on a bridge, and whether the axle-group loads should be in phase. In the latter connection, however, it may be observed that the presence of

Table 7.4

Test	*Maximum I + 1*	*Mean I + 1*
Pritchard (1)	2.32	1.380
Pritchard (2)	2.31	1.716
Chan	2.25	1.509
Overall mean		1.471

a fixed bump, or a series of bumps, constitutes a disturbance that is in phase with location on the bridge. The other major factor that has not been addressed here, although it has been, to some extent, considered by Chan (1988), is the statistical variation of these loads. It is not really sufficient to determine the maximum effect that can occur in a few vehicle-crossings; rather one must consider the probability of an extreme event that may occur within the life of a bridge.

Many others have considered highway bridge impact, such as Nowak (1994) and Wang, Shahawy and Huang (1993). Miller (1984), Miller and Swannell (1983) and Swannell and Miller (1987) have described another Queensland analysis for bridge-vehicle interaction with some other test results for Six Mile Creek Bridge. Gupta and Traill-Nash (1980) (see also Mulcahy, Pulmano and Traill-Nash 1983) have studied impact effects due to a vehicle braking on a bridge; Sweatman (1983, 1987) has also studied suspension units for heavy vehicles. Heywood (1994) measured high impact values on another Australian bridge.

The problem of impactive vertical loads on railway bridges is essentially different – in regard to the dynamic characteristics of vehicle suspensions, the geometry of the vehicles, and the nature of rail and wheel roughness. One useful reference is IEAust (1981); another is the more recent paper by Kalay and Reinschmidt (1989); a third is the more general study contained in Frýba (1996).

7.5 Code provisions for horizontal dynamic loads

This section will look at some code provisions, with highway loads treated first, followed by railway loads.

The British Departmental standard BD37/88, used recently for bridge design in the United Kingdom, makes the following provision for centrifugal loads. For highway bridges carrying carriageways with a horizontal radius of curvature less than 1000 m, the centrifugal loads are specified (Cl.6.9) as a series of point loads, 50 m apart, in any two notional lanes of each carriageway, with each load given by

$$F_C = 40{,}000/(R + 150)\ \text{kN}$$

where R is the radius of curvature of the lane in metres.

With each centrifugal load there shall also be considered a vertical live load of 400 kN, distributed over the notional lane for a length of 6 m (Cl.6.9). For comparison, the normal HA loading consists of a knife edge load of 120 kN, together with a distributed load whose intensity varies with the loaded length, L. If L equals 50 m, this distributed load is 24.4 kN/m, totalling 488 kN. When the knife edge load is added, the total load is 608 kN. A value of L equal to 6 m gives a load intensity of 101.2 kN/m, a total distributed load of 607 kN, and a total load of 727 kN. It follows that the vertical load specified for combination with the centrifugal load is somewhat less than the HA loading, for L equal to or less than 50 m.

The basis for the centrifugal force is not given, but can be deduced as follows. The 400 kN vertical load has a mass, m, of 40,770 kg. The corresponding centrifugal force is given by mV^2/R. If this is equated to the specified force,

$$40770\ V^2/R = 40 \times 10^6/(R + 150)$$

The assumed velocity, V (m/sec), is therefore given by

$$V^2 = 981R/(R + 150)$$

These values of V, together with the ratios of horizontal/vertical force, are listed below for various values of R in Table 7.5.

Two conclusions may be drawn:

(a) the combination of load and velocity has been controlled to some extent by the radius, R; and
(b) the ratio of horizontal to vertical force is large for values of R less than 100 m; it approaches as a limit the value 0.667.

The coefficient of friction for a pneumatic tyre sliding from rest on a clean, dry road is typically of the order of 0.81, but reduces if the tyre slides at speed. Gillespie (1992: 346, 7) plots values reducing to about 0.73, 0.56 and 0.36 at 60 kph, and then 0.65, 0.47 and 0.19 at 100 kph, on dry asphalt, wet concrete and wet asphalt, respectively. The transverse slip of a rotating tyre is, however, a complex problem, as also is overturning due to a centrifugal force (see also Gim 1988; Moore 1975). It must be remembered that a curved roadway will typically be given a transverse slope or cant, and this will affect both sliding and overturning actions. One concludes, therefore, that the problem is complex, and the figures given by BD37/88 form a useful guide.

The earlier BS 5400: Part 2: 1978 specified the somewhat smaller value of $F_C = 30{,}000/(R + 150)$, with an associated vertical load of 300 kN.

BD37/88 also specifies the following longitudinal loads resulting from traction or braking (Cl.6.10):

(a) for the HA loading, 250 kN plus 8 kN per metre of loaded length, but not more than 750 kN, acting in one lane only;
(b) 25% of the weight of an HB vehicle; or

Table 7.5

R (m)	*V (m/sec)*	*V (kph)*	*Horizontal/vertical force*
30	12.8	46.0	0.555
50	15.7	56.4	0.500
100	19.8	71.3	0.400
300	25.6	92.1	0.222
500	27.5	98.9	0.154
1000	29.2	105.1	0.087

(c) a horizontal load of 300 kN, acting in any direction, in association with the HA loading.

For all these loads, taken in association with the appropriate vertical loads, a load factor of 1.25 is applied in checking an ultimate limit state, or 1.00 for a serviceability limit state. The load factors for the vertical loads taken alone are 1.50 for HA and 1.30 for the HB loading.

The weight of the HB vehicle depends on the specified number of HB units, which may vary between 30 and 45 for public highway bridges in Great Britain. The number of axles is four, with 10 kN per axle for one unit of loading. The total vehicle weight varies from 1200 to 1800 kN; 25% of this gives a range of braking forces from 300 to 450 kN.

The HA loading in one lane consists of a 120 kN vertical knife edge load together with a uniformly distributed load, W, kN per metre, expressed in terms of the loaded length, L.

For $L \leq 50$ m,

$$W = 336(L)^{-0.67}$$

For L between 50 and 1600 m,

$$W = 36(L)^{-0.1}$$

A comparison of the braking load defined in (a) above, and the vertical HA loading, is shown in Fig. 7.25. Two discontinuities occur: at $L = 50$ m, when the formula for the vertical load changes; and at $L = 62.5$ m, when the total horizontal load reaches the limiting value of 750 kN. Despite these discontinuities, it is evident that the ratio of horizontal to vertical load lies between 0.4 and 0.5 for much of the range shown; it reduces for longer lengths (beyond 50 m) to 0.25 (the value used for the HB loading) at about $L = 130$ m. All lie below the longitudinal friction coefficients quoted previously for a tyre skidding at speed on a dry asphalt surface. One would expect, therefore, that higher braking loads may occur in practice, at least for the shorter lengths. It must be remembered, however, that the specified loads are equivalent static loads; the problem is a dynamic one, with loads of short duration applied to a deck of large inertia or mass.

The earlier BS 5400 (Cl.6.7) specified a single horizontal load 'due to skidding' equal to 250 kN, which may readily be compared with the BD37/88 values shown in Fig. 7.25.

Part 3 of Eurocode 1 (ENV 1991-3: 1995: Cl.4.4.2, Table 4.2) specifies characteristic values of the centrifugal force, F_C, in terms of the sum, W, of the vertical loads acting on two axles.

For the radius, R, less than 200 m,

$$F_C = 0.2W$$

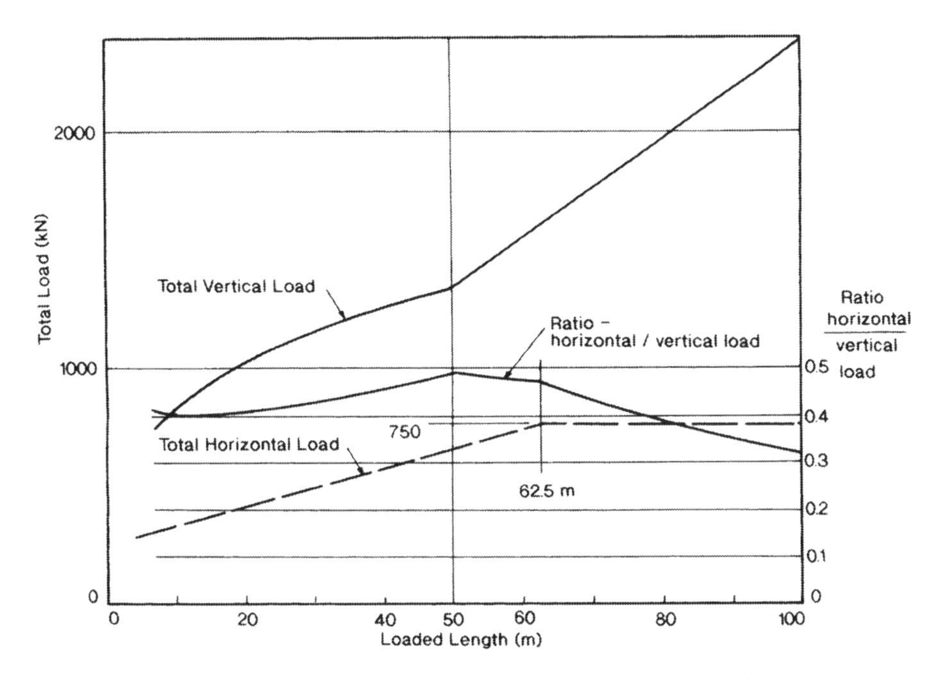

Fig. 7.25 A comparison of braking and vertical loads specified by BD37/88

For R between 200 and 500 m,

$$F_C = 40W/R$$

For $R > 1500$ m,

$$F_C = 0$$

The basic vertical load on each of the two axles is 300 kN in lane 1, 200 kN in lane 2 and 100 kN in lane 3. These are modified by an adjustment factor which may vary with the class of route. However, assuming this factor is unity, the total vertical loads are 600, 1000 and 1200 kN for one, two or three lanes. If the corresponding mass were m, then $m = W/g = 0.102W$ (kg), and the true expression for centrifugal load is

$$mV^2/R = 0.102WV^2/R$$

where

V is the velocity

As with BD37/88, this can be used to estimate the velocities implicit in the code formulae. For R between 200 and 500 m, the implied velocity is constant at 19.8 m/sec (71.3 kph). For smaller values of R, the velocity, V, reduces; some

Table 7.6 Relative values of centrifugal loads (kN)

Radius (m)	*BD37/88*	*Eurocode 1*		
		1 lane	*2 lanes*	*3 lanes*
30	220	120	200	240
50	200	120	200	240
100	160	120	200	240
200	114	120	200	240
300	89	80	133	160
500	62	48	80	96
1000	35	24	40	48

typical values are 14.0 m/sec (50.4 kph) at R = 100 m, and 9.9 m/sec (35.6 kph) at R = 50 m; all are somewhat less than the values calculated previously for BD37/88. It is also instructive to compare the relative values of the centrifugal loads themselves. Some of these are listed in Table 7.6.

For the most part, the BD37/88 values lie between 1 and 2 lanes of Eurocode 1.

Eurocode 1 (Cl.4.4.1) also specifies values for the braking force, F_B, including an upper bound of 800 kN for the full width of the bridge. For a single 3 m wide lane, with an adjustment factor, α, taken as unity, and a loaded length, L, greater than 1.2 m,

$$F_B = 360 + 2.7L \text{ (kN)}$$

For this case, the upper bound corresponds to L = 163 m, with F_B varying linearly from 360 to 630 kN over the range L = 0 to 100 m, or 360–529 kN for L from 0 to 62.5 m. For BD37/88, as shown in Fig. 7.25, the specified braking force varies linearly from 250 to 750 kN over this range, but is applied only to a single lane.

Turning to the United States, the 1973 AASHO Code specified a longitudinal or braking force of 5% of the live load in all lanes carrying traffic headed in the same direction, or in all lanes for bridges likely to become one-directional in the future. For this purpose, the live load was taken as the lane load, without impact. This was made up of a concentrated load of 80 kN, together with a uniformly distributed load of 9.34 kN/m for one lane, with both multiplied by a reduction factor for multiple lanes in excess of two, with 100% applied in each of one or two lanes, 90% in each of three lanes and 75% in each of four lanes or more.

Buckland's committee (Buckland 1981: 1164f, 1178f) reviewed this loading and recommended that the longitudinal load due to braking in one lane should be 80% of the weight of the design truck, and to this should be added 5% of the lane load in additional lanes, where the reduction factors for these lane loads were 0.7 for the second lane and 0.4 for each additional lane. This paper offers an interesting commentary, together with a comparison of longitudinal loads specified in various codes. It suggests that 'there is reasonable evidence that the braking force which can be applied by one vehicle is at least equal to the weight of the vehicle', but added:

> it is arguable that the design truck may not be able to exert 100% friction because of jack-knifing or other restraints.... It is unlikely that a large number of vehicles will all be exerting the maximum braking force on the bridge simultaneously, so some reduction in load intensity must be made when considering more than one vehicle. The recommended loading is a compromise ... based on 'common sense' rather than on any firm data.

The chief reference for some of this is a study published by the Transportation Research Board in 1975, with the title 'Effects of Studded Tires'. Lesser coefficients of friction may occur with un-studded tyres.

Buckland's code comparisons for braking force for one lane, 30.5 m long, are:

AASHTO	18.2 kN
British	445
Canadian	449
French	294
Ontario	105
ASCE (Buckland 1981)	256

It is evident that the 1973 AASHTO code provision is relatively very low.

The 1994 AASHTO *LRFD Bridge Design Specifications* (Cl.3.6.3) specify the use of a factor, C, for centrifugal force, where

$$C = 4V^2/(3gR)$$

where

V is the highway design speed and
R is the radius of curvature

The factor, C, is applied to axle weights of the design truck (or tandem) described in Section 4.3, with V not less than the value specified in AASHTO (1990). The form of this expression is as in the correct theoretical expression for centrifugal force, with the gravitational acceleration, g, included to make the adjustment from vehicle weight to mass. The vertical loading specified in this code requires the truck load to be applied with the lane load. The combination represents a single vehicle that is heavier than the design truck; this is allowed for in the numerical factor, 4/3. The clause allows also for a correct variation of vehicle speed with radius.

The 1994 AASHTO code (Cl.3.6.4) specifies a braking force taken as 25% of the axle weights of the design truck or tandem, and gives a theoretical basis for this expression. For a vehicle of weight, W, the mass is W/g and the kinetic energy, $WV^2/(2g)$. If the deceleration force, W, is assumed to be constant over a braking length, a, then the work done in stopping the vehicle is bWa. Equating the two,

$$bWa = WV^2/(2g)$$

that is,

$$b = V^2/(2ga)$$

The code assumes $a = 122$ m, and $V = 90$ kph (25 m/sec). Then, for $g = 9.807$ m/sec^2, b becomes 0.26, close to the specified value of 0.25. The corresponding time for the vehicle to come to rest is 2.71 sec.

Concerning railway bridges, Section 4.8 drew attention to the early recognition of the need to provide for dynamic effects, and quoted rules specified in many codes for the amplification of vertical loads. For many reasons, the nature of dynamic railway loads differs from that for highway bridges, and they must be considered separately.

Despite its title, *Loads for Highway Bridges*, the British Department of Transport Standard BD37/88 includes also detailed specifications for railway bridge live loads (Section 8 and Appendix D); these are identical with those in BS 5400: Part 2: 1978 (see Section 8 and Appendix D). The basic design RU and RL vertical loads and the corresponding dynamic factors were included in Section 4.8, with the specified provision for lurching. It was noted there that the RU loading was for main lines, both in Great Britain and on mainland Europe, while the RL loading was for lines carrying only rapid transit passenger stock and light works trains. There is also a specified nosing load (Cl.8.2.8), taken as a single lateral load of 100 kN, applied at track level.

The basic expression for centrifugal force is:

$$F_C = fP(v + 10)^2/(127R)$$

where

P is the static equivalent uniformly distributed load (80 kN/m) for RU loading, or 40 kN/m for RL loading,
v is the greatest speed envisaged on the curve in question in kph,
R is the radius, and
f is a factor defined as follows

$$f = 1 - \left[\frac{v - 120}{1000}\right] \times \left[\frac{814}{v} + 1.75\right] \times \left[1 - \sqrt{\frac{2.88}{L}}\right]$$

where

L is the loaded length

For L less than 2.88 m, or v less than 120 kph, $f = 1.0$.

If the velocity, V, is expressed in m/sec, then, for $f = 1$,

$$F_C = \frac{P(V + 2.78)^2}{9.80} = \frac{P(V + 2.78)^2}{g}$$

consistent with the correct theoretical expression for centrifugal force, but with the maximum envisaged speed increased by 10 kph, or 2.78 m/sec. If P is expressed in kN/m, then F_C will have the same units. The basis of the factor, f, is not given, but in relation to dynamic vertical effects, Appendix D of the code mentions that passenger locomotives may travel at speeds up to 250 kph, and light, high speed trains up to 300 kph. For example, $v = 250$ kph and $L = 28.8$ m,

$$f = 1 - (0.13 \times 5.01 \times 0.68) = 0.55$$

For smaller values of v or L, f increases, reaching unity at either $v = 120$ kph, or $L = 2.88$ m.

Longitudinal loads are specified in Clause 8.2.10 and Table 18 of BD37/88, separately for traction and braking, as percentages of the vertical loads on the driving or braked wheels, where these percentages are 30% for traction and 25% for braking. The specified vertical loads are shown in Fig. 4.15. For the RU loading, both traction and braking forces may be calculated directly from the loadings shown; for example, for a loaded length of 7 m, the total vertical load is 1000 kN, giving traction and braking loads of 300 and 250 kN. For the RL loading, the traction and braking loads are specified for particular loading lengths; some examples are: up to 8 m, 80 and 64 kN; and from 8 to 30 m, 10 and 8 kN/m, respectively. An upper bound of 750 kN is specified for the traction force; the maximum braking force is unbounded, and expressed as $20(L - 7) + 250$ kN, for $L > 7$ m. It is noticeable that the traction loads are large. For bridges supporting ballasted track, up to one-third of the longitudinal loads may be assumed to be transmitted by the track to reactions outside the bridge, provided that no rail discontinuities exist within 18 m of either end of the bridge.

Eurocode 1 (ENV 1991-33: 1995: Cl.6.3.2) includes a Load Model 71 for vertical loads that is identical with the British RU loading, together with load models SW/0 and SW/2 for use on designated lines that carry heavy rail traffic. The specified centrifugal forces for Load Model 71 are the same as in BD37/88. The longitudinal forces differ, and are specified as follows:

traction 33 kN/m, but $\not>$ 1000 kN;
braking 20 kN/m, but $\not>$ 6000 kN.

Although the maximum braking force is large, it corresponds to a length of 300 m. For shorter lengths, the values specified in BD37/88 are larger, but with a reduced upper bound of 750 kN for the traction force. For a length of 300 m, BD37/88 specifies a braking force of 6110 kN.

Although not strictly relevant, it is of some interest to observe that Eurocode 1 does discuss comfort levels for passengers in coaches during travel (Cl.G.3.1.3) and links this to vertical accelerations as follows: for a very good level of comfort, the vertical acceleration should not exceed 1.0 m/sec^2; for good comfort, 1.3 m/sec^2; and for acceptable comfort, 2.0 m/sec^2. For this reason, the code also limits the maximum vertical deflection of the structure; for example, for simply-supported girders with a span of 100 m, the following limits apply: for $V \leq 200$ kph, $L/600$; for $V = 200$–280 kph, $L/1400$.

One example of American practice will be given. For centrifugal force, the AREA1990 code (Cl.1.3.6) specifies a horizontal load, applied 1.83 m above the top of the rails, equal to $0.00117S^2D$ of the specified axle load without impact, where S is the speed in miles per hour and D is the 'degree of curve'. The longitudinal load from trains (Cl.1.3.12) shall be assumed as 15% of the live load (without impact), for the case where the rails are not continuous. If they are continuous (with either welded or bolted joints) across the entire bridge from embankment to embankment, the effective longitudinal load is specified as $L/36.6$ (for L in metres) multiplied by the full longitudinal load, but with this factor ($L/36.6$) taken as not more than 0.80.

Chapter 8

Light rail

Light Rail Transit (LRT) has seen something of a revival in recent times with many cities introducing it in an effort to curb private motor vehicle use and improve urban environments. The definition of light rail is a matter of debate. The Light Rail Transit Subcommittee of the US Transportation Research Board (ECMT 1994: 14) defined LRT as

> a metropolitan electric railway system characterised by its ability to operate single cars or short trains along exclusive rights-of-way at ground level, on aerial structures, in subways, or occasionally, in streets, and to board and discharge passengers at track or car floor level.

Under this definition tramways are defined as running primarily in mixed traffic while LRT has an exclusive right-of-way.

The European Conference of Ministers of Transport (ECMT 1994: 15) defined LRT as

> a rail-borne form of transportation which can be developed in stages from a modern tramway to a rapid transit system operating in its own right-of-way, underground, at ground level or elevated. Each stage of development can be the final stage, but it should also permit development to the next stage.

The latter definition allows a wider range of systems to be included in LRT. To the bridge engineer there is little to distinguish the vehicles which run in the various LRT systems, and hence the latter definition will be used here. The vehicles themselves, including trams, will be referred to as Light Rail Vehicles (LRV).

LRVs consist of a single, rigid or articulated car. However, two cars may be coupled together and run as a single, longer unit during peak periods. They

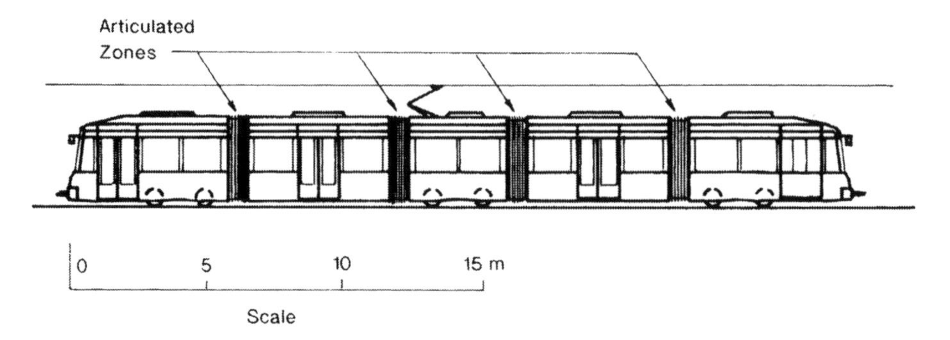

Fig. 8.1 Typical Light Rail Vehicle

have steel wheels on bogies running on steel rails and are electrically powered (usually DC). Modern LRVs often have low floors to improve accessibility and a typical LRV is shown in Fig. 8.1. LRVs differ from heavy rail in a number of ways. Heavy rail utilises longer trains (series of cars or carriages), and the stations are usually spaced further apart along the route and have higher platforms. Heavy rail is not as flexible as light rail and requires larger radius turns. LRT is not as flexible as bus transit and this has implications for loads imposed on structures by LRVs.

Light Rail Vehicle loading typically has not been specified by bridge design codes. The AASHTO (1994) LRFD Bridge Design Specifications state that 'where a bridge also carries rail-transit vehicles, the Owner shall specify the transit load characteristics and the expected interaction between transit and highway traffic'. If the rail transit occupies exclusive lanes the bridge is required to have sufficient strength to carry highway loading, assuming that the transit lanes are not present.

AUSTROADS (1996) simply states that 'the operating Authority for the utility shall be consulted to determine the appropriate design loads and load factors'. BS 5400: 1978 and BD37/88 (British Department of Transport 1988) do not address light rail explicitly. Eurocode 1 (ENV 1991-3) refers the designer to the relevant authority for the loading and characteristic values of actions for narrow-gauge railways, tramways and other light rails, preservation railways, rack and pinion railways and funicular railways.

The Swiss Standard SIA 160 (1989) does not refer explicitly to light rail but has a load model for rail traffic on narrow-gauge lines for urban and suburban traffic. The load model is reproduced in Fig. 8.2. The load may be applied to any two tracks but additional tracks are unloaded. The dynamic coefficient is defined by the following equation:

$$\Phi = 1.44/(\sqrt{l_\Phi} - 0.2) + 0.82 \qquad \text{with } 1.0 \leq \Phi \leq 1.67$$

where

Φ = dynamic coefficient, and
l_Φ = influence length (or span) in metres

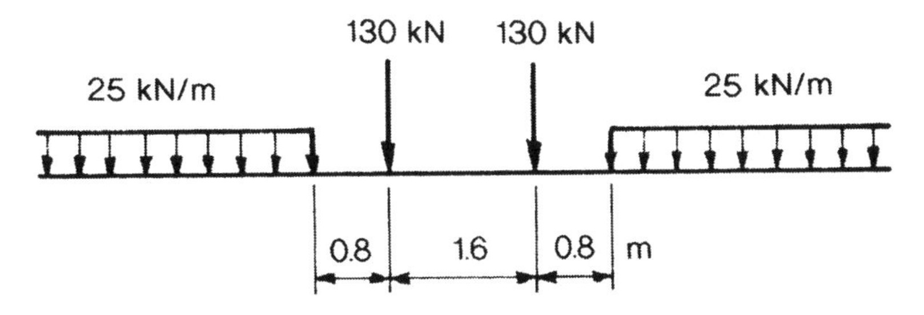

Fig. 8.2 Load Model 3 – SIA Standard 160 (1989)

The accelerating force is defined as 0.3 × the vertical load (without the dynamic allowance) up to a maximum of 250 kN, and the braking force is 0.25 × the vertical load (without the dynamic allowance) up to a maximum of 500 kN. When two tracks are loaded, the starting force is assumed to act on one track, while the braking force is assumed to act simultaneously on the other track in the same direction as the starting force. It is a characteristic of rail traffic that lurching, or the transfer of load from one rail to the other and back again, will occur. This changes the vertical loads on a rail, but not the total load on the two rails. Oscillation will also lead to lateral loads on the rails. SIA 160 specifies that the lateral force (nosing force) generated by this action is to be taken as 50 kN. The centrifugal force is assumed to act 1.8 m above the top of the rails and is given by the following formula:

$$F_c = fV^2Q_r/(gR)$$

where

F_c = representative value of the centrifugal force,
f = reduction coefficient (usually 1.0 for LRVs),
V = design speed in m/sec,
g = acceleration due to gravity,
R = radius of curvature in m, and
Q_r = representative value of vertical load due to rail traffic without dynamic component

This formula is similar to that given in BS 5400 and ENV 1991-3: 1995 for centrifugal forces on rail (see Section 7.5).

The lack of guidance given by codes causes some difficulty to designers of bridges that carry light rail. Brisbane is currently designing a light rail network. The proposed light rail route crosses the Victoria Bridge over the Brisbane River. This has three pre-cast, post-tensioned, segmental box girders joined by cast-in-situ slabs, and cast-in-situ cantilever footpaths. The bridge has three spans and a total length of 313 m. The central span consists of a suspended unit, 27.6 m long, supported by two cantilevers, each of 57.3 m, giving a total span of 142.2 m. The two anchor spans are each 85.4 m. The bridge is 15.24 m wide between kerbs, sufficient for five traffic lanes. Construction of the bridge was completed in 1969 and hence the design loads were about three-quarters of the loads required by Austroads (1996), which is the current Australian code.

Shaw and Stevens investigated the need to strengthen the bridge for light rail and this work was reported by Stevens (1998). The capacity of the bridge is about 7.2 kN/m per lane, whereas the current code loads work out at about 9.6 kN/m per lane. A lane of buses (1 m spacing) imposes 13.3 kN/m per lane; a line of design trucks (T44s at 1 m spacing) imposes 24 kN/m and a lane of fully loaded family cars imposes 3.3 kN/m. Of historical interest is the fact that another bridge in Brisbane, opened in 1932, was designed for trams using a lane load of 24.3 kN/m, and a design for the Victoria Bridge in the 1950s used a lane load of 17.2 kN/m for trams.

Data on the masses and weights of current LRVs were assembled for the recent study and some of these are included in Table 8.1. This shows the actual lengths of LRVs and their crush loads, where the crush load of an LRV is typically its own weight plus the weight of passengers, calculated as the number of passengers multiplied by an average weight of 700 N. The number of passengers is generally assessed as the number of seats plus 8 × the standing area (in m^2), although some manufacturers of LRVs use 6 persons per m^2 to calculate the crush loads. These weights have been converted to linear loads with vehicles end-to-end and with a 1 m spacing between vehicles. Maximum axle loads have been shown in the last column where available. For comparative purposes the loads defined by SIA 160 (1989) and those quoted by some other researchers have been included.

Table 8.1 indicates that LRVs vary in weight from 16.1 kN/m to 27.25 kN/m, although some of the variation may be due to differences in the assumption of the density of the crush load. This has implications for the bridge designer, interested in a design life of 100 or 120 years. During this period, a light rail transit operator may change the type of vehicle being used. One philosophy would be to design for a particular vehicle type and to strengthen the bridge when a heavier vehicle is added to the fleet. This philosophy has two drawbacks. First, it would require the LRT operator to refer all decisions regarding fleet purchases to the bridge owner; this may or may not happen and would certainly be seen by the LRT operator as a serious constraint on his ability to operate the system efficiently. Second, it is usually more expensive to add strength later to a bridge, rather than during its initial construction. More seriously, it may not be possible to strengthen a bridge at all. There may also be other limitations restricting the type of vehicle which can be used in the future.

The alternative strategy is to design the bridge so that it will carry all vehicles likely to be purchased by the LRT operator over the life of the bridge. This philosophy is consistent with that adopted for highway loads. In the case of Victoria Bridge it was decided to use 24 kN/m as the design load for lanes carrying light rail. Austroads (1996) specifies a load factor of 2.0 for normal traffic loads and 1.6 for railway loads for the ultimate limit state. The factor of 1.6 reflects the lower probability of overloading. Since the probability of overloading on a LRV is also low, a factor of 1.6 was used. Austroads (1996) specifies a dynamic load allowance of 0.2 for general traffic and 0 for railway traffic on a bridge with a natural frequency of 0.6 Hz. A factor of 0.05 was selected for the LRVs whose characteristics differ from heavy rail, and 0.2 for general traffic.

Table 8.1 Comparison of Light Rail Vehicle loads

Source	*Length (m)*	*Crush load mass (kg)*	*Linear weight (kN/m)*	*Linear weight 1 m Headway*	*Maximum axle load (kg)*
Alsthorne, France	28.5	60850	20.9	20.2	10410
ASEA AB, Sweden	22.15	43070	19.1	18.2	8180
Bombardler, Canada	27.17	60520	21.8	21.1	10890
Fiat, Italy	29.73	67470	22.3	21.5	
Kawasaki, Japan	26.3	61270	22.9	22.0	
UTDC, Canada	23.63	52260	21.7	20.8	9670
Waggon-Union, FRG	28.4	62290	21.5	20.8	11090
Tokyo, Japan	20.37	49320	23.7	22.6	12330
Simmering, Austria	26.75	52570	19.3	18.6	8150
MAN AG, FRG	38.58	95370	24.2	23.6	
DUEWAG, FRG	24.28	52020	21.0	20.2	9750
Breda, Italy	24.35	56220	22.6	21.8	22490
ABB/Comeng	16.56	27250	16.1	15.2	8100
Adtranz, Australia	28.28	51600	17.9	17.3	
PTC, Australia	23.53	46700	19.5	18.7	8650
SIA 160 (1989)			25.0		
Buckland (1981)	16.15	27020	16.4	15.5	
Buckland (1981)	26.2	49960	18.7	18.0	
TTC, Canada*	22.86	63490	27.25	26.1	
WMATA, US*	22.86	54420	23.35	22.37	
MATA, US*	22.86	55330	23.74	22.74	
UTDC, Canada*	12.8	22430	17.19	15.94	
DCRT, US*	22.86	52380	22.47	21.53	

*Nowak and Grouni 1983.

Contact was made with authorities in Melbourne and Sydney. Although there was some variation in their approach to loading, it was generally agreed that the LRV lanes should be fully loaded over the critical length of the bridge, and that, if two LRV lanes were present, then both lanes should be fully loaded. This is consistent with the section of the Australian code on railway loading and other international codes. Having determined a loading for two LRV lanes on Victoria Bridge, the question that remained was what portion of the traffic loading on the remaining three lanes should be considered to act with the LRV loads. For five traffic lanes, Austroads (1996) specifies that the normal traffic loading in all five lanes is to be multiplied by a factor of 0.6 and for three lanes the factor is 0.8. The total live loading on the bridge was calculated initially using two lanes of LRVs with no lane reductions, together with three traffic lanes each with a lane modification factor of 0.8. This was considered to be the upper bound of the most likely loading. Similarly, a loading calculated by applying a factor of 0.6 to all five lanes was chosen initially as a lower bound. A load of two lanes of LRVs and three lanes of average traffic as determined by survey was found to be higher than the latter case and became the new lower bound. Investigation of the bridge was eventually carried out using lane modification factors of 1.0 for the two LRV lanes and 0.6 for the remaining three traffic lanes. The serviceability limit state was determined using LRV loads equal to 5/6 of the crush loads to represent more normal loads. Local effects were checked using an axle load of 12000 kg, a dynamic load allowance of 0.4 and an ultimate load factor of 1.6.

In order to maximise its existing strength, the bridge was measured in detail and the results of concrete testing carried out at the time of construction were obtained. This revealed that the bridge was about 2% heavier than the design target. The anchor spans were found to be the critical sections of the bridge. However, as the bridge is an anchor-cantilever type, it is the difference between the dead loads on the cantilever and the anchor spans which dominates. Based on the detailed measurements, it was determined there was negligible imbalance between the percentage errors in the self-weight of the bridge on either side of the piers. Therefore the code figures of 1.0 for favourable dead loads and 1.2 for unfavourable dead loads were reduced to 1.0 and 1.05 respectively. The 1.05 allowed for some load imbalance, unforeseen stress distributions and modelling errors. An analysis of Victoria Bridge based on the preceding assumptions showed that the bridge required strengthening in order to carry light rail transit and its accompanying track slab. This strengthening is anticipated to be carried out by further prestressing the bridge with external tendons between the box girders and by installing vertical prestressing bars to increase shear resistance. It should be noted that a limiting factor in the load-carrying capacity of the bridge is the movement capacity of the central joint.

The above example demonstrates the issues facing a bridge designer in the absence of guidance from design codes. This lack of guidance leads to differing approaches by designers and results in designs with differing reliability indices. Nowak and Grouni (1983) investigated the impact of these differing approaches by examining five transit systems, as noted in Table 8.1. They observed that transit guideways were generally designed using highway and railway bridge design codes. This was the procedure adopted in the example above. They noted further that there are significant differences between bridges supporting highway or railway loads and those supporting transit loads. Their work was based on elevated transit guideways (bridges) for the exclusive use of LRVs and hence the question of traffic mix did not arise.

Nowak and Grouni (1983) considered ultimate and serviceability limit states, including fatigue and cracking. They assumed a load and resistance factor code format, and calculated reliability indices for the guideway structures. The reliability index β for the ultimate limit state was defined as follows:

$$\beta = (R_M - L_M)/\sqrt{\sigma_R^2 + \sigma_L^2}$$

where

R_M and σ_R are the mean resistance and its standard deviation, and
L_M and σ_L are the mean load effect and its standard deviation

The reliability index is described more fully in Section 3.3. Reliability indices were also calculated for the concrete cracking and fatigue serviceability limit states.

The mean-to-nominal value and the coefficient of variation were determined for each load parameter based on previous research or engineering judgement. The nominal load capacities of the structures were calculated using AASHTO, and the mean-to-nominal values and coefficients of variation were determined for resistance in flexure and shear using basic data about the uncertainties in material properties, dimensions, fabrication and analysis. For the serviceability cases a mean cracking stress and tensile stress and corresponding coefficients of variation for the concrete were nominated.

Sets of load factors for the ultimate limit state were determined such that for each load component the probability of exceedence was the same. It was also assumed that load factors for similar load components were the same. Three load combinations were considered for the ultimate limit state:

1. the operational condition, with dead and live loads and the largest effect of support settlements, temperature, wind, earthquake, ice and snow;
2. the non-operational condition, with an empty vehicle plus wind; and
3. an operational condition with a derailed vehicle.

The first two combinations were also used for the serviceability limit cases.

Thirty-three guideway structures in the five transit systems were examined. The values of L_M and σ_L were determined for each structure for each load combination. For flexure in the ultimate limit state, the resistance factor ϕ was taken as 0.95. The nominal resistance was determined by summing the factored loads in each combination and dividing by ϕ. The mean resistances were then calculated by using the nominal-to-mean ratios. The reliability indices were subsequently calculated using the mean resistances and their corresponding standard deviations, and L_M and σ_L determined previously. The reliability index was also calculated based on AASHTO load factors and the actual capacities of the structures.

A target reliability index of $\beta = 4$ was chosen, compared with a value of 3.5 for highway bridges. The reliability indices based on the actual capacities varied substantially – between about 5 and 8. The load factors for AASHTO and the recommended set are shown in Table 8.2. The reliability indices based on the AASHTO load factors varied between 5.68 and 6.8, whereas the recommended set resulted in reliability indices varying from 3.69 to 4.26. The recommended set of load factors was used to calculate the reliability indices for shear; $\phi = 0.75$ was found to give values closest to the target value of 4.

Table 8.2 Load factors for ultimate limit state

Load component	*Nowak and Grouni (1983)*	*AASHTO*	*Dolan (1986)*
Dead loads: Factory-produced elements and weight of trackwork	1.20	1.30	1.30
Dead loads: Cast-in-place elements and ballast	1.30	1.30	1.30
Live loads: LRV and passengers	1.30	2.17	1.7
Prestressing effect	1.1	1.0	1.0

A similar exercise was carried out for the serviceability limit state using load factors of 1.0. An upper limit on the tensile stress in the concrete of the form $k\sqrt{f'_c}$ was assumed, where k is a constant. A value of $k = 2$ was found to give reliability indices closest to the target values of $\beta = 2.5$ for cracking and $\beta = 2.0$ for fatigue. Nowak and Grouni concluded that the AASHTO specification is conservative for the design of transit guideway structures and that the set of recommended load and resistance factors, although less conservative, provided satisfactory levels of safety.

A committee of the American Concrete Institute (Dolan 1986) has also published a guide with recommendations for the analysis and design of reinforced and prestressed concrete structures supporting transit guideways (light rail). Estimates for most load parameters, load and resistance factors and load combinations are given in the guide. One ultimate strength combination is included in Table 8.2 for comparative purposes.

No standard vehicle load is given, but the dynamic load allowance is given by:

$$I = \frac{VCF}{f} - 0.1$$

For continuous span structures,

$$I = \tfrac{1}{2}\frac{VCF}{f} - 0.1$$

where

$$VCF = \frac{\text{vehicle speed (m/sec)}}{\text{span length (m)}}$$

f = natural frequency

I must be ≥ 0.1 for continuously welded rail and ≥ 0.3 for jointed rail.

Longitudinal forces (acceleration and braking) are given as 30% of the vertical live load for emergency braking and 15% for normal braking. The nosing loads are specified as 8% of the vertical live load for non-steerable axles and 6% for steerable axles.

To sum up, light rail transit is experiencing a revival and many cities are introducing new systems in an effort to improve their environments and discourage private car usage. Bridges supporting the trackwork for these systems will most likely be designed on the basis of existing highway and railway bridge design codes. It has been common to design these bridges on the basis of a particular light rail vehicle. Light rail vehicles have an expected economic life of 10 to 15 years, hence replacement vehicles can be expected during the lives of the bridges. Therefore the practice of designing bridges for specific vehicles seems unnecessarily limiting for future expansion of the network. Nowak and Grouni (1983) have further demonstrated that the application of current highway and bridge codes to transit guideways or bridges supporting light rail transit leads to inconsistent levels of reliability. In view of the resurgence in light rail transit it is felt that there is a need to codify the design of bridges for these systems if communities are to have economical systems of acceptable reliability.

Chapter 9

Pedestrian and cycle loads

9.1 Nature of pedestrian loading

It is usual to treat cycle loads and pedestrians together. The pedestrian loading dominates due to its more mobile and compact nature. The difference between cycle and pedestrian bridges lies in the geometrical requirements, mainly in terms of width and barrier heights. Hence this Chapter will refer to pedestrian loadings only, but the cases presented cover cycle bridges too.

Pedestrian loading encompasses a number of forms. The static loading due to a particular density of pedestrians is well known. The loads placed on handrails by groups of people or by a crush of people require consideration. Loads on handrails will be considered in Chapter 10. Less obvious loads are due to a single person. A vandal, or groups of vandals, could apply unusual loads to the structure, which could damage it. Due to the flexibility of footbridges, vibration of the bridge due to a single person walking over it or due to deliberate activity by some people must also be considered. Vibration may occur in either the vertical or horizontal directions; it can cause discomfort for some sections of the community and this must be guarded against.

Consideration must be given to the likely usage of a footbridge and its maintenance. If a footbridge is in a remote site, e.g. a park, it is unlikely to suffer loading from a crowd of people. However, it may be used by a vehicle maintaining a park, to access all areas of the park. Either this must be considered in the design or special steps must be taken to prevent use of the footbridge by vehicles. In the case of a major, large span footbridge, crossing a river in a central city area, higher usage of the bridge should be expected. The bridge may be used for special events such as viewing fireworks or regattas on a river. Maintenance of this type of bridge is also likely to be a major consideration as it is unlikely that the bridge can be maintained by any means other than from the bridge itself.

Various densities of pedestrian loading are shown pictorially in Figs. 9.1 to 9.5. Each figure shows people standing in an area 2 × 2 m square and the densi-

Fig. 9.1 1.5 kPa pedestrian load: 8 persons in a space, 4 m^2

Fig. 9.2 2 kPa pedestrian load: 11 persons in a space, 4 m^2

Fig. 9.3 3 kPa pedestrian load: 17 persons in a space, 4 m^2

Fig. 9.4 4 kPa pedestrian load: 22 persons in a space, 4 m^2

Fig. 9.5 5 kPa pedestrian load: 28 persons in a space, 4 m^2 (Courtesy Brisbane City Council staff)

ties assume that each person weighs an average 700 N. A load of 1.5 to 2 kPa is regarded as a domestic loading; 3 to 4 kPa is usually a commercial loading and 5 kPa is a 'crowd' loading. A crowd loading usually occurs in special circumstances, such as for access to a sports stadium or a bridge, which can be used as a vantage point by spectators. It is usual to specify a lower static loading for footbridges of longer length, due the lower probability of having the bridge fully loaded over its full length.

The Australian Bridge Design Code (Austroads 1996) requires footbridges and footways independent of the road bridge superstructures to be designed for 5 kPa for loaded areas up to 85 m^2, and 4 kPa for loaded areas above 100 m^2. A linear variation is assumed for loaded areas between 85 m^2 and 100 m^2. For footways attached to road bridge superstructure, the code specifies 5 kPa for loaded areas up to 10 m^2 and 2 kPa for loaded areas over 100 m^2. A linear variation is assumed for loaded areas between 10 m^2 and 100 m^2. Where it is possible for a vehicle to mount the footway or a footbridge the code requires the facility to be designed for a concentrated load of 20 kN.

The AASHTO LRFD Bridge Design Specifications (1994) specifies a pedestrian load of 3.6 kPa on sidewalks greater than 600 mm in width. Bridges exclusively for pedestrians or bicycles are to be designed for 4.1 kPa; access by maintenance vehicles is to be considered.

BS 5400: 1988 specifies a design load of 5 kPa for spans up to 30 m.

This load is reduced for spans greater than 30 m according to the following formula:

$$5.0 \times \frac{\text{nominal HA UDL for appropriate loaded length (kN/m)}}{30 \text{ kN/m}} \text{ (kPa)}$$

where HA UDL is the uniformly distributed component of the HA traffic loading described in Section 4.2. The code contains the proviso that when the loaded lengths exceed 30 m, special consideration should be given to cases where exceptional crowds may be expected. A footbridge serving a sports stadium is given as an example. Where the footbridge or cycle bridge is not protected from access by vehicles, the code requires that the bridge be designed for any four wheels of 25 units of HB loading acting in any position. It is understood that BD37/88 represents the design practice in the UK today. It specifies 5 kPa for loaded lengths (L) up to 36 m; for loaded lengths greater than this the load is reduced in accordance with the following formula:

$$5.0 \times \frac{\text{nominal HA UDL for appropriate loaded length (kN/m)} \times 10}{L + 270} \text{ (kPa)}$$

Part 3 of Eurocode 1 (ENV 1991-3) specifies a design load of 5 kPa for foot or cycle bridge spans up to 10 m. For spans, L_{sj}, greater than 10 m, the loading, q_{fk}, is reduced in according with the following formula:

$$2.5 \text{ kPa} \leq q_{fk} = 2.0 + \frac{120}{L_{sj} + 30} \leq 5.0 \text{ kPa}$$

For a footway on a road bridge the design load is 2.5 kPa in combination with other loads, or 5 kPa when applied on its own or in combination with crowd loading on the carriageway. A concentrated load of 10 kN acting on a square of sides of 0.1 m is specified for local effects. If the bridge is to be designed for a maintenance vehicle then this is used for the design rather than the concentrated load.

The *Bridge Manual* published by Transit New Zealand (1994) nominates a load of between 2 kPa and 5 kPa, as defined by the expression 6.2 – *S*/25, where *S* is the loaded length. This translates to bridges with spans up to 30 m being designed for 5 kPa, and bridges with spans over 105 m being designed for 2 kPa. If a crowd loading can be expected, such as on a bridge leading to a sports stadium, or where a bridge can be used as a vantage point to view a public event, then 5 kPa should be used.

The loads specified by the various codes for pedestrian or cycle bridges are summarised in Table 9.1.

Table 9.1 Comparison of code pedestrian loads for bridges

Code	*Lowest load (kPa)*	*Lowest load length (m)*	*Transition*	*Highest load (kPa)*	*Highest load length (m)*
Austroads	4	Area $\geq$ 100 m^2 $\geq$33*	linear	5	Area $\leq$ 85 m^2 $\leq$28*
AASHTO	4.1			4.1	
BS 5400	1.5	$\geq$380	$L^{-0.475}$	5	$\leq$30
BD37/88	0.46	1600	$\approx L^{-1.1} - L^{-1.67}$	5	$\leq$36
Transit New Zealand 1994	2	$\geq$105	linear	5	$\leq$30
ENV 1991-3	2.5	$\geq$210	120/(L − 30)	5	$\leq$10

*Assuming width of footbridge is 3 m.

9.2 Periodic and impulsive loads

For road bridges, periodic or impulsive loads are likely to be due to dynamic loading by vehicles. The dynamic response of the bridge should be determined and the vibration assessed for acceptability (see Chapter 7 for the dynamic vibration of bridges due to vehicles).

Footbridges, however, are usually of light construction. This brings with it special considerations, in that the natural frequencies of footbridges may fall in the range excited by a single pedestrian. Codes approach this problem in two ways. The simplest way is to determine the natural frequencies of the footbridge and ensure that they lie outside the range of pedestrian movement. The second way is to model a pedestrian walking or running across a bridge or jumping on the bridge and determine the maximum response of the bridge.

BS 5400: 1978 requires no further checks on vibration at serviceability if the fundamental natural frequency of the unloaded bridge exceeds 5 Hz. Where the natural frequency f_0 is $\leq$5 Hz, the maximum acceleration shall be limited to $0.5\sqrt{f_0}$ m/sec^2. The maximum acceleration may be calculated by assuming that a pedestrian exerts a static load of 700 N and has a dynamic component equal to 180 sin $(2\pi f_0 T)$, in N. T is the time in seconds and the pedestrian is assumed to be moving at a speed equal to 0.9 f_0 m/sec. For values of f_0 greater than 4 Hz the calculated maximum acceleration is to be reduced by an amount varying linearly from 0.0 at 4 Hz to 70% at 5 Hz.

BS 5400 gives an alternative method for deriving the maximum vertical acceleration (m/sec^2):

$$a = 4\pi^2 f_0^2 y_s K\psi$$

where

f_o is the fundamental natural frequency in Hz,
y_s is the static deflection in metres due to a pedestrian imposing a static force of 700 N at the centre of the main or centre span,

K is the configuration factor reflecting the number of continuous spans and relative span lengths, and
ψ is the dynamic response factor dependent on the degree of damping present in the bridge

BD37/88 contains the same provisions for the assessment of vibration as a serviceability limit state. At the time of writing, Eurocode 1 had not published appropriate models for walking, running and jumping pedestrians. The AASHTO (1994) is silent on vibrations due to pedestrians.

Austroads (1996) specifies that road bridges with footways shall be investigated by calculating the maximum static deflection due to a T44 vehicle with the dynamic load allowance. This deflection is not to exceed the specified limit indicated by a figure. This limit depends on the first mode flexural frequency of the bridge and reduces as the frequency increases. Footbridges are to be investigated for a pedestrian weighing 700 N crossing the structure at 1.75 to 2.5 footfalls per second. The commentary indicates that no further checks are required if the fundamental natural frequency of the footbridge exceeds 5 Hz. Footbridges with a fundamental natural frequency in the range of 1.5 to 3.5 Hz are noted as being susceptible to significant vibrations. If a detailed assessment of the dynamic response of the footbridge is required, the reader is referred to work by J.E. Wheeler.

Wheeler (1980, 1982) modelled 21 footbridges in Perth, with analytical predictions that compared favourably with experimental results. Pedestrian loads were based on a static load of 700 N and were modelled by a Fourier series at walking speeds. The shapes of the force-time histories were based on recorded pedestrian loads. At jogging and running speeds the pedestrian load was modelled as a half sine wave, the parameters of which were given graphically and depend on the number of footfalls per second. His structures were modelled using lumped masses and analysed using modal analysis methods.

Rainer, Pernica and Allen (1988) measured the dynamic responses due to pedestrians walking, running and jumping and determined equivalent Fourier series. The peak acceleration response, a (in m/sec^2), is determined as follows:

$$a = (2\pi f)^2 \frac{\alpha P}{k} \Phi$$

where

k is the stiffness of the span at the centre,
P is the static load due to a pedestrian,
f is the fundamental natural frequency of the span,
α is the Fourier coefficient, and
Φ is the dynamic amplification factor, which depends on the damping in the structure and the number of cycles of loading occurring on the span.

Values of α are given as 0.41 for walking, 1.1 for running and 1.7 for jumping. The above equation was found to give good agreement with experimental results

9.3 Perception of vibration

The human response to vibration is difficult to quantify since different people react in different ways and the context of the vibration has a significant influence. For example, small vibrations in a laboratory or operating theatre are likely to be intolerable, whereas quite large vibrations in a manufacturing plant may be acceptable. The time of day as well as the duration of exposure to vibration influences people's reactions. The severity of vibration is usually indicated by the measured or calculated, peak or root mean square (rms) accelerations. Sometimes acceleration is converted to displacement for ease of calculation, using the following formula:

$$a = (2\pi f)^2 \delta$$

where a is acceleration,
f is the frequency of vibration in Hz, and
δ is the displacement

The international standard ISO 2631-1 1990 sets out the general requirements for the evaluation of human exposure to whole-body vibration. The three axes of the human body are treated separately, with the human response to 'longitudinal' (head to foot vibration) being different from the other two forms of vibration. The primary quantity used to describe the intensity of vibration is acceleration and usually the rms value is used. Root mean square values can be converted to peak values by multiplying by $\sqrt{2}$. Limitations are placed on the type of vibrations covered by the standard.

Three levels of exposure limits are set. These are the 'fatigue-decreased proficiency boundary' (preservation of working efficiency), the 'exposure limit' (preservation of health and safety) and the 'reduced comfort boundary' (preservation of comfort). The reduced comfort boundary relates to the difficulty of eating, reading etc. and does not apply to situations where the socio-psychological impact of vibration may be more important, such as in buildings.

ISO 2631-2 sets out the special requirements for continuous and shock-induced vibration in buildings. This Part of the standard does not apply to transient or impulse vibrations characterised by a rapid build-up to a peak with a damped decay of vibrations. The limits specified in this part of the standard are generally lower than those specified in Part 1. Neither Part of the standard is directly applicable to bridges where a degree of vibration is expected and tolerated. A higher limit could be expected since a person is usually on a bridge for a short time only.

Irwin (1978) discussed the human response to vibrations in structures, and the limits for vibrations in buildings presented in his paper are consistent with those in ISO 2631-2. However, he also discusses vibrations in bridges and offshore structures. He nominates two vertical vibration limits: one is for frequent events and the other is for 'storm' conditions. The limit for 'storm' conditions is higher than that for frequent conditions, since it is expected that people are

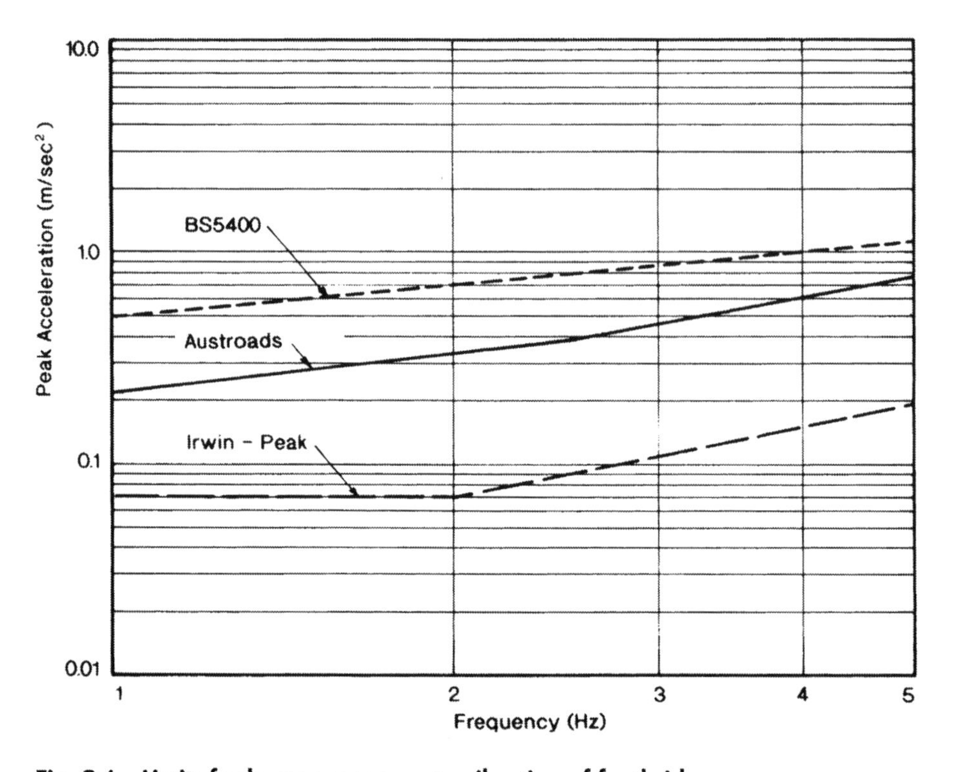

Fig. 9.6 Limits for human response to vibration of footbridges

unlikely to be working or standing on a bridge during a storm. The base curve for vertical vibrations (frequent events) has been converted from rms acceleration to peak acceleration and plotted in Fig. 9.6.

Austroads (1996) specifies a peak displacement limit for a given natural frequency of vibration and is depicted graphically. The limits appear to be based on work by Kobori and Kajikawa (1974), who specify 'unpleasant' limits of $y\omega^{1.4} = 73$ mm/sec for standing pedestrians, and $y\omega = 24$ mm/sec for walking pedestrians. The symbol y represents the peak displacement and ω is the natural frequency of vibration in radians per second. These peak displacements have been converted to peak accelerations and plotted in Fig. 9.6. For comparative purposes the peak acceleration limit specified by BS 5400 ($0.5\sqrt{f_0}$ m/sec^2) has also been plotted in Fig. 9.6. The spread shown in Fig. 9.6 indicates the subjectivity of the limits, which must be applied cautiously.

If a footbridge should behave unsatisfactorily under dynamic excitation the problem can be addressed by modifying the mass, stiffness or damping. Reduced mass or increased stiffness will increase the natural frequencies of the structure, while increased damping will reduce its response. The use of a tuned damper, i.e. a mass attached to the structure through a spring of a particular stiffness with or without a shock absorber, has proved successful in addressing dynamic problems. Goodger and Kohoutek (1990) and Tilly, Cullington and Eyre (1984) have discussed the use of tuned mass dampers (TMDs) to reduce the impact of vibrations in footbridges. TMDs can be used to split the natural frequencies of the footbridge into two smaller frequencies, or to absorb the energy applied to the structure by the pedestrian. Tilly et al. (1984) suggest that an additional mass equal to 5% of the superstructure mass is required for a frequency splitter, whereas a mass of only 0.5% is required to dampen vibrations. The use of cables attached to the middle of the span of the footbridge has also been proposed to increase the frequency of vibration and increase damping of the structure.

Although the previous discussion has concentrated on vertical accelerations due to pedestrians, consideration should also be given to other modes of vibrations. For example, a pedestrian overpass incorporating ramps on cantilever landings has been observed to exhibit unacceptable horizontal movements of the bridge. Resolution of the dynamic problem in this case involved modelling the structure as an inverted pendulum and increasing the stiffness of the columns.

Chapter 10

Barrier and railing loads

10.1 Vehicle barriers

Vehicle barriers are an essential part of a bridge, providing safety both for vehicles, as well as for people in the case of an errant vehicle. Barriers may be required to protect passengers in the vehicle, other users of the bridge or people in the near vicinity, and structures adjacent to the bridge from collapse that may cause injury or death. In detailing barriers, attention must be directed to ensuring they do not become hazards themselves during a crash. The objectives of a vehicle barrier are as follows.

1. Barriers should offer a level of protection commensurate with the level of risk involved.
2. Barriers should minimise the deceleration of an errant vehicle and its occupants. Colosimo (1996) suggests a limit of $5g$ or 50 m/sec^2.
3. Barriers should redirect an errant vehicle in a direction parallel with the barrier or intended direction of travel. Rebounding of the vehicle back into traffic must be avoided in order to prevent a second perhaps more lethal accident. Colosimo (1996) suggests the exit angle should be less than half the impact angle.
4. Barriers should have sufficient structural strength to prevent parts becoming separated as debris, which may penetrate the vehicle, causing injury or death to its occupants.
5. Barriers should be capable of rapid repair or replacement.
6. Vehicle barriers should not damage the bridge structure when impacted by an errant vehicle.
7. The barrier should not detrimentally affect sight lines to and from vehicles and pedestrians.
8. Barriers should be compatible with the bridge, allowing the required thermal, live load and long-term movements of the bridge.

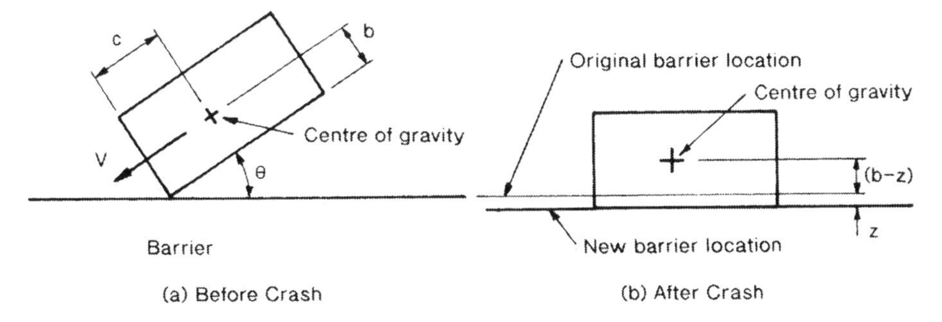

Fig. 10.1 Vehicle impact on a crash barrier

9. Barriers should make a minimal contribution to the hydrodynamic forces acting on a bridge should it become submerged during a flood.
10. Barriers on bridges should be compatible with those on the bridge approaches and departures.
11. Barriers should desirably contribute to the aesthetics of the bridge.

Designing a barrier is predominantly an exercise in absorbing the kinetic energy of an errant vehicle. Figure 10.1 shows such a vehicle (a) just prior to, and (b) just after impacting a barrier. The vehicle is assumed to be travelling at a speed V m/sec and at an angle θ to the barrier. The centre of the mass of the vehicle is assumed to be at a distance of c metres along the vehicle and b metres across the vehicle from the point of impact. The energy of the vehicle parallel to the barrier is of no interest as the objective is to halt movement of the vehicle in a lateral direction. The kinetic energy of the vehicle in the lateral direction is

$$\tfrac{1}{2}m(V \sin \theta)^2$$

where m is the mass of the vehicle in kg.

During impact it is assumed that the barrier and vehicle will suffer a combined deformation (crumple) of z metres and that the vehicle will be redirected to a direction parallel to the barrier, as shown in Fig. 10.1(b). The energy absorbed by deformation of the vehicle and barrier is given by the product of the average force of the impact, F, and the distance that the centre of mass travels before being brought to rest laterally. The distance travelled by the centre of mass is

$$c \sin \theta + b \cos \theta + z - b = c \sin \theta + b\,(\cos \theta - 1) + z$$

Equating the initial lateral kinetic energy with work done by the barrier and vehicle during impact (laterally) gives

$$F\,(c \sin \theta + b\,(\cos \theta - 1) + z) = \tfrac{1}{2}m(V \sin \theta)^2$$

Therefore,

$$F\,(\mathrm{N}) = m(V \sin\theta)^2/(2(c \sin\theta + b(\cos\theta - 1) + z))$$

and the average deceleration, a (m/sec^2), is given by

$$a = (V \sin\theta)^2/(2\,(c \sin\theta + b\,(\cos\theta - 1) + z))$$

These formulae are identical to those given in BS 6779: Part 1: 1992. It should be noted that the formulae give the average impact force and deceleration during the crash. The code indicates that the actual impact force for a vehicle impacting on a metal barrier is likely to have high initial and final peaks. Using $b = 0.76$ m, $c = 2.44$ m, $z = 0.64$ m and $\theta = 20°$, which are presented as possible values in the Standard, and assuming a vehicle of 2000 kg mass travelling at a speed of 120 kph (33.3 m/sec), the average impact force is found to be 91 kN. The corresponding deceleration is 45.5 m/sec^2 or 4.6 g.

Figures 10.2 to 10.5 show the effects of varying parameters in the equation for vehicle impact force based on the above basic values. Figure 10.2 shows that the impact force increases linearly, as expected, with increasing mass of the vehicle. The forces generated by large vehicles are quite large, although it should be borne in mind that the dimensions of the vehicle and the likely greater crumple distance would change the output of the formula. Figure 10.3 shows that the impact force increases with the square of the vehicle speed. Figure 10.4 shows vehicle impact increasing in a non-linear fashion with the angle of impact. In the limit, as θ becomes 90°,

$$F = mV^2/(2(c - b + z))$$

For the example given in Fig. 10.4 this value is 479 kN. It should be noted, however, that this value assumes that the vehicle could still spin around and end up parallel with the barrier. If this is not the case, the average impact force is given by

$$F = mV^2/(2z)$$

For the example given, this force becomes 1736 kN.

Finally, Fig. 10.5 shows the reduction in impact force with increasing crumple depth. This demonstrates the value of flexible barriers in reducing impact forces, and hence vehicle damage and personal injury. There is also a finite impact force for a crumple depth of zero metres, this being the mean force required to re-orientate the vehicle parallel to the barrier without damage to the vehicle or barrier. This is a theoretical value only, as some damage to the vehicle and barrier would be expected, due to the force itself. It does, however, indicate that energy is absorbed in rotating the vehicle as well as in the deformation (crumple) of the barrier and vehicle.

The preceding analysis of a vehicle crash on a barrier is most applicable to a metal barrier, usually consisting of a series of posts linked by a number of hor-

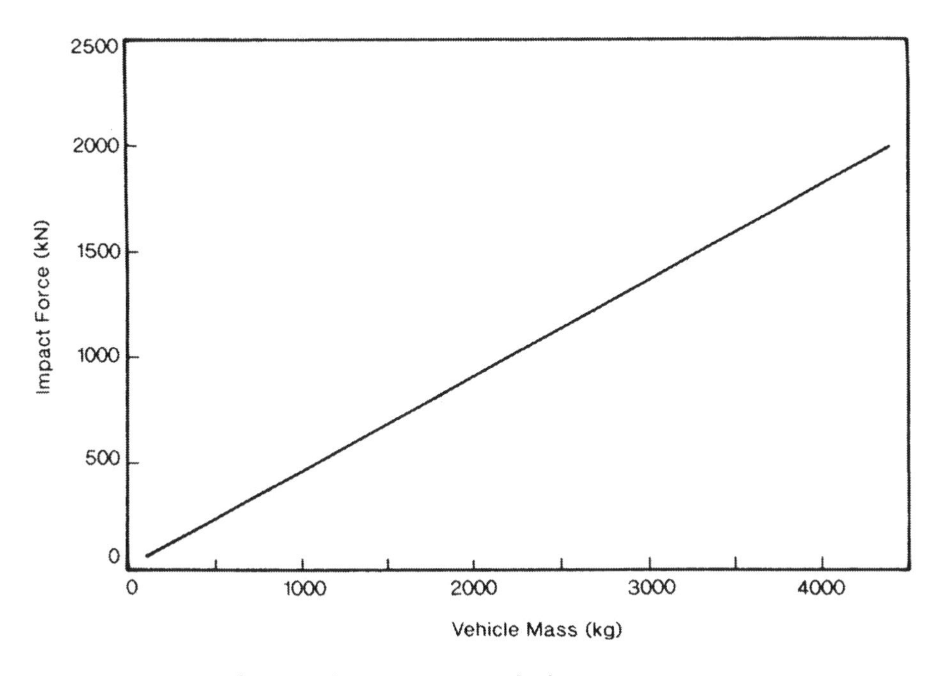

Fig. 10.2 Impact force on barrier versus vehicle mass

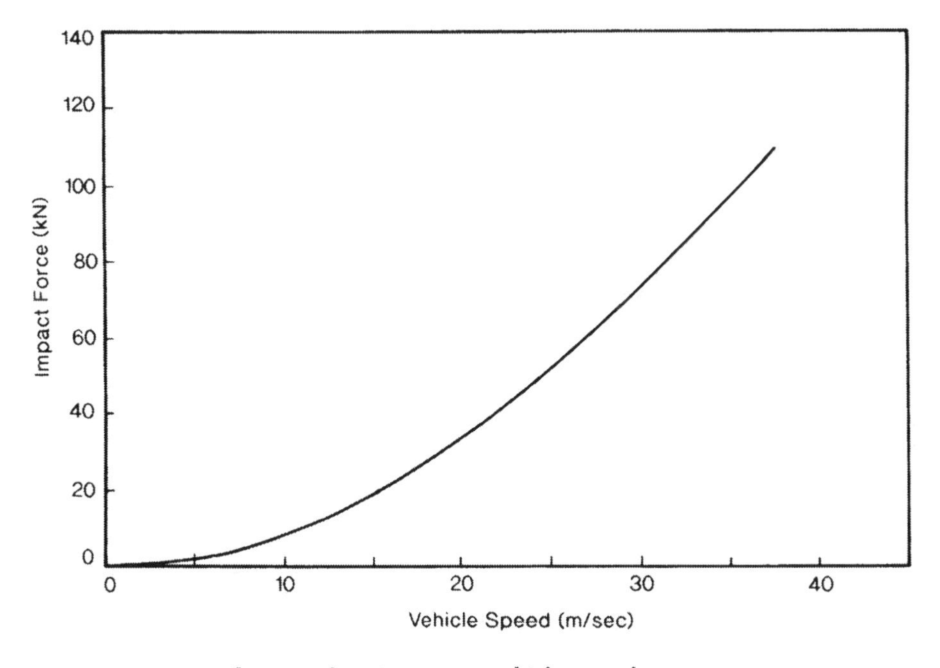

Fig. 10.3 Impact force on barrier versus vehicle speed

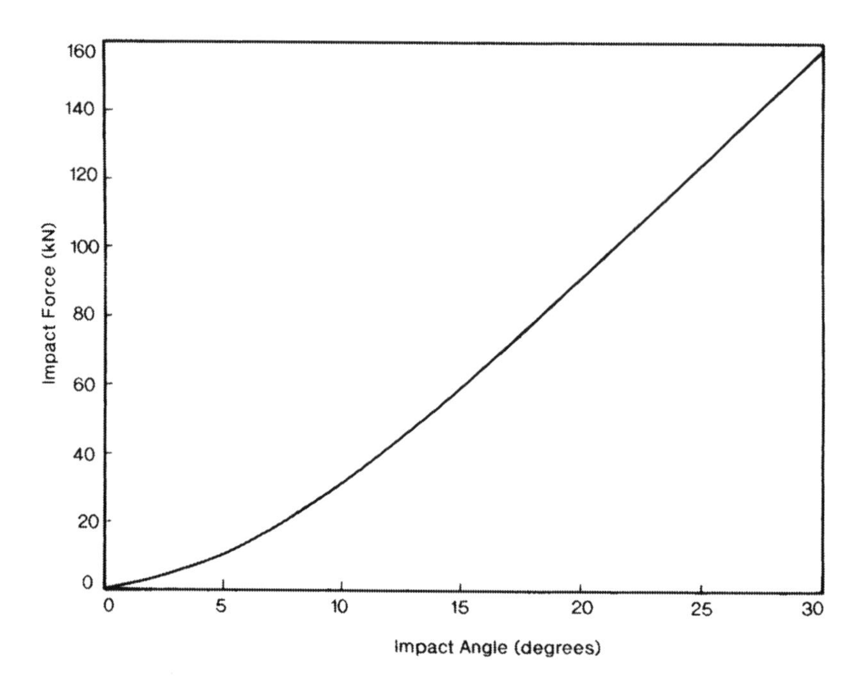

Fig. 10.4 Impact force on barrier versus vehicle impact angle

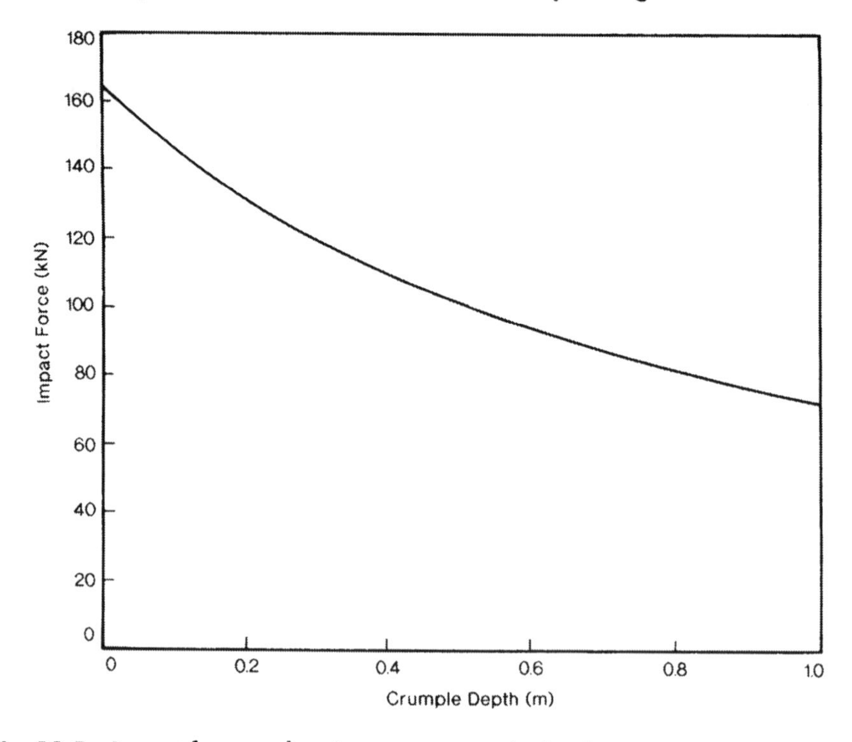

Fig. 10.5 Impact force on barrier versus crumple depth

izontal rails. They are normally damaged during impact and require repair. Damage to the errant vehicle would also be expected. The lateral displacements of barriers can present a problem in places of limited room. These constraints led highway authorities to experiment with rigid concrete barriers. It is not known exactly when concrete barriers were first used but such a barrier was reported to be in use in California in the mid-1940s. It was developed to minimise the number of vehicles penetrating the barrier and to minimise the cost and need for barrier replacement in locations with high accident records. The presence of narrow medians was also a consideration.

The following brief history of the New Jersey barrier is based on information in a booklet, *The Safety Shape (G39)*, produced by the Cement and Concrete Association of Australia. Highway authorities in New Jersey first installed a rigid, concrete barrier in 1955 specifically to address the problem of narrow medians. The first barrier was basically a low wall with kerbs on either side. Based on experience and crash testing, the shape depicted in Fig. 10.6(a) was developed. In 1963, General Motors USA developed a barrier very similar to the New Jersey barrier in shape, for use on bridge parapets. Testing was also directed at developing a metal rail on top of the barrier to contain vehicles during high-speed, wide-angle impacts. Unlike flexible barriers, which absorb energy through deformation of the barrier and vehicle, rigid barriers absorb energy by compression of a vehicle's suspension system. Crabbing of the wheels and friction at the tyre–road interface dissipates additional energy. It is a feature of these barriers that quite often the errant vehicle does not sustain damage to its bodywork. Vehicles tend to hug the barriers rather than rebound back into the traffic stream, and overturning is reduced. The barrier has been successful in redirecting large tractor-trailers. The New Jersey success has seen similar barriers, as shown in Fig. 10.6(b), adopted in codes such as Austroads (1996). The

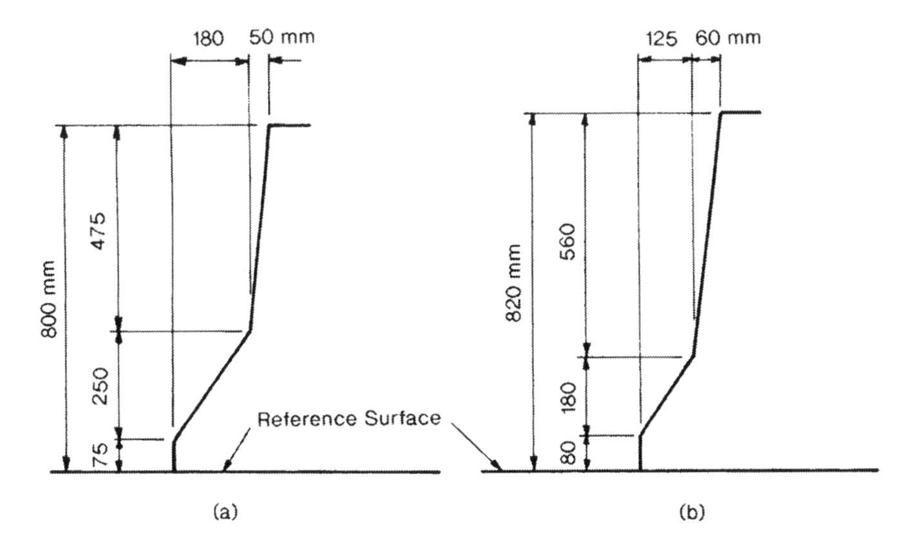

Fig. 10.6 Rigid barrier profiles

unifying features of the two shapes are that the lower sloped face is at 55° to the horizontal and the upper sloped face at 84°.

Barriers are normally classified by the level of containment (low, medium and high) that they offer. Low-level barriers are designed to contain light vehicles such as saloon cars, medium-level to contain medium-sized trucks and buses, and high-level barriers to contain large trucks and tractor-trailers. The level of containment required usually depends on the level of risk pertinent to a particular situation. Although codes of practice may give guidance on the design of barriers, most require crash testing of barriers before acceptance.

Austroads (1996) advises that barriers need not be provided at low-level bridges subject to flooding, where

(a) the provision of barriers would prevent the passage of debris and/or be frequently damaged by heavy debris;
(b) traffic volumes are less than 150 vehicles per day;
(c) the bridge is over water normally less than 1.2 m deep, or is less than 1.5 m above ground;
(d) there is an essentially straight approach road and bridge alignment (radius > 1500 m), and where the approaches have sufficient sight distance (>stopping distance);
(e) the width between kerbs is not less than 6.6 m for a two-lane bridge or 4.2 m for a single-lane bridge;
(f) appropriate signposting is erected, warning that the bridge is subject to frequent flooding; and
(g) continuous kerbs not less than 250 mm high are provided on the deck. Castellated kerbs may be used where afflux is critical, where permitted by the Authority.

Colosimo (1996) adds the following conditions:

(a) in rural locations without anticipated pedestrian traffic;
(b) with a barrier offset at least 1.2 m from the edge of the traffic lanes; and
(c) where the land beneath the structure is low-risk; that is, if conditions under the bridge must not increase the level of risk to occupants of a vehicle leaving the bridge.

All the above conditions must be satisfied if a barrier is to be omitted.

AASHTO (1989) and Colosimo (1996) indicate methods for selecting the appropriate level of barrier, based primarily on traffic volume, numbers of commercial vehicles and level of risk.

10.1.1 High-level containment barriers

High-level barriers should be provided in high-risk locations. Austroads (1996) refers to these as Level 1 barriers and suggests they should be provided for:

(a) bridges over other major roadways;
(b) bridges over railways;
(c) bridges over houses, factories, etc;
(d) very high bridges;
(e) bridges over deep water;

or in other situations as directed by the Authority.

Assessment of the level of hazard is to take into account the volume and types of vehicles, total traffic volumes, road alignment, bridge width, and the consequences of barrier failure. No design criteria are specified, but the barrier should be able to contain vehicles identified by the assessment at the prevailing speed of the road for impact angles up to 15°. The safety shape shown in Fig. 10.6(b) is suggested as a basis.

Colosimo (1996) suggests that a high-level barrier should be capable of containing a van-type tractor-trailer vehicle with a mass of 44 tonnes at speeds of 90 kph, at impact angles up to 15°. An ultimate design containment force of 900 to 1200 kN, and a minimum barrier height of 1300 to 1500 mm are nominated. Since the high-level barrier is designed for heavy vehicles, Colosimo suggests such a barrier may be required when commercial vehicle volumes on busy, major roadways exceed 500 per day in rural areas and 2000 per day in urban areas. If these volumes are exceeded, high-level barriers should be specified in the following situations:

(a) bridges over major roadways with an AADT of 8000 or more vehicles per day;
(b) bridges over electrified railways or goods lines carrying large quantities of either noxious or flammable substances;
(c) bridges over water more than 2 m deep;
(d) bridges over houses, factories, etc;
(e) bridges more than 10 m high;
(f) bridges on horizontal curves with a radius of 600 m or less.

BS 6779: Part 2: 1991 specifies that a high-level barrier is to resist penetration from a 30-tonne rigid tanker with centre of mass 1.8 m above the ground, travelling at 64 kph at an impact angle of 20°. Parts 1 and 3 of the Standard specify a similar vehicle except that the height of the centre of mass is 1.65 m. The British Standard is divided into three parts – BS 6779: Part 1: 1992 covers metal barriers, and requires that metal barrier construction be tested dynamically prior to acceptance for use; BS 6779: Part 2: 1991 specifies the design of barriers of concrete construction; and BS 6779: Part 3: 1994 specifies the design of combined metal and concrete barriers.

For high-level concrete barriers where shear transfer between panels is provided, BS 6779: Part 2 specifies that barriers be designed for a nominal bending moment, generated by applying a force of $(180 + 40L)$ kN per panel at the top of the panel, where L is the length of the panel (1.5 to 3.5 m). Where shear transfer between panels is not provided, the force is $(210 + 40L)$ kN. The force is to be applied uniformly along the top of the panel in a horizontal direction.

For shear design, the relevant forces are $(90 + 50H)L$ kN per panel and $(110 + 50H)L$ kN per panel respectively, where H is the height of the barrier and the shear force is to be resisted by any transverse section of the panel. The minimum height of the barrier is 1.5 m. BS 6779: Part 3: 1994 specifies a nominal load of 320 kN for combined concrete and metal panels, and the design of a number of elements of the barrier is based on this load. The nominal shear force per panel is 380 kN where shear transfer takes place between panels, and 440 kN where it does not. Load factors have to be applied to the nominal values.

AASHTO (1989) designates the high-level barrier as PL-3, capable of containing a 22-tonne tractor-trailer at 80 kph at an impact angle of 15°. AASHTO (1994) indicates that PL-3 barriers are normally used for freeways of variable cross slopes, reduced radii of curvature, high heavy-vehicle content and high speeds. There are two additional, higher performance barriers, the PL-4 and PL-4T, capable of containing 36-tonne tractor-trailers and 36-tonne tankers respectively. These barriers are required only at sites where such vehicles are prevalent.

PL-3 barriers are designed for a transverse force of 516 kN, a longitudinal load of 173 kN and a vertical load of 222 kN downward. The transverse and longitudinal loads need not be considered simultaneously with the vertical load, which represents the weight of the vehicle resting on the barrier. The transverse (impact) force and longitudinal (frictional) loads act over a distance of 2.44 m; the length of vehicle resting on the barrier is assumed to be 12.2 m. The code gives guidance for the design of railing-type barriers and the design of concrete barriers using yield line analysis, but requires all barriers to be tested prior to acceptance.

Hirsch and Fairbanks (1985) have described the design and testing of a rigid concrete barrier to contain 36-tonne tankers at speeds of 80 kph at an impact angle of 15°.

10.1.2 Medium-level containment barriers

Medium-level barriers should be provided in medium-risk locations. Austroads (1996) refers to these as Level 2 barriers and suggests they be provided where Level 1 barriers are not required but conditions exceed those applicable for Level 3 barriers (see Section 10.1.3). The standard accepts the concrete barrier, shown in Fig. 10.6(b), and a steel post and rail system (2 or 3 rails) as conforming to this level. The barrier is to be designed for an ultimate horizontal design force of 90 kN for heights up to 850 mm. For heights greater than 850 mm, the load is increased by multiplying by $(1 + (H - 850)/450)$, where H is the height of the concrete parapet or the top rail. Colosimo (1996) advises that 'this requirement is basically the traditional specification to direct a 2-tonne passenger vehicle (car) at 100 kph and 25° impact angle'. There is an inconsistency between this specification and the target vehicles for Level 2 barriers.

Colosimo (1996) nominates the medium-level containment barrier as capable of containing a 33-tonne tractor-trailer (van type) at a speed of 80 kph and an impact angle of 15°. An ultimate design containment force of 500 kN and a minimum barrier height of 1000 mm are nominated. This barrier satisfies the base requirements for highways and freeways.

BS 6779: Parts 1 and 2 specify that a normal-level barrier must be capable of resisting penetration by a 1.5-tonne saloon whose centre of mass is 0.6 m above the ground, travelling at 113 kph at an impact angle of 20°. BS 6779: Part 3 specifies a similar vehicle, except that the height of the centre of mass is 0.53 m. Part 2 specifies a nominal load of 50 kN applied horizontally over a length of 1.0 m at the top of the wall for bending, and a nominal force of 80 kN/m for shear. Part 3 specifies a nominal load of 50 kN for the design of combined metal and concrete panels and a nominal shear force of 50 kN per panel, with the minimum height of the barrier 1000 mm.

AASHTO (1989) designates its medium barrier as PL-2, capable of containing a 2.4-tonne vehicle at 100 kph at an impact angle of 20°, and an 8-tonne rigid vehicle at 80 kph at an impact angle of 15°. It is current practice in the USA to use this as the base level barrier on all highways, though some authorities recommend the PL-2 barrier as it is only marginally more expensive than the PL-1. PL-2 barriers are designed for a transverse (impact) load of 240 kN, a longitudinal (frictional) load of 80 kN and a vertical downward load (vehicle weight) of 80 kN. Requirements for the design and testing of these barriers are similar to the high-level PL-3 barriers, with the exception that the transverse and longitudinal loads are applied over a length of 1.07 m and the length of the vehicle resting on the barrier is 5.5 m.

10.1.3 Low-level containment barriers

Low-level barriers should be provided in low-risk locations. Austroads (1996) refers to these as Level 3 barriers and suggests they be provided in the following situations:

(a) bridges on roads with low traffic volumes (<500 vehicles per day);
(b) bridges on roads with low volumes of vehicles with a high centre of mass (<100 vehicles per day);
(c) bridges over shallow water (<1.2 m deep);
(d) bridges with low height above ground or water (<4 m);
(e) bridges with an essentially straight alignment (radii > 1500 m);
(f) bridges with width between barriers not less than 6.6 m for a two-lane or 4.2 m for a single-lane bridge.

Austroads (1996) nominates the flexible steel W-beam guardrail as an example of a Level 3 barrier, capable of containing cars at the prevailing speed at an impact angle of 15°.

Colosimo (1996) refers to the low-level barrier as a regular barrier capable of containing a 2-tonne vehicle travelling at a speed of 100 kph at an impact angle of 25°. The critical vehicle, however, is a 2.4-tonne vehicle travelling at

100 kph at an impact angle of 20°. An ultimate design containment force of 200 kN and a minimum barrier height of 800 mm are nominated. A typical concrete barrier would be as shown in Fig. 10.6, or the steel post and rail type, generally used for rural highways and major urban roads. Colosimo also refers to a 'low'-level barrier providing less containment than the regular level. It corresponds approximately to the AASHTO PL-1 level and could be a standard roadside barrier modified in stiffness for bridge use.

BS 6779: Part 1: 1992 specifies that a low-level barrier is to resist penetration from a 1.5-tonne saloon car with centre of mass 0.6 m above the ground, travelling at 80 kph at an impact angle of 20°.

AASHTO (1989) designates its low barrier as PL-1, capable of containing a 2.4-tonne vehicle at 70 kph at an impact angle of 20°. PL-1 barriers are designed for a transverse (impact) load of 120 kN, a longitudinal (frictional) load of 40 kN and a vertical downward load (vehicle weight) of 20 kN. The transverse and longitudinal loads are applied over a length of 1.22 m and the length of the vehicle resting on the barrier is 5.5 m. Requirements for the design and testing of PL-1 barriers are similar to those for high-level PL-3 barriers.

10.2 Pedestrian and bicycle barriers

The purpose of a pedestrian barrier is obviously to restrain people from falling off a bridge which may be high or above a water hazard. The need might arise as a result of children playing or fishing on the bridge, or pedestrians being knocked from their feet while the bridge is in flood, or from pressure from a crowd crossing the bridge. The design of a pedestrian barrier must be a compromise. An open design may reduce loading on the bridge due to stream and debris forces during a flood, but too large an opening will fail to prevent children from falling off the bridge. The size of openings must be carefully chosen to prevent entrapment of body parts.

The most common form of barrier is post-and-rail, with two or three rails specified in order to prevent people falling through gaps in the barrier. This type has the disadvantage that it is climbable by children, who may perch on the top rail and overbalance. An alternative type is a balustrade consisting of many vertical bars closely spaced and supported between a top and bottom rail, offering a greater degree of safety, but at an increased cost. A solid barrier offers a safe solution, but would impose large stream forces on the bridge during a flood.

Bicycle barriers are usually designed for loads similar to those for pedestrian barriers. The difference is usually one of geometry, the height of the barrier being increased to cope with the higher centre of gravity of the cyclist.

BS 5400: 1978 requires a pedestrian parapet to be designed for a nominal load of 1.4 kN/m, acting at a height of 1.0 m above the footway. BD37/88, British Department of Transport (1988), retains the loading specified in BS 5400, noting that the load is to be applied concurrently with the pedestrian loading. ENV 1991-3: 1995 specifies a line load of 1.0 kN/m, acting vertically or horizontally on the top of the parapet.

AASHTO (1994) nominates the minimum height of the rail as 1060 mm for pedestrians and 1370 mm for cyclists, and allows both horizontal and vertical elements. The maximum spacing of the rails is 150 mm, except where vertical elements are also present, in which case the maximum spacing may be increased to 380 mm for that part of the railing above 685 mm. Each rail is to be designed for loads of 0.73 kN/m, acting vertically and transversely simultaneously. Each rail also has a concentrated load of 0.89 kN, acting in any direction, at any point, simultaneously with the former loads.

Austroads (1996) specifies a serviceability design load of 0.75 kN/m acting vertically and horizontally on each rail. Posts are to be designed to resist the load from one bay (i.e. the line load times the spacing of the posts). The deflection limits are given as $L/800$ for rails where L is the span of the rail between posts and $H/500$ for posts, where H is the height of the top rail. Geometric properties of the rails are as follows:

(a) a preferred height of 1.1 m (1.0 m minimum);
(b) balusters shall be spaced not more than 130 mm clear for the safety of children;
(c) vertical balusters without climbing footholds shall be provided (in preference to a series of horizontal rails); and
(d) where cyclists may use the walkway, the preferred railing height is 1.2 m (1.1 m minimum).

Austroads (1996) raises two other issues to be considered by designers. Any rail attached to a vehicle barrier to act as a pedestrian rail is to be detailed so that it cannot become loose during a vehicle crash and spear into the vehicle. Second, where the pedestrian barrier is above an electrified railway line, the Railway Authority may require a special, possibly higher, barrier, which may also need to be electrically isolated to avoid the possibility of its becoming charged.

10.3 Light rail barriers

Since Light Rail Vehicles (LRVs) are guided vehicles, the probability of errant vehicles is reduced. However, derailment can occur due to a steering mechanism failure, a fault or wear in the rail, or vehicle collision. In the latter case, collision can occur between LRVs or, in the case of 'street' running, with a road vehicle. The support structure needs to be designed for the vertical load imposed by the LRV in its derailed location, and barriers to contain the LRVs need to be designed for the appropriate collision forces. Dolan (1986) advises:

> The magnitude and line of action of a horizontal derailment load on a barrier wall is a function of a number of variables, among which are the distance of the tracks from the barrier wall, the vehicle weight and speed at derailment, the flexibility of the wall, and the frictional resistance between vehicle and wall. In lieu of a detailed analysis, the barrier wall

should be designed to resist a lateral force equivalent to 50 per cent of a standard vehicle weight distributed over a length of 15 ft (5 m) along the wall and acting at the axle height.

An impact factor of 1.0 (i.e. static load should be doubled) should be included in the wheel loads when assessing the strength of the deck slab. The location of the derailed vehicle should be as close to the barrier wall as possible.

10.4 Railway barriers

Discussion of railway barriers must address the associated questions of derailment and collision loads on bridge supports. The issues are similar to those for light rail, except that the forces are greater.

Austroads (1996) states that the likelihood of a train being derailed depends on the track and the following components in decreasing order of probability:

1. catch points and derailers;
2. slips (track component);
3. facing turnouts;
4. diamonds;
5. trailing turnouts;
6. curved track (particularly on descents and especially at the bottom of a long descent); and
7. straight track on a descent (especially at the bottom of a long descent).

If a train leaves the track completely, such as in a *major derailment*, it will travel in a tangent to the track. If only one or two bogies leave the track (*minor derailment*), the train will travel parallel to the track.

Consider first the case of a bridge supporting a railway. In the event of a minor derailment, the train will leave the track but remain close to the tracks. Therefore the bridge must be checked for the loads being applied adjacent to the tracks. In the event of a major derailment, the train will continue towards the edge of the bridge and remain perched on the bridge. Damage to the bridge may be accepted but total collapse should normally be avoided.

BS 5400: 1978 requires that railway bridges be designed so as not to 'suffer excessive damage or become unstable in the event of a derailment'. Three conditions are specified. Condition (a) is where a train suffers a minor derailment. The code requires that no permanent damage be caused for the serviceability limit state. Condition (b) is also where the train suffers minor derailment, and in this case collapse of any major elements is avoided and local damage accepted as an ultimate limit state. Condition (c) is a major derailment of a locomotive and one wagon. The locomotive and wagon are assumed to be located on the parapet of the bridge, and no overturning or instability is accepted, but damage may occur.

For designs using the RU loading, BS 5400: 1978 specifies the following:

Condition (a): A pair of parallel vertical line loads of 20 kN/m each, 1.4 m apart, parallel to the track and applied anywhere within 2 m either side of the track centreline; or an individual concentrated vertical load of 100 kN anywhere within the width limits specified for the former loading;
Condition (b): Eight individual concentrated loads each of 180 kN, arranged on two lines 1.4 m apart, with each of the four loads 1.6 m apart on line, applied anywhere on the deck;
Condition (c): A single vertical load of 80 kN/m applied along the parapet or outermost edge of the bridge, limited to a length of 20 m anywhere along the span.

For designs using the RL loading, BS 5400: 1978 specifies the following:

Condition (a): A pair of parallel vertical line loads of 15 kN/m each, 1.4 m apart, parallel to the track and applied anywhere within 2 m either side of the track centreline (or within 1.4 m either side of the track centreline where the track includes a substantial centre rail for electric traction or other purposes), or an individual concentrated vertical load of 75 kN anywhere within the width limits specified for the former loading;
Condition (b): Four individual concentrated loads each of 120 kN, arranged at the corner of a rectangle of length 2.0 m and width 1.4 m, applied anywhere on the deck;
Condition (c): A single vertical load of 30 kN/m, applied along the parapet or outermost edge of the bridge, limited to a length of 20 m anywhere on the span.

ENV 1991-3: 1995 has a general rider that, in the event of a derailment, damage to the bridge is to be limited and that overturning or collapse of the structure is to be avoided. In the event of a minor derailment, collapse of major elements is to be avoided but local damage may be tolerated. The load for the ultimate state consists of two vertical line loads of 50 kN/m each over a length of 6.4 m and spaced 1.4 m apart. The outer extremity of the load is located a maximum of 1.5 × the track spacing on either side of the centreline of the track. If a wall is located closer than 1.5 × the track spacing, the load is to be located against the wall. In the event of a major derailment, the bridge must not overturn or collapse. The load for this case consists of a single, vertical line load of 80 kN/m over a length of 20 m and located a maximum of 1.5 × the track spacing on either side of the centreline of the track. The load is to be located against a wall or at the edge of the bridge if these elements are closer to the track centreline.

Austroads (1996) specifies that, in the event of a minor derailment, the strength of the bridge is to be checked for two alternative loads, using an ultimate load factor of 1.2. The first alternative is the 300-A-12 loading, applied as wheel loads parallel to the track and located within a distance, G_B, from the track centreline. The second alternative is a single point load of 200 kN located within a distance, G_B, from the track centreline. The distance G_B is taken to be

2.0 m for standard gauge lines and 1.5 m for narrow gauge lines. A minor derailment is considered to be when the train does not move further than 1.3 m laterally from the track. It is expected that major structural elements will not suffer more than superficial damage. In the case of a major derailment, the strength and stability of the bridge is to be checked for a vertical line load of 100 kN/m, over a length of 20 m, located on the edge of the bridge. The ultimate load factor is 1.0. It is expected the bridge will not be damaged beyond economic repair and that it will remain stable.

Now consider the case of a rail over-bridge, where the bridge carries some other traffic over the railway. The objective in this case is to prevent the collapse of the over-bridge on to the derailed train, with a high potential for loss of life. There are many documented cases where loss of life has occurred under these circumstances, for example, at Granville (Section 2.4). The first approach is to avoid the use of support columns close to trackwork; the second is to use barriers to protect the columns; the third is to design the columns to resist the impact forces. The column should behave in a ductile way and a blade column may offer the best solution. Ability to repair the column should be considered. Finally, consideration may be given to designing the bridge to support its own weight and a portion of the live load with the column removed.

BS 5400: 1978 offers no guidance on design loads for collisions against columns and refers the designer to the relevant authority. A similar approach is adopted by ENV 1991-3: 1995. Austroads (1996) advises that bridges should have clear spans between abutments. However, if columns are necessary and located within 5.5 m of the centreline of the track, they must be designed for 2000 kN parallel to the rails, acting simultaneously with 1000 kN normal to the rails. These are ultimate loads, applied horizontally at a height of 2.0 m above the rails. The code advises that 'a solid concrete full height pier, 600 mm or more thick and 4 m or more wide, with 1% vertical steel and with adequate connection to the superstructure should satisfy these load conditions in most cases'.

The final topic under this heading concerns the requirements for railway barriers. As would be expected, given the low probability of a tracked vehicle colliding with obstacles, there is little guidance given. BS 5400: 1978 and ENV 1991-3: 1995 are silent in this regard. Austroads (1996) states that the loads specified for the design of piers adjacent to railways are to be used for the design of crash resistant walls. A load factor of 1.5 is applied for the ultimate limit state.

Chapter 11

Stream loads

11.1 Introduction

Stream flow is a complex subject, whose proper study extends well beyond the scope of this book. The interaction between a bridge and stream may be divided into two parts.

There is, first, the effect of the bridge on the stream. The bridge typically causes some constriction of the waterway, resulting in *afflux*, or the elevation of flood levels upstream from the bridge. This afflux must be controlled, and the selection of bridge spans and the division of the bridge between embankments and open waterway is often determined by this factor. In addition, the reduction in waterway brings about an increase in the velocity of the stream beneath the bridge; this may cause scour, which is undesirable in itself and also leads to the deposition of material in the bed downstream. Moreover, the course of the stream may be diverted. To quote from *Waterway Design* (Austroads 1994: 14), '... modification to perennial rivers' [those with a permanent flow] 'can be quite extensive, however the dynamic effects of rivers in arid and semi-arid regions ... can be even more extensive and dramatic'. The disturbance caused by a bridge may cause major variations in the course of a river, such as one with meanders.

Second, there is the effect of the stream on the bridge, and this includes such matters as flow loads on bridge piers and abutments, the forces applied to the bridge by floating objects and debris, scour of the kind that may undermine and weaken the foundations, and possible inundation.

It is probably fair to say that these effects of stream flow have been the most common cause of bridge failure. Some examples may be given. The Roman bridge across the Guadiana River at Mérida in central Spain still serves as the city bridge (O'Connor 1993: 106f; Section 2.1). It has had, however, a chequered history. When first built, possibly by Augustus about 25BC, it had 36 arched spans, with a total length of about 260 m, of which 80.5 m were taken up by the piers, leaving 179.5 m, or 69%, as waterway. This proved insufficient,

Fig. 11.1 The two-span Roman Pons Fabricius in Rome, showing rock protection around central pier (O'Connor photograph)

and a flood cut through the northern approach, leading to the addition of a further 10 spans, with a total length of about 129 m, of which 81.8 m, or 64% was waterway. This again was insufficient, for a later flood cut through the southern approach. The bridge was further extended, possibly about the period AD98–117, to become the longest of all Roman stone bridges, with about 62 spans and a total length of about 755 m (with 64% waterway). Casado (n.d.: 33) has a photograph of this bridge, in the flood of 1942, with the water reaching in places to the deck. Another major Roman bridge carried the Via Flaminia over the Nera River at Narni: O'Connor (1993: Frontispiece, 78ff) has a reconstruction. It was originally about 160 m long, with a maximum height above normal water level of 30 m and four spans, 19.6, 32.1, 18 and 16 m. Of these, only span 1, of 19.6 m, is now intact, but pier 3 (between spans 3 and 4) also stands. Pier 2 is also shown as standing in some early photographs, but has now fallen. It is of some importance to note here the nature of its failure, for it fell upstream, clearly as a result of local scour beneath its upstream face. A third Roman bridge is the Pons Fabricius in Rome, crossing the left branch of the Tiber to the Isola Tiberina, or the island in the Tiber (O'Connor 1993: 65). It still stands as an impressive example of Roman achievement, with its completion date, 62BC, carved into its stones. It is shown in Fig. 11.1. A glance is sufficient to reveal the additional stonework deposited around the single central pier: evidence of a continuing battle against scour, and also of the concurrent increased risk of damage due to larger stream velocities in the impeded waterway.

Two further examples have been given in Section 2.4: the failure of part of the Tasman bridge due to ship impact, and the effects of the extraordinary debris mat deposited by flood against the Ngalimbiu River in the Solomon Islands, 'a solid wall of timber some 200–300 m wide, standing 3–4 m above the water surface' (Boyce 1987, 1989), that demolished certain spans and caused the stream to bypass the bridge and cut through its approaches.

Only some of these matters will be discussed here: a brief introduction into the basic phenomena, a brief look at scour, and then a slightly fuller account of some of the loads applied to a bridge. A thorough treatment of bridge hydraulics may be found in the recent book of this name by Hamill (1999). Basic early work in this area was that by Bradley (1960, 1978). Other books include Austroads (1994), Farraday and Charlton (1983), Neill (1973) and Robinson (1964). A classic early study of the hydraulics of open channels was by Bakhmeteff (1932); other works are by Chanson (1999), Chow (1959), French (1985), Hamill (1995), Henderson (1966) and Montes (1998). The estimation of stream flows, or hydrology, has been the subject of many books, such as those by Linsley et al. (1988) and Wilson (1983). In Australia, the classic reference work is *Australian Rainfall and Runoff* (IEAust 1987), dependent on rainfall data published by the various states (for example, 'Queensland Stream Flow Records for 1979', Queensland Water Resources Commission 1980). Similar data should be available for other geographic regions. 'The Rivers Handbook', edited by Calow and Petts (1992–) presents a wider view. An emphasis on scour at bridge sites may be found in Andreski (1987), Breusers and Raudkivi (1991), Hopkins et al. (1975), Laursen (1960, 1963), Neill (1964), Simons and Lewis (1971), US Federal Highway Administration (1993) *Circular 18* and Melville (B.W., 'Scour at Bridge Sites', 327–62 in Cheremisinoff et al. 1987–8 Vol.2). Further references on specific subjects will be cited in the following sections.

Finally, it should be noted that the French engineer, Perronet, was one of the first to recognise the need to minimise harmful interactions between a bridge and a stream, as illustrated initially by his bridge at Neuilly, of 1772, built in the days of stone, before wrought iron and steel came into widespread use. The steps he took were:

(a) to keep the pier thickness to a minimum, with a ratio of thickness to clear span of 0.104 at Neuilly;
(b) to use slender, segmental stone arches so as to keep the level of their springings as high as possible; and
(c) to bevel the underside of the arches near the springings, so as to improve flow beneath the bridge.

Descriptions of his bridges may be found in general works on the history of bridges, and also in his own work, *Construire des Ponts au XVIIIe Siècle*, published originally in 1782, and reissued in 1987 with this title by l'École Nationale des Ponts et Chaussées.

11.2 Stream flows

The basic formula used to describe normal stream flow in a channel of constant cross-section is that given by Manning:

$$Q = (AR^{\frac{2}{3}}S^{\frac{1}{2}})/n$$

where

Q is flow (m^3/sec),
A is cross-sectional area (m^2) of the flow,
R is hydraulic radius (m),
$= A/WP$,
WP is wetted perimeter (m),
S is hydraulic slope (m/m), and,
n is the Manning roughness coefficient

Typical values of the Manning coefficient, n, are:

regulated waterway	0.035
cultivated land	0.040
scattered trees and brush	0.070
wooded areas	0.100–0.120
wooded areas with flood reaching branches	0.120–0.160

Figure 11.2(a) shows the cross-section of a stream in flood (taken from Austroads 1994: 50). With respect to some datum, the surface level is at an elevation of 35 m; the flood stage is said to be 35 m. The total cross-sectional area is 261.4 m^2, with a wetted perimeter (lower surface) of 104.4 m. The overall hydraulic radius is 2.50 m. However, the nature of the flooded surface varies, and with it, the Manning coefficient. It is assumed, therefore, that the total cross-section may be divided up, as shown, into seven parts. For each of these, the wetted perimeter is the length of the underlying surface. The hydraulic slope is assumed in this exercise to be 0.00042 m/m, where this figure is the same for all segments of the cross-section. The flow for each segment may then be calculated, and these added to give the total flow. For example, for segment 4, the area is 47.3 m^2, with a wetted perimeter of 11.0 m. The hydraulic radius is 4.30 m, and the flow,

$$\begin{aligned} Q_4 &= (47.3 \times 4.30^{\frac{2}{3}} \times 0.00042^{\frac{1}{2}})/0.035 \\ &= 73.2 \text{ m}^3/\text{sec} \end{aligned}$$

The total flow, Q, for all seven segments, is 220 m^3/sec.

Conversely, the mean velocity, V_1, is given by

$$\begin{aligned} V_1 &= Q/A \\ &= 220/261.4 \\ &= 0.84 \text{ m/sec} \end{aligned}$$

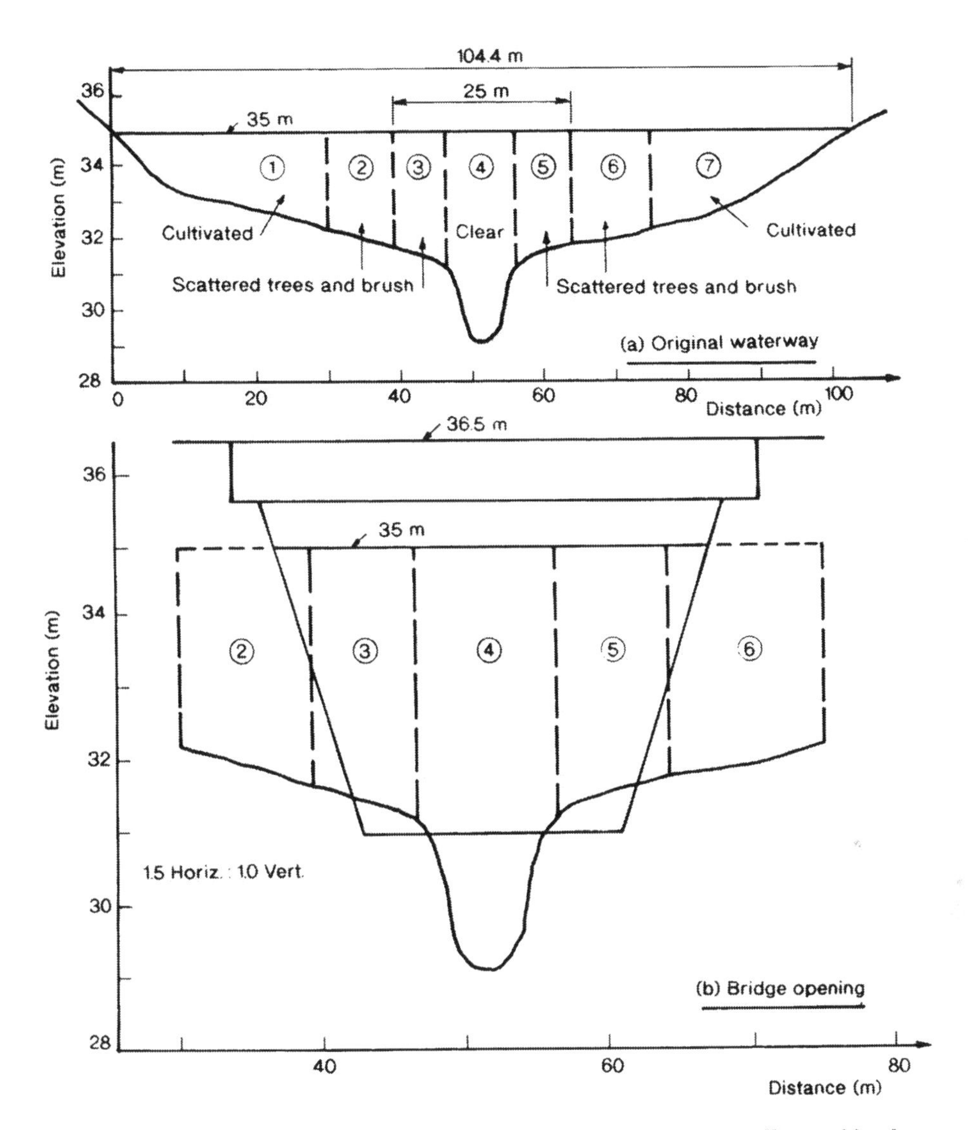

Fig. 11.2 Cross-section of stream in flood: (a) natural flow; (b) effects of bridge (Redrawn from Austroads 1994)

If the total flow, Q, is specified, then the flood height may be determined by trial and error; or, for various values of flood height, the discharge, Q, may be plotted against height or stage.

Suppose that it is necessary to pass the flood flow of 220 m^3/sec through the bridge opening shown in Fig. 11.2(b). If the surface level remains at 35 m, the cross-sectional area is 98.9 m^2, with an average water velocity, $V_2 = 2.22$ m/sec. Suppose that the bridge has no piers. The bridge opening corresponds approximately to regions 3, 4 and 5 of the unobstructed waterway. These regions have a total area of 101.5 m^2, and their contribution to the total uninterrupted flow is 109.8 m³/sec. If the original flow is subdivided into Q_a, through regions 1 and 2; Q_b, through regions 3–5; and Q_c, through regions 6 and 7, then the bridge opening ratio, M, is defined as

$$M = \frac{Q_b}{Q_a + Q_b + Q_c} = \frac{Q_b}{Q}$$

In this case, $M = 109.8/220 = 0.50$.

The kinetic energy corresponding to a stream velocity, V, has the form $\rho Q V^2/2$, where ρ is the water density (mass per unit volume). The corresponding potential energy is $\rho Q h g$, such that a pressure head, h, would cause the velocity, V. Equating the two,

$$h = V^2/2g$$

More truly, the velocity will vary across the section. The previous calculation gave Q_4, the initial flow in region 4, as 73.2 m³/sec, through a regional area of 47.3 m². The average velocity in this region is 1.548 m/sec., compared with the overall average of 0.84 m/sec. It is appropriate, therefore, to multiply the expression for equivalent head by a factor, α:

$$\alpha = \frac{\Sigma(qv^2)}{QV^2}$$

where

q = flow through a segment,
v = velocity through a segment,
Q = total flow,
V = total average velocity

In this case, $\alpha = \alpha_1 = 1.59$.

A similar factor, α_2, exists for flow through the constricted area beneath a bridge. Here, because of the constriction, the velocity is also non-uniform. Austroads (1994), based on the observation of flows beneath existing bridges, suggests the use of a graph linking α_2 to α_1 and M. An equivalent expression is

$$\alpha_2 = 1.0 + M(\alpha_1 - 1)$$

In the present case, for $\alpha_1 = 1.59$, and $M = 0.5$, $\alpha_2 = 1.30$.

Figure 11.3(a) shows a plan of the flow, and identifies four locations or sections. Sections 2 and 3 are at the upstream and downstream faces of the bridge embankments, section 1 is upstream, and section 4 well downstream. Above section 1, the stream flow is uniform, with parallel flow lines. Between section 1 and the constricted section 2, the outer flows of the stream are directed inwards to the opening. These sectional locations are also shown in Fig. 11.3(b), which is a longitudinal section through the centre of the stream. It shows the sloping bed, and a line marked NWS (Normal Water Surface), indicating that

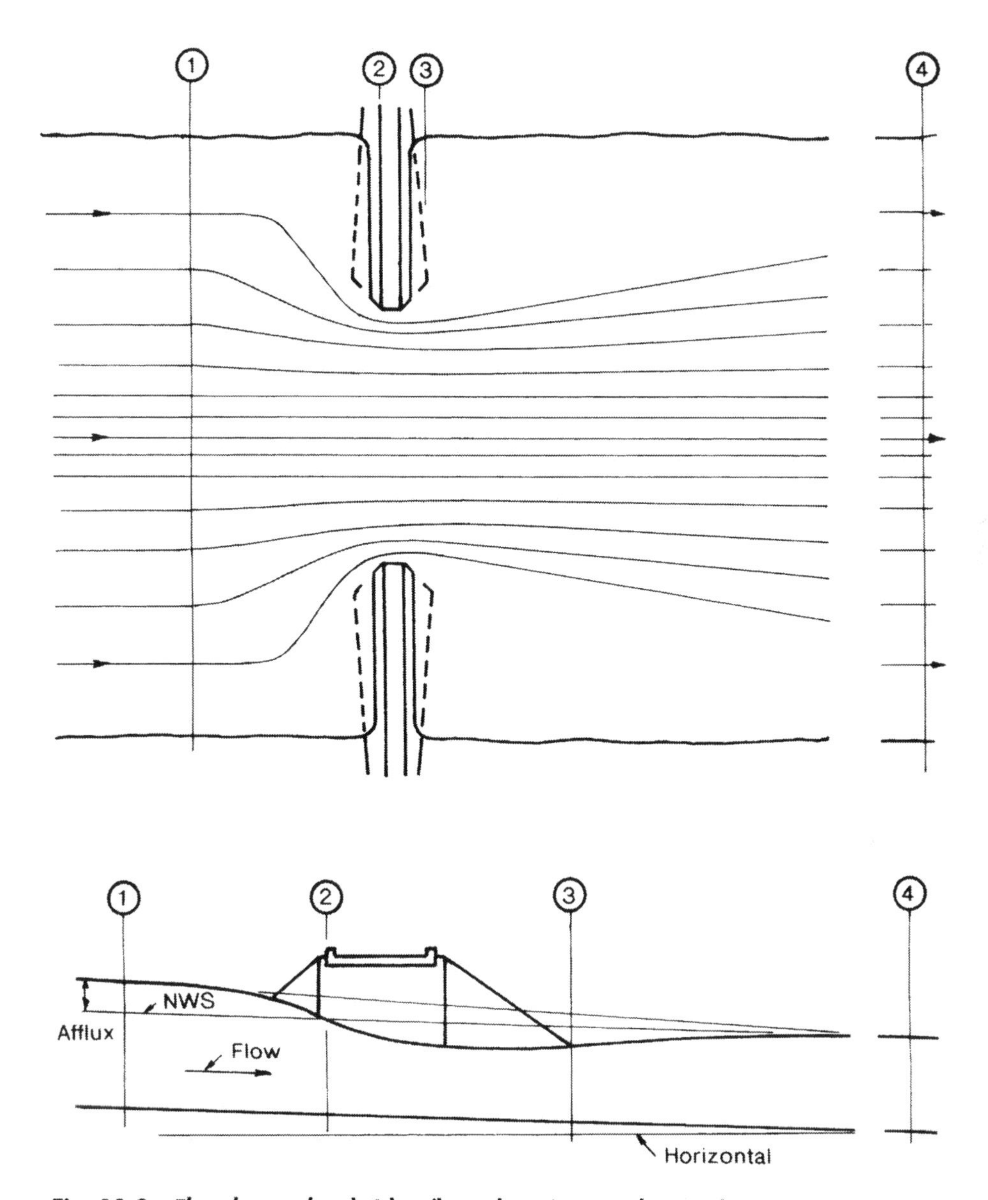

Fig. 11.3 Flow beneath a bridge (based on Austroads 1994)

below the bridge, at section 4, the flood depth is the same as it would have been if the bridge were not there. The drawing assumes that the design of the bridge is such that, at section 2, the flood depth also reaches NWS; but at section 1, there is an elevation in water depth, by the afflux, h_1. Bradley (1960, 1978) developed methods for the estimation of h_1 (see also Austroads 1994 and Hamill (1999). A first approximation is:

$$h_1 = K_b \alpha_2 V^2{}_2/2g$$

where

> V_2 is the average velocity in the constricted area,
> α_2 is the corresponding energy correction term, and
> K_b is the backwater coefficient.

Values of the coefficient, K_b, were determined experimentally by Bradley (1960, 1978) as functions of the bridge opening ratio, M, and are presented in Fig. 11.4 (see also Austroads 1994: Fig. 5.13; Hamill 1999: 145). For the present case, $M = 0.50$. The resulting value of K_b is about 1.2. It follows that a first approximation to h_1 is

$$h_1 = 1.2 \times 1.3 \times 2.22^2/(2 \times 9.81)$$
$$= 0.39 \text{ m}$$

This estimate must be corrected for a number of reasons: for the change in kinetic energy between sections 4 and 1 (Fig. 11.3), for energy losses due to piers in the flow, for eccentricity in the flow ($Q_a \neq Q_b$), and for skew, where the crossing is not normal to the stream flow. Methods to calculate these corrections may be found in the cited references, but in the present case, the effect on h_1 is small.

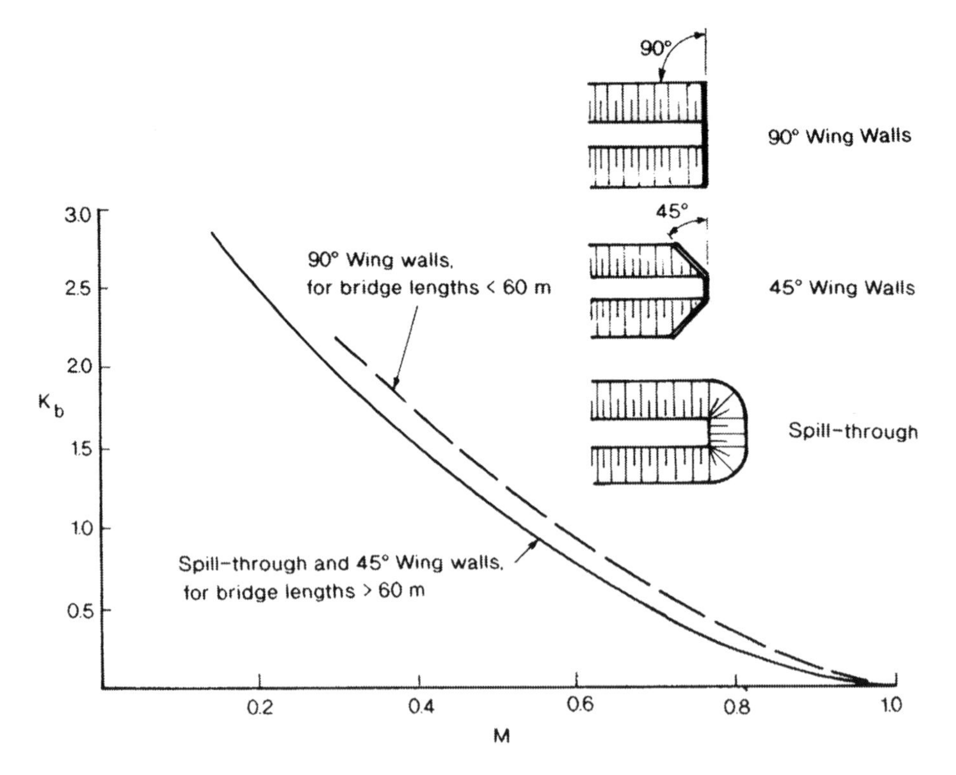

Fig. 11.4 Backwater coefficient (Bradley 1978)

There is, however, another assumption used in this illustration. Flows described by the Manning Formula are called subcritical flows. For steeper bed slopes, another type of flow, supercritical flow, may develop, with larger stream velocities and smaller depths for the same discharge. Transition from subcritical to supercritical flow occurs at a critical depth or critical velocity, where either correspond to a unit value of the dimensionless parameter, F, called the Froude number,

$$F = V/(gY)^{\frac{1}{2}}$$

V is the velocity that is regarded as 'characteristic' for the cross-section and Y is the characteristic depth. In the present case, V may be taken (with some approximation) as the average velocity, and Y, called also the hydraulic depth, as the average depth.

For the constricted flow shown in Fig. 11.2(b), the total area is 98.9 m^2, with a surface width of 30.6 m. The corresponding value of Y is:

$$\begin{aligned} Y &= 98.9/30.6 = 3.23 \text{ m} \\ \therefore F &= 2.22/(9.81 \times 3.23)^{\frac{1}{2}} \\ &= 0.39 \end{aligned}$$

This is less than 1.0, and in this case all regions of the flow (sections 1–4 in Fig. 11.3) are subcritical.

To take a simpler case, consider a rectangular channel, of width, b, with a water depth, d. Then,

$$\begin{aligned} Y &= d \\ Q &= Vbd \end{aligned}$$

For

$$\begin{aligned} F &= 1 \\ V &= Q/bd = (gY)^{\frac{1}{2}} = \sqrt{gd} \end{aligned}$$

Rearranging,

$$d = \left(\frac{Q^2}{b^2g}\right)^{\frac{1}{3}}$$

This value of d is, therefore, the critical depth. At this depth,

$$V = Q/bd = (Qg/b)^{\frac{1}{3}}$$

which is the critical velocity for a discharge Q through a rectangular channel of width b.

For small slopes, the flow is as described by the Manning Formula. For either a rectangular or complex cross-section it is possible to determine both the

Fig. 11.5 Macrossan bridges over Burdekin River, Queensland (O'Connor photograph)

average velocities, V, and depths, d, for a particular discharge as the stream slope is increased, with the corresponding values of F. Critical flow will occur (at least approximately) when $F = 1$. It is not proposed to consider here the problem of supercritical flow.

This treatment has considered the effects of a particular discharge, Q. The further problem is to select an appropriate value of Q for bridge design. One must recognise, first, that stream discharges or flows may vary widely, both from month to month, and from year to year. Figure 11.5 is a distant view of two bridges across the Burdekin River at Macrossan in North Queensland. Both have six main spans, each of 76.2 m; the first was placed in service in 1899; it was replaced by the second in 1964 but still stands. The design of the first was described in a paper by Jackson (1900). The clear height is about 26 m above the river bed. As can be seen, the flow at the time of the photograph is small, but the design assumed a flood depth of 23.5 m. Jackson states:

> The highest recorded flood occurred in the year 1870 with a rainfall of 15.8 inches [0.401 m], which gave a cross-sectional area of 81,244 square feet [7548 m^2] occupied by flood-water, the observed surface velocity being 9 miles per hour [4.02 m/sec].

If 4.02 m/sec is assumed to be the average velocity, then multiplying this by the observed cross-sectional area gives a discharge of 30,000 m^3/sec. For comparison, modern stream flow records for this site, over the period 1947–69 (Qld W.R.C 1980–), give a maximum flood of 23,300 m^3/sec., in March 1950. This is not claimed to be the maximum flood ever recorded, but the order of magnitude lends support to the flow given by Jackson. Similarly, Jackson quotes a drainage area of '14,000 square miles' (36,260 km^2), compared with the more recent figure of 36,390 km^2, effectively the same. These similarities lend support to Jackson's estimates of flood height and flood velocity of 4.02 m/sec.

Referring again to the more recent records, the figure of 23,300 m^3/sec is a 'monthly maximum instantaneous flow'. The second largest flow in the recent records is 19,200 m^3/sec in March 1956. To assess flow variability, it is appropriate to look for the minimum value of the monthly-maximum flow that occurred between these two maxima. For the four-month period from July to October 1952, the flow varied between zero and 0.18 m^3/sec. Two conclusions may be reached:

1. A flow through a waterway, 7500 m^2 in cross-sectional area, 23.5 m deep, with a recorded velocity of 4.02 m/sec, is impressive.
2. The flow through this site is most variable.

This site at Macrossan (or Sellheim) is 297 km from the river mouth. Figure 11.6 is a photograph of the bridge between Home Hill and Ayr, built in 1955 to cross the Burdekin 23 km from its mouth. It has 10 major spans, each of 76.2 m. This total length of 762 m corresponds to the distance between the vegetated banks of the river. Between is a major sandy bed, with the stream flow carried by narrow, meandering channels. In times of flood, the river rises above these banks; the current bridge provides a total waterway width of 1091 m. The

Fig. 11.6 Burdekin Bridge between Home Hill and Ayr, Queensland (O'Connor photograph)

greatest recorded flood depth in 1950 was 12.7 m above the average dry weather low-water level.

Here, records for the period 1921 to 1958 give a maximum value of the monthly instantaneous flow of 40,390 m^3/sec, somewhat more than at Macrossan. It occurred in March 1946. There was another large flow, of 30,400 m^3/sec in March 1950. Between these was a period of three months, from October to December 1948, with the maximum monthly flows recorded as 0.09, 0.00 and 0.00 m^3/sec. Again there is evidence of an extreme variation in the rate of flow.

It is noticeable that the two sites differ greatly. Some natural rock is visible in the river bed at Macrossan, and this should provide a safeguard against erosion. On the other hand, the bridge at Home Hill is founded in deep sand, clearly vulnerable to scour. Its foundations consist of caissons carried down to a depth of about 30 m below the river bed. Just visible at the right of Fig. 11.6 is the low-level bridge, placed in service in 1913. It is typical of low-level bridges built in Queensland about that time and earlier (Brady 1896). Although these were commonly put out of service by flooding in the wet season, it was found better, with the funds then available, to place them low, to reduce the risk of damage caused by floating debris that was picked up and then passed over them at higher stages of the flood.

Section 5.1 introduced and defined 'recurrence interval' and quoted from hydrological practice; the term 'average recurrence interval' (ARI) is now preferred ('Australian Rainfall and Runoff', IEAust 1987: 7). In Australian bridge design (Section 4.7), an ultimate limit state has been defined as one with a 5% chance of occurrence in a bridge life of 100 years. The corresponding notional Average Recurrence Interval is 2000 years. It needs to be seen if the event with this ARI can be predicted using current statistical procedures. Section 3.3 introduced the Normal Distribution and also discussed the Lognormal Distribution, where the logarithm $\ell n(x)$ of a function, x, is normally distributed.

The normal probability density distribution (Fig. 3.1) is symmetrical. If x is the variable whose probability is plotted, then the mean, $\bar{x}$, is also the value with the highest probability of occurrence. The median or mode of x is the value $\tilde{x}$ such that the total number of occurrences less than $\tilde{x}$ equals the total number greater than $\tilde{x}$. For a symmetrical distribution, $\bar{x} = \tilde{x}$. Similarly, for a lognormal distribution, either $y = \ell n(x)$ or $y = \log_{10}(x)$, the graph of p against y is symmetrical, with $\bar{y} = \tilde{y}$. However, in this case, the graph of p versus x is not symmetrical, but is skewed, as shown in Fig. 3.4, with the right hand tail $(x > x_{max})$ longer than that to the left. This skew is said to be positive. It is generally found (see, for example, Wilson 1983: 220) that histograms of flood records, such as the maximum annual floods, are skewed – it is impossible, for example, in most streams (but not all) to have a negative flood. As a result, a lognormal plot will give a better approximation to recorded floods than a normal distribution. However, such a plot is still generally unacceptable. The so-called log-Pearson Type III distribution was adopted in 1977 by American agencies and has come to be used generally ('Australian Rainfall and Runoff', IEAust 1987; Linsley et al. 1988: 347f; Wilson 1983: 220f).

The use of this method can be described as follows.

1. Suppose that the maximum annual flows, Q, have been recorded. Then, for each Q, calculate $X = \log_{10}(Q)$.
2. Calculate the mean, X_M, and standard deviation, σ_x, of these values of X.
3. An estimate of the degree of skewness of X is found by calculating the skew coefficient,

$$G = \frac{n\Sigma(X - X_M)^3}{(n-1)(n-2)(\sigma_x)^3}$$

where

n is the number of values

For n large, the value $(n - 1)(n - 2)/n$ approaches n.

4. For a particular average recurrence interval (ARI), the annual probability of exceedence is (1/ARI). For this value of ARI, the corresponding flood is given by

$$X = X_M + K\sigma_x$$

where K is a function of the distribution, and depends on ARI and G. Values of K were calculated by Pearson (1930). More recent tables may be found in the sources quoted above, in Harter (1969) and guidelines published by the US Interagency Advisory Committee on Water Data (1982). Some values of K are listed in Table 11.1.

The values of K for $G = 0$ are as for a normal distribution. For example, a value of ARI = 50 years corresponds to a probability of exceedence of 0.02, or of non-exceedence, 0.98. Tabulated areas of the standard normal distribution list a value of 0.98 at a point that is 2.054 × the standard deviation above the mean, the same as that listed. The nature of the log-Pearson III distribution can be gauged by comparing the rows in the above tabulation with that for $G = 0$ (lognormal) – (see Fig. 3.4).

It is of some interest to observe the ratio of the predicted 500-year and 100-year events. IEAust (1987) lists flows for an Australian stream on the

Table 11.1 Values of K

G	*ARI (years)*					
	2	*5*	*10*	*50*	*100*	*500*
2.0	−0.307	0.609	1.303	2.912	3.605	5.215
1.0	−0.164	0.758	1.340	2.542	3.023	4.088
0	0	0.842	1.282	2.054	2.326	2.878
−1.0	0.164	0.852	1.128	1.492	1.588	1.741
−2.0	0.307	0.777	0.895	0.980	0.990	0.998

mid-north coast of New South Wales (the Styx River). The maximum annual floods, Q, between 1954 and 1966 varied from 8.2 to 878 m^3/sec. The mean value of X (or $\log_{10}Q$) was 2.07, with a standard deviation of 0.48. The coefficient of variation, σ_x, was, therefore 0.231. For the present purposes, consider a stream with $\sigma_x = 0.25$ and $X_M = 1$. For G equal to 2.0 and an ARI of 500 years, K is 5.215, giving $X = 2.304$, and $Q = 201.3$. The corresponding mean flow, Q, is 10. A similar calculation for an ARI of 100 years gives $Q = 79.7$; that is, the predicted 500-year flood is 2.53 × the 100-year flood. The corresponding ratios for other values of G are: (1.0, 1.85), (0, 1.37), (−1.0, 1.09) and (−2.0, 1.00), with each pair given in the order, G, then the ratio of the 500-year to the 100-year flood (Q). In Australian practice, it is found that negative values of G are common. Nevertheless, it is of some concern that for values of G from 0 to +1.0, this ratio varies from 1.37 to as much as 2.53. Australian practice is to avoid extrapolations beyond 100 years.

There are other ways of estimating flood flows. The simplest is by direct observation of the maximum historical flood height, either from physical traces such as debris heights, or from contemporary records. The British Standard, BS 3680: Part 5: 1992 (the same as ISO 1070: 1992) describes the 'slope area method of estimation', essentially using the Manning Formula in the way described previously, with suitable estimates of the coefficient n and the slope estimated 'from flood marks on the channel banks'.

Flood flows may also be estimated from rainfall records using hydrological methods. One application of these methods allows the use of regional rainfall and storm records to estimate the Probable Maximum Flood (PMF: see IEAust 1987: 271ff; Linsley et al. 1988: 367ff; Wilson 1983: 228ff). This may be used to estimate the 1 in 2000-year flood in the manner shown in Fig. 11.7; the vertical scale represents the discharge, Q, on a natural scale, with the Average Recurrence Interval (ARI, years) plotted horizontally to a $\log_{10}$ scale. The procedure is as follows.

1. Use the log-Pearson III procedure to estimate flows for ARI = 20, 50 and 100 years.
2. Estimate the PMF.
3. For this PMF, assume a return period (here shown as 10^6 years).
4. By linear interpolation, estimate the discharge Q for ARI = 2000 years.

11.3 Scour

Some general references on the subject of scour were listed in Section 11.1. Scour caused by a bridge is generally classed as (a) constriction scour, caused by increased stream velocities resulting from a reduced waterway area, and (b) local scour, near piers and abutments. The approach conditions also affect the scour. *Clear-water* scour occurs when there is no movement of the bed material upstream from the bridge. Conversely, *live-bed* scour happens when the bed material upstream is moving. An example of the latter is that illustrated in

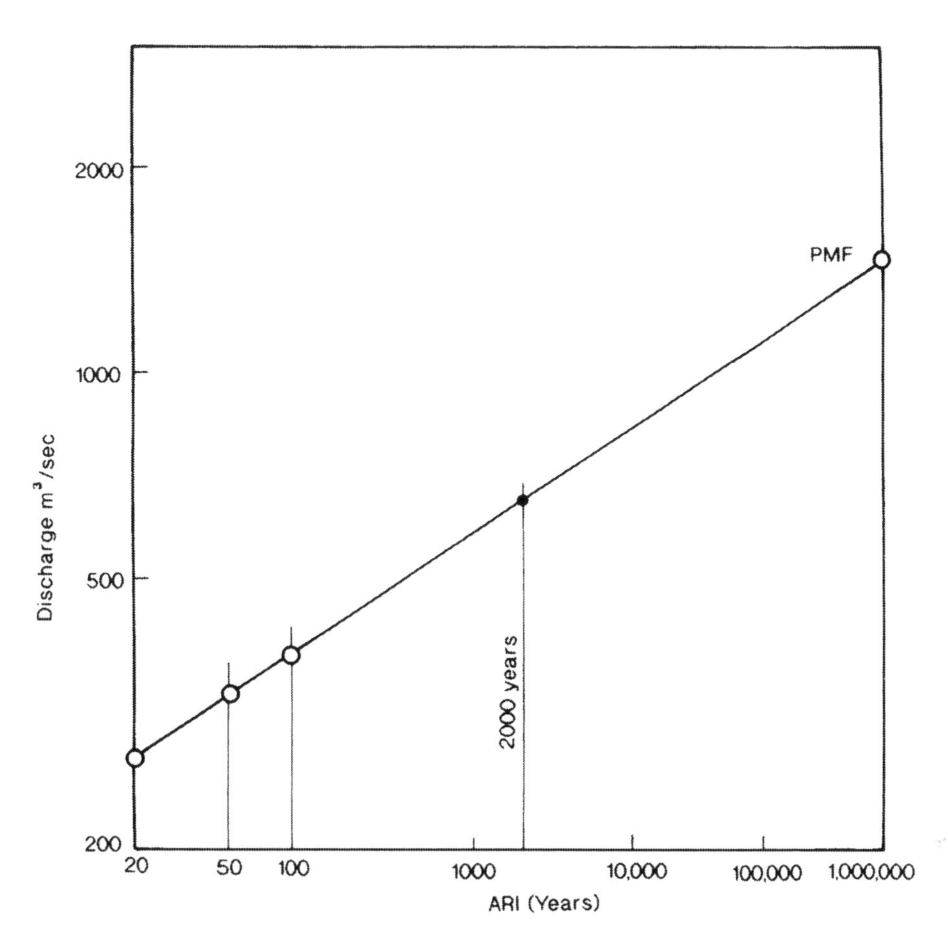

Fig. 11.7 Use of the Probable Maximum Flood (PMF) to estimate flood flows for particular recurrence intervals

Fig. 11.6, in the Burdekin River at Home Hill, where in times of flood the sandy bed tends to move downstream.

Laursen (1960) gave the following equation for constriction scour:

$$\frac{y_2}{y_1} = \left[\frac{Q_2}{Q_1}\right]^{0.86}\left[\frac{W_1}{W_2}\right]^k$$

where

the second term is omitted for clear-water scour,
y_1, Q_1 and W_1 are the average depth, flow and bottom width in the upstream main channel, and
y_2, Q_2 and W_2 are the corresponding terms at the bridge opening

It follows that $(y_2 - y_1)$ is the average scour depth. It is important to observe that, upstream, only the terms from the main channel should be included. For example, in the upstream cross-section shown in Fig. 11.2(a), the main channel corresponds only to region 4 and the banks are vegetated; for the bridge cross-section shown in (b), there is no constriction of flow. On the other hand, for the Burdekin River crossing at Home Hill (Fig. 11.6), if the bridge were carried on embankments that entered the sandy bed, there would be constriction. For many bridges, therefore, there is no constriction scour predicted by Laursen's equation.

Consider the case where there is; then it is necessary to assume whether there is live-bed or clear-water scour. The fall velocity, w(m/sec), of the bed material in water can be estimated from its median diameter, D_{MED} (Austroads 1994: Figure 6.10). Some examples are:

$D_{MED} = 0.01$ mm, $w = 1.4 \times 10^{-4}$ m/sec
$D_{MED} = 0.1$ mm, $w = 8.5 \times 10^{-3}$
$D_{MED} = 1$ mm, $w = 0.16$ m/sec

where all are approximate.

The shear velocity in the upstream section is defined as

$$V_S = (gy_1S)^{\frac{1}{2}}$$

where

S is the hydraulic slope

Then,

for V_S/w less than 0.5 $k = 0.59$;
for V_S/w between 0.5 and 2.0 $k = 0.64$;
for V_S/w greater than 2.0 $k = 0.69$.

For example, if $D_{MED} = 1$ mm, then $V_S/w = 12.5$, and the second term in the Laursen equation becomes $(W_1/W_2)^{0.69}$. For $W_1/W_2 =$ (say) 0.5 (Fig. 11.2), this equals 0.62, and inclusion of the second term reduces the estimated scour. Omission of the second term is, therefore, a conservative assumption and may be done on the basis of observations of the stream and its bed. To continue with Laursen's equation, for $k = 0.62$ and, for example, $Q_2 = 2 \times Q_1$, then $y_2 = 1.13 \times y$; or $(y_2 - y_1) = 0.13y_1$.

However, it is possible to estimate the conditions that cause initial sediment movement (Breusers and Raudkivi 1991: 13; Farraday and Charlton 1983: 35). The Reynold's Number (a dimensionless parameter) is

$$RN_P = V_S\, D_{MED}/\nu$$

where ν is the kinematic viscosity of water, equal to 1.14×10^{-6} m²/sec for water at 15°C.

The dimensionless particle size, D^*, is defined as

$$D^* = [(\rho_P/\rho - 1)g/\nu^2]^{\frac{1}{3}} . D_{MED}$$, where ρ_P is the particle density.

Then particle movement will occur:

for $RN_P < 1$, when $D^* \leq 2.15\ (RN_P)$;
for RN_P between 1 and 10, when $D^* \leq 2.5(RN_P)^{4/5}$;
for $RN_P > 10$, when $D^* \leq 3.8\ (RN_P)^{5/8}$.

These conditions may be solved to give the critical particle size, D_{CMED} for a particular velocity, V_S. For example, for $\rho_P/\rho = 2.7$, $g = 9.81$ m/sec^2 and $\nu = 1.14 \times 10^{-6}$ m^2/sec,

$$RN_P = 877190\ V_S\ D_{MED}$$
$$D^* = 23412\ D_{MED}$$

If

$$D^* = 3.8(RN_P)^{5/8} \text{ (i.e. } RN_P > 10)$$
$$= 19690\ (V_S\ D_{MED})^{5/8}$$

Then

$$D_{MED}{}^{3/8} = 0.841\ V_S{}^{5/8}$$
$$D_{MED} = D_{CMED} = 0.6301\ V_S{}^{5/3}.$$

Alternatively, the critical shear velocity is

$$V_S = V_{SC} = 1.319\ D_{MED}{}^{3/5}$$

For

$V_S = 0.05$ m/sec
$D_{CMED} = 0.00427$ m $= 4.27$ mm
with $RN_P = 188$

Similarly, for

$1 < RN_P < 10$
$D_{CMED} = 8220 V_S{}^4$
$V_{SC} = 0.105\ D_{MED}{}^{1/4}$

For

$V_S = 0.015$ m/sec
$D_{CMED} = 0.000416$ m $= 0.415$ mm
with $RN_P = 5.47$

For a medium sand, D_{MED} is of the order of 0.4 mm (Breusers and Raudkivi 1991: 8). It follows that a sand of this size may be lifted by a critical velocity, $V_{SC} = 0.015$ m/sec. This velocity is quite low. At the commencing phase of a flood, velocities will increase with time, and a previously stable bed may be lifted. At a particular location there may be new bed material coming down from upstream to replace that which is lost, but this will depend on the supply. Scour in a flood is a non-steady state phenomenon and needs to be treated as such. Nevertheless, the foundations of a bridge may be threatened by conditions that would arise even if the bridge were not there.

There are other natural conditions that may threaten. Figure 11.8(a) is a cross-section of the Brisbane River at Indooroopilly, at the site of the Albert railway bridge. The first bridge here was opened in 1876 and washed away in 1893 (Stanley 1898). The site is at a marked bend in the river (Fig. 11.8(b)), with a steep rocky bank to the north and a sandy beach formed commonly at the end of the peninsula at the south. During large floods, a major eddy is caused to the stream flow by the bend, with large velocities at the north bank, and upstream flows formed near the south. In 1893, these upstream flows caused scour to a maximum depth 'over 40 feet' (12.2 m; Stanley 1898: 289), leading to the failure of one of the piers. The bridge was replaced in 1895 by the Second Albert bridge, with two 100.7 m spans and a single river pier.

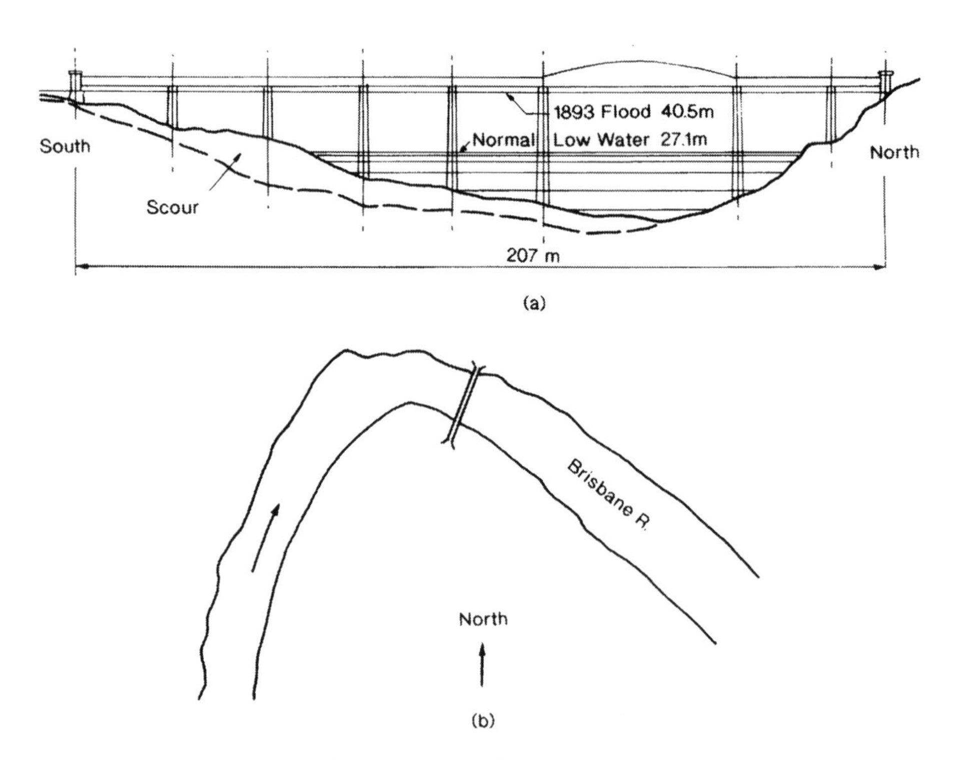

Fig. 11.8 Scour at First Albert bridge, Indooroopilly, Brisbane, as influenced by the bend in the river

At a more local scale, it is the formation of eddies or vortices that contributes greatly to scour at a bridge pier. The cylindrical pier forms the simplest case, and is shown in Fig. 11.9(a). Three distinct vortex types occur. At the leading edge, there is a downflow leading to vortices with a horizontal axis, and the locus of these axes forms what may be seen as a horseshoe around the pier – they are called horseshoe vortices. A vertical shaft in a horizontal flow will shed vortices with a vertical axis from alternative sides of the trailing edge. There are also vortex components with a horizontal axis in this area, acting as extensions of the horseshoe vortices.

Scour due to these effects tends to be greatest in front of the leading edge of the pier, where a deep hole may be formed, as in Fig. 11.9(b). The depth of scour depends not only on the stream velocity and particle size, but also on the shape and grading of the particles (Breusers and Raudkivi 1991: 61f). The figure (also taken from this source) shows typical scour depths, for various values of the average velocity V divided by the critical shear velocity V_{SC} for entrainment of particles, based on laboratory tests done by Raudkivi at the University of Auckland, using uniform sediments; where a particle size, D, less than about 0.7 mm is classed as a 'ripple-forming sediment', and for D greater than 0.7 mm, as a 'coarse sediment'. The coarse sediments will have a larger critical velocity but, as shown in the figure, may have greater depths of scour. For $V < V_{SC}$, clear-water scour will occur, with live-bed scour at $V \geq V_{SC}$. For the coarse sediment, the maximum scour depth occurs at a particular range of velocities, about $V = V_{SC}$, and may be as much as 2.3 × the pier diameter. For the smaller sediments, the scour depth is less, but tends to increase with V to about 2 × the pier diameter at large velocities.

The scour depth varies also with the pier shape. Although formulae have been proposed for predicting depth, it appears that these predictions may not be reliable (Breusers and Raudkivi 1991: 65). Neill (1973) suggested the figures for scour depth, y_S, shown in Table 11.2 for flow parallel to a pier of width b.

The value of 1.5 for a cylindrical pier is somewhat less than some of the values shown in Fig. 11.9. Scour is also affected by pier alignment, and may be significantly increased for piers with a significant angle of attack; for example, by a factor as much as 2×, for an angle of 15° from the pier to the stream flow (Breusers and Raudkivi 1991: 72, 3). Other cases are also discussed in the literature; for example, pile groups, and flow around embankment training walls.

The most common method of protecting against scour is to dump stones on the river bed around a pier or in other critical locations. Breusers and Raudkivi (1991: 91) suggest that the width of protection should be about three or

Table 11.2

Pier shape in plan	y_s/b
oblong with rounded nose	1.5
cylindrical	1.5
rectangular with square nose	2.0
ogival (sharp nose)	1.2

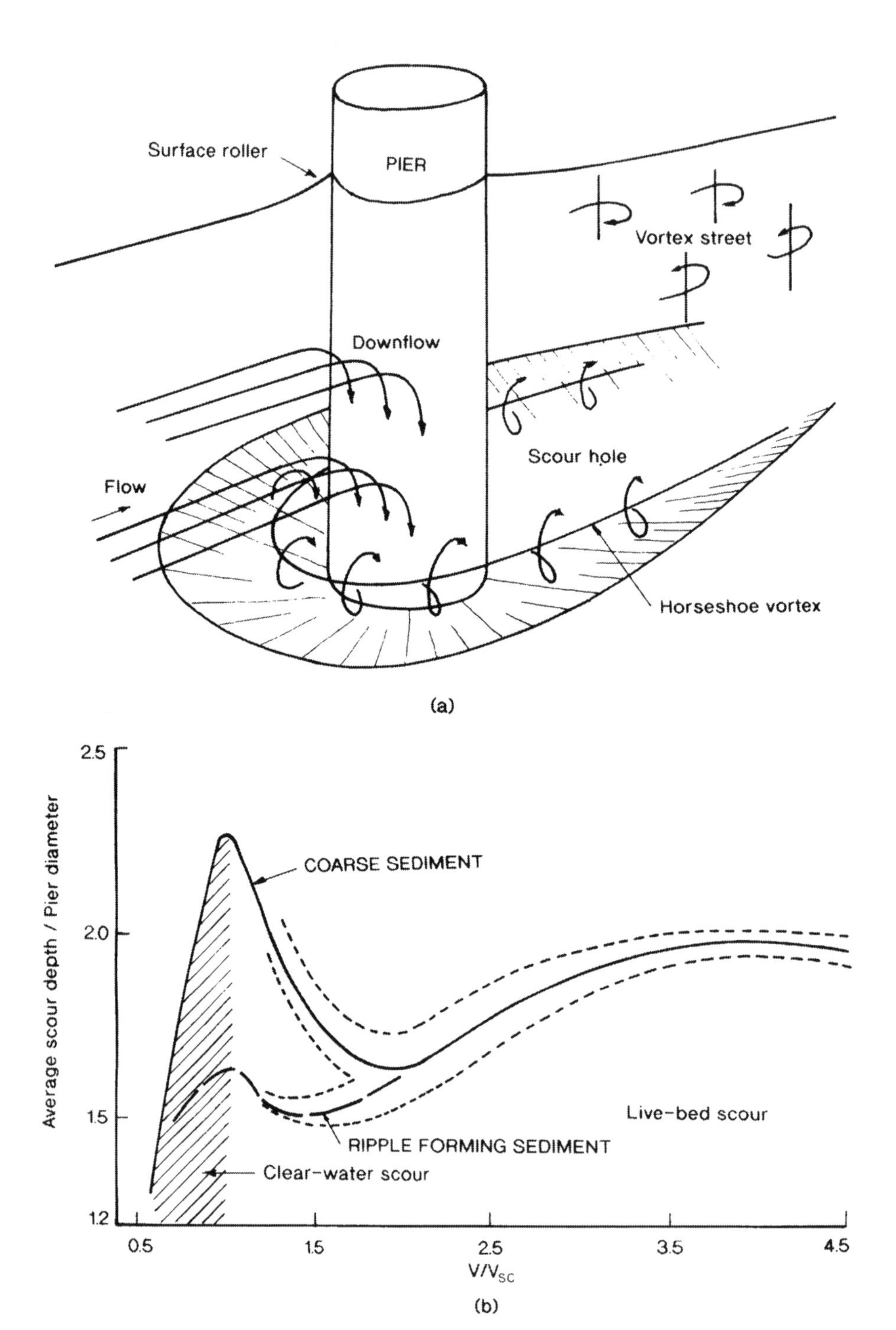

Fig. 11.9 (a) Eddy formation around a cylindrical pier; (b) scour depths (Redrawn from Breusers and Raudkivi 1991)

four times the pier thickness. They give the following empirical relationship between critical mean velocity, V_C, and rock size:

$$V_C = 4.92\sqrt{d}$$

where d is the diameter of the equivalent sphere in metres. The assumed relative density of the rock is 2.6.

11.4 Flow loads on piers and bridges

Much of the work on flow loads on bridges has been done by Apelt and others (see Apelt 1960 and 1986, Apelt and Isaacs 1968 and Apelt and Piorewicz 1986; also Jempson et al. 1997a, b) at The University of Queensland: it is cited, for example, by Farraday and Charlton (1983: 44) and Austroads (1994: 26). It commenced with work done by McKay and Apelt (personal communication: C.J. Apelt) following the failure of the railway bridge across the Pioneer River at Mirani in 1956 due to flood loads on a pier. This bridge was completed in 1913 with truss spans of 18.6, 36.9 and 18.6 m. A further flood in February 1958 destroyed piers of the replacement bridge, one of which had been completed to full height and others were close to completion. Initial model studies on a new pier design were carried out by G.R. McKay, and then Apelt. However, some less substantial earlier work had been done at this University in 1951 by J. O'Connor on hydraulic model studies on a proposed central river pier for another railway bridge at Indooroopilly.

With regard to flood loads on bridge piers, two simple cases may readily be visualised. A single cylindrical pier will clearly experience large *drag* forces applied downstream in the direction of the stream flow. On the other hand, consider a flat, plate-like pier, rectangular in cross-section, with either triangular or rounded ends. Then, if the axis of the pier is inclined in plan to the direction of the pier flow, observation may disclose that the level of the water in contact with one face of the pier is higher than at the other face. If this height difference is h, then one would expect an out-of-balance uniform pressure applied transversely to the pier (a *lift* force) of the order of ρgh, where ρ is the water density.

Figure 11.10 shows cross-sections of piers tested by Apelt and Isaacs (1968) in hydraulic model studies. For the first case, with two separate cylindrical piers, values of the lift and drag coefficients, C_L and C_D, are plotted in Fig. 11.11(a) against the angle of attack, α. The lift and drag forces, L and D, can be calculated from,

$$L \text{ (or } D) = C_L \text{ (or } C_D) \tfrac{1}{2}\rho V^2 hd$$

where

ρ is the water density,
V is the average water velocity over the pier height,
h is the depth of flow, and
d is the cylinder diameter

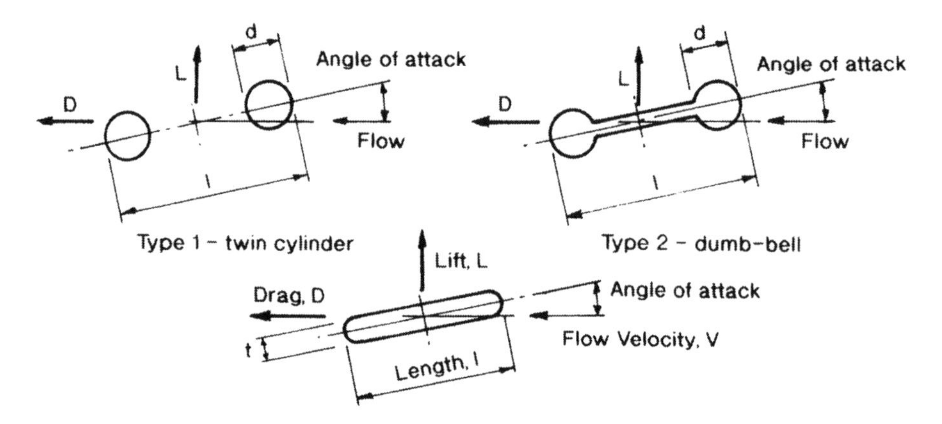

Fig. 11.10 Pier cross-sections (Redrawn from Apelt and Isaacs 1968)

The lift coefficient does not vary greatly with the cylinder spacing and reaches a maximum of about 0.6 at $\alpha = 10°$. A greater variation is apparent in C_D, which, as the overall pier length, ℓ is increased from 2.625 to 5 times the cylinder diameter, increases from about 0.9 to 2.0 at $\alpha = 10°$. For comparison, for a single circular cylinder, C_D is given as 1.17. A widely spaced pair of unconnected cylinders has values of a combined C_D that approach twice this value for large values of α. On the other hand, for closely spaced cylinders ($\ell = 2.625d$) at small α, the combined C_D is as little as 0.7.

Figure 11.11(b), with a different vertical scale, plots C_L and C_D for dumb-bell piers, but expressed in terms of the overall length, ℓ. That is,

$$L \text{ (or } D) = C_L \text{ (or } C_D) \tfrac{1}{2}\rho V^2 h\ell$$

Consider the case with the centre-to-centre spacing of the cylinders equal to $1.625d$, or $\ell = 2.625d$. Then, at $\alpha = 10°$, $C_L = 0.6$ and $C_D = 0.16$. The revised values, calculated in terms of the cylinder diameter instead of ℓ, are $C_L = 1.6$, $C_D = 0.42$. As compared with the pair of separate cylinders, C_L is increased and C_D reduced.

Lift and drag coefficients for the plate pier (in terms of the overall length, ℓ; $\ell/t = 6.52$) are shown in Fig. 11.11(c), with, for comparison, values for a flat plate (of small thickness). They are about equal, and the same may be said if a comparison is made with values for the longest dumb-bell pier ($\ell/d = 5$). These lift and drag coefficients are based on model tests with the following values of the Froude and Reynold's Numbers:

$$F = V/\sqrt{gh} = 0.19$$
$$RN = Vd/\nu = 12{,}700$$

Apelt and Isaacs (1968: 28f) suggest that the coefficients quoted here should be sufficiently accurate to apply to full scale flow, at least for the purpose of comparing pier types, and for making 'a preliminary estimate ... of the

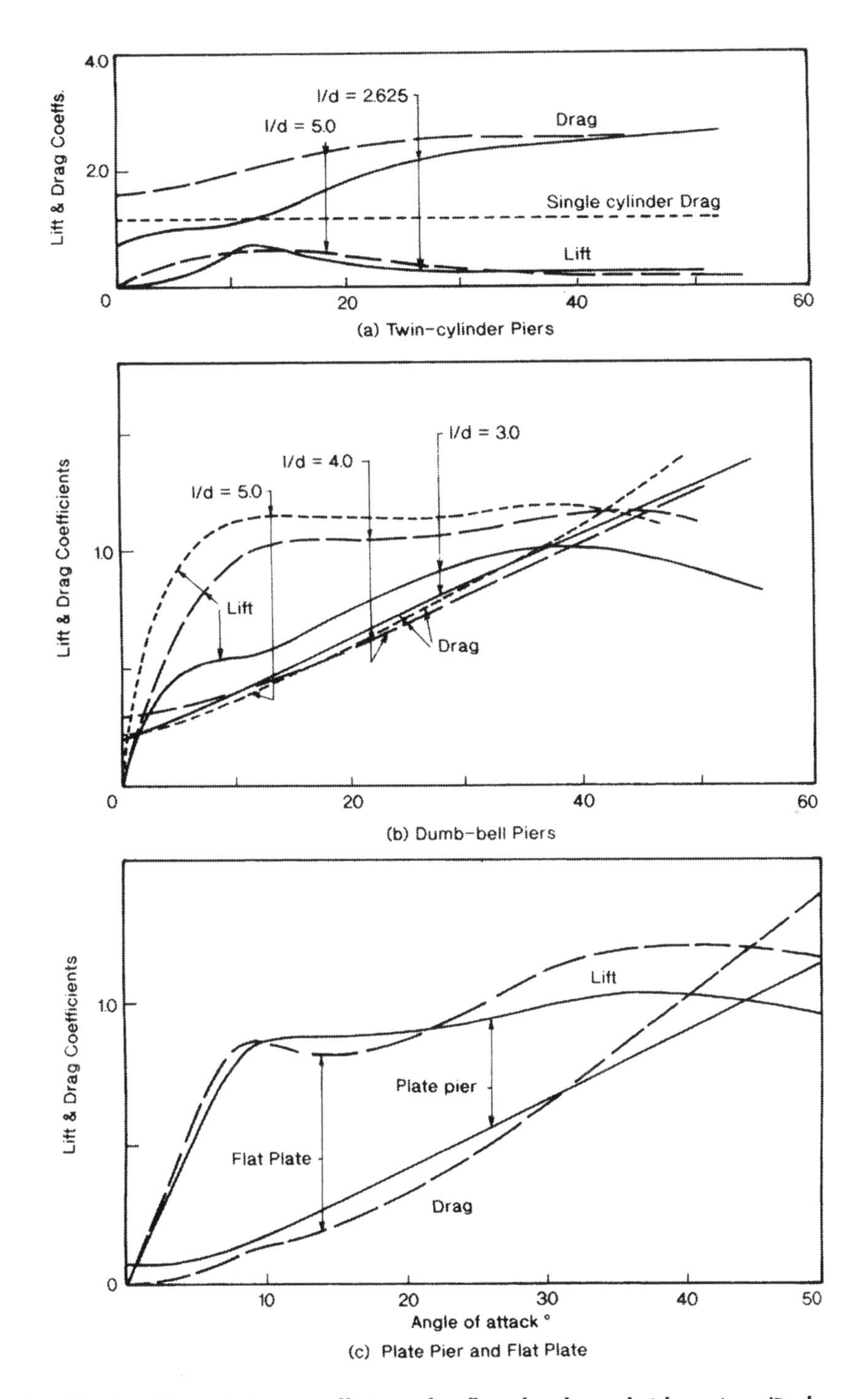

Fig. 11.11 Lift and drag coefficients for flow loads on bridge piers (Redrawn from Apelt and Isaacs 1968; used with permission)

hydrodynamic forces expected in the full scale situation', in order to determine whether, in a particular case, 'a detailed model study should be carried out'. There were also 'fluctuating (force) components superimposed on the average values (observed in these tests). These fluctuating components were least for the plate pier. The fluctuating components experienced by the twin cylinder and dumb-bell piers were generally quite large and complex.'

The AASHTO LRFD Code (1994) considers flow loads on bridge piers, expressed in the form

$$p_L \text{ (or } p_D) = 5.14 \times 10^{-4}\, C_L \text{ (or } C_D)\, V^2$$

where

p_L and p_D are pressures in MPa, and
V is the 'design velocity of water for the design flood'

The pressure, p_L, is applied to the face area of the pier (ℓh for the plate pier of Fig. 11.11(a)) and p_D to 'the projected area exposed ... to longitudinal stream pressure' (th in Fig. 11.11(a)).

For plate type piers, the following values of C_D are specified:

semi-circular nosed piers	0.7
square ended piers	1.4
wedge nosed pier, with nose angle 90° or less	0.8

The values of C_D given by Apelt and Isaacs for this case and shown in Fig. 11.11(c) are based on the pier length, $\ell = 6.52t$. For comparison, they must be multiplied by 6.52, or, alternatively, the AASHTO value of C_D for a pier width with a semi-circular nose, must be divided by 6.52, giving 0.7/6.52 = 0.11. The AASHTO figure is, therefore, consistent for the case with $\alpha = 0°$, but Apelt and Isaacs give values of C_D also for $\alpha \neq 0°$, and these may be considerably higher than the AASHTO value.

For the lift force, AASHTO (1994) lists the values of the coefficient C_L given in Table 11.3.

These agree with the values by Apelt and Isaacs and lend support to the use of their values in prototype structures.

Apelt and Piorewicz (1986) have also reported on 'breaking wave forces

Table 11.3

α	C_L
0°	0
5°	0.5
10°	0.7
20°	0.9
≥30°	1.0

on vertical cylinders', which may be relevant in some cases. There has also been work reported by Apelt (1986) and Jempson, (see Jempson and Apelt 1995 and Jempson et al. 1997a) on flood loads on submerged and semi-submerged bridge superstructures. This is important, but will not be included here. Inundation of a bridge affects not only bridge loads but also afflux (Austroads 1994: 45ff).

11.5 Debris and ship-impact loads

The AASHTO LRFD Code (1994) also refers to debris loading on piers, with a pressure, $p = 5.14 \times 10^{-4} 1.4V^2$, applied to the area of a debris raft lodged against the pier, as seen from upstream. It quotes, for guidance, a draft New Zealand Highway Bridge Design Specification which specifies a triangular area, of width B and depth A, where B is the base and A the height of an inverted triangle placed symmetrically on the pier centreline. The width B is 14 m, or half the sum of the two adjacent spans if this is less. Height H is 3 m, or one half of the water depth, when this is less. The major width of the raft is assumed to be at the water surface, with the dimension H extending downwards. The area is, therefore, $AB/2$. The New Zealand code specifies $C_D = 0.5$, instead of the figure of 1.4 in AASHTO (1994), presumably to allow for some porosity in the debris mat.

Boyce (1987, 1989) described the debris mat that caused a flood to damage and bypass the Ngalimbui bridge in the Solomon Islands in 1986, as mentioned in Section 2.4. The bridge had five 21 m spans, with a concrete deck acting compositely with steel joists, 0.76 m deep. It was designed for a debris mat 1.2 m deep by 10 m long on each pier. The assumed average stream velocity was 3.3 m/sec for the 100-year flood. The mid-stream velocity was assumed to be 1.5 times this, leading to a design debris load of 152 kN acting at the headstock level of each pier. The actual debris was described as 'a solid wall of timber some 200–300 m wide standing 3–4 m above the water surface and bearing down at about 20 km/hr' (5.5 m/sec). Two of the spans were washed away and the flood also cut through both bridge approaches.

Jempson et al. (1997b) carried out model studies to estimate forces applied to a bridge pier by assumed debris mats. These were assumed to take the form either of a rectangular, flat plate (face vertical) or a segment of a cone with a rough surface, centred on the pier with its apex downwards, so that its shape as viewed from the upstream side was as in the New Zealand model. Flow around the debris mat was found to cause not only a drag force in the direction of the current, but a vertical force with a resultant some distance in front of the face of the bridge, leading to a moment about the bridge face. Data given in the paper allows these forces to be estimated.

Another aspect of the problem is that logs, or other floating objects, may be brought down at considerable velocity and strike either the bridge or its piers. Such moving objects possess considerable kinetic energy, part or all of which is dissipated in the impact. If two bodies that are completely rigid strike each other, then the impact force is theoretically infinite. More truly, yield will occur in one or both of the bodies and, also, the point of impact may not lie on a line parallel to the relative velocity and passing through the centres of gravity of the bodies. The impact forces may cause angular accelerations to one or both of the bodies, which, in the present context, would be about a vertical axis.

The Australian NAASRA (1965) *Highway Bridge Design Specifications*, as quoted by Farraday and Charlton (1983: 47), suggested that the designer allow for the force resulting from a 2 t log, travelling at the normal stream velocity, arrested within a distance of 75 mm for a solid pier. If M is the moving mass, V its velocity, P an assumed constant contact force, and d the relative movement, then

$$MV^2/2 = Pd$$

or, for example, for

$$V = 3.0 \text{ m/sec}$$
$$P = MV^2/2d = 2000 \times 3^2/(2 \times 0.075) \text{ N} = 120 \text{ kN}$$

Farraday and Charlton quote some UK firms as designing for a mass of 10 t, arrested in 75 mm. The problem is considered more fully by Andreski (1987: Part II).

Farraday and Charlton also discuss the problem of ship impact, quoting from Frandsen and Langsø (1980), Minorsky (1959), Ostenfeld (1965) and Saul and Svensson (1982). There is also an AASHTO *Guide Specification and Commentary on Vessel Collision Design of Highway Bridges* (1991). Some of this work resulted from the collapse of part of the Tasman bridge, in Tasmania, in 1975, from ship impact, as described in Section 2.4. The problem has two parts: (1) the estimation of the probability of ship impact at a particular site, and (2) estimation of the impact forces.

Essentially the aim is to ensure that the ship fails, rather than the bridge. The deformation of a ship absorbs energy that is closely related to the volume of deformed steel, R (m^3), in the damaged area (Saul and Svennson 1982: 31; quoting from Minorsky 1959):

$$\Delta w \text{ (MNm)} = 47R + 32$$

where in many cases, with large Δw, the constant term (32) on the right hand side may be ignored. The kinetic energy, E, associated with a moving ship is

$$E = kMV^2/2$$

where

V is the ship's velocity,
M is its mass (or displacement), and
k is a correction to include a mass of water moving with the ship, estimated to be about 1.05

Frandsen and Langsø (1980: 87) report work by Woisin, who carried out tests with models of ship bows run down inclines and then horizontally into stationary reactive systems. He found that the average impact force, $\overline{P}$, was approximately constant for most of the deformation, but with short-term maximum values, equal to about twice the average value, shortly after impact. He also found that these resistance forces tended to be independent of the impact velocity, but varied with the tonnage and hence the details of construction of the equivalent ship. As an approximation to this observed behaviour he gave

$$P_{max}(\text{MN}) \approx 0.88\sqrt{\text{DWT}}, \pm 50\%$$

where

DWT is the ship's carrying capacity in dead weight tons
(1DWT = 2240 lb = 1016 kg = 1.016 t)

This can be used to estimate the extent of damage to a ship. Consider the case where a ship hits a rigid pier, with the approach at right angles to the pier face. Let a be the length of ship, from its bow, that is crushed in the impact. The following is an example taken from Saul and Svennson (1982: 36f), who give, first, graphs relating to the displacement or mass of modern ships, in tonnes to the DWT. For example, for bulk carriers, approximately

$$M(t) = 1.25 \text{ DWT}$$

They take the case of a tanker, with a displacement of 45,000 t and a DWT of 38,000; that is, with $M = 1.18$ DWT. From the above expression,

$$P_{max}(\text{MN}) = 172 \pm 86 \text{ MN}$$

The average impact force is

$$\overline{P} = \tfrac{1}{2}P_{max} = 86 \pm 43 \text{ MN}$$

For an approach velocity of 3 m/sec, the kinetic energy of the approaching ship is

$$E = 1.05MV^2/2 = 1.05(45 \times 3^2) = 213 \text{ MNm}$$

The damage length is

$$a = 213/(86 \pm 43) = 1.65 \text{ to } 4.95 \text{ m}$$

The data shown in this example corresponds to that for an actual ship impact, when a tanker of 45,000 t displacement hit a massive main pier of the Newport Bridge, Rhode Island in February 1981. The pier was essentially undamaged, but the ship's bow was flattened for a length of 3.5 m.

The time, t, over which the force $\overline{P}$ is applied to the pier may also be estimated. The mean deceleration equals $M/\overline{P}$; i.e. 45/(86 ± 43) = 0.35 to 1.05 m/sec^2. The time t is, therefore, 3/(0.35 to 1.05) = 8.6 to 2.9 secs. The maximum force, P_{max}, is transient, and occurs only for a short part of this time, quoted as 0.1 to 0.2 secs. Analyses are available also for the case of oblique impact (Saul and Svennson 1982).

Forces due to ship impact may be very large, and may require special pier protection (see, for example, Frandsen and Langsø 1980). Figure 2.18 shows the Bowen bridge, Hobart, built upstream from the Tasman bridge in the period following its failure in 1975 (Section 2.4). Bowen bridge was completed in 1984 with ten major spans, including eight of 109 m. The pier base has pointed ends, designed so as to deflect or sink the ship.

Estimation of the probability of ship impact may be difficult. Frandsen and Langsø (1980) list fourteen cases of bridges struck by a ship between 1965 and 1967, when sometimes the bridge superstructure was struck, rather than the piers. Woisin's work on ship impact forces included a proposed probability density distribution that may be used in statistical analyses.

Other forms of impact are possible, such as between a derailed locomotive and carriages, and a railway bridge overpass. The consequences of this may be major, such as in the Granville bridge disaster in Sydney in 1977, where the destruction of a pier by a derailed locomotive caused the heavy concrete superstructure to fall onto crowded carriages, with great loss of life. In such cases, the provision of deflecting structures to protect the bridge piers is essential.

11.6 Effects of ice

Farraday and Charlton (1983: 46f, Appendix B) provide a useful introduction to this problem and quote work done in Canada and the Soviet Union. The thorough paper by Neill (1976) reflects Canadian experience; it has an extensive range of references (see also Neill 1981). Hobbs (1974) has the title *Ice Physics*.

Montgomery and Lipsett (1980) have described both experimental and analytical studies of ice forces on a pier of the Athabasca River bridge at Hondo in Alberta. A typical ice force history is presented for a time of 24 sec and an ice thickness of 0.91 m. It shows peak forces up to 800 kN repeated at intervals of about 3 to 6 sec, superimposed on a relative low base level ranging from about 0 to 50 or 80 kN. The picture is clearly one of ice fracturing against the inclined upstream face of the pier, which was about 22° from the vertical, and it appears that this is often the case. Indeed the correct modern approach to design appears to be to shape the piers so as to fracture the ice.

An example of this, and an illustration of the magnitude of the problem, may be seen in the Confederation Bridge across Northumberland Strait to Prince Edward Island, part of eastern Canada, opened in 1997 with a total length of 12.9 km, over a waterway that is commonly choked with ice in winter. Thurston (1998) speaks of it as a 'Canadian-built-and-designed marvel of cold ocean engineering', and describes the ice ridges that may be formed of 'large masses of compressed ice rubble', up to 6 or 7 m high above the surface, and extending up to 20 m deep below. The bridge was required not only to withstand forces from this ice, but also to avoid delays in the movement of ice out of the Strait in spring. The bridge has 44 main spans totalling 11,080 m, with a typical distance of 250 m between pier centrelines. Each pier has a conical base, with faces inclined at 30° from the vertical. As the ice moves against the pier, two factors are present. The circular shape of the pier in plan tends to split the ice; and in addition, the ice floe tends to ride up the pier face, with vertical forces that cause the ice to fracture in tension in its upper face. The two effects combine to minimise forces applied to the pier, and to allow the ice to move past. Similar strategies can also be adopted in smaller bridges.

Farraday and Charlton (1983) and Neill (1976) quote values from the 1978 and 1974 editions of the Canadian Standards Association's standard CSA-S6. The horizontal force, F, exerted on a pier by moving ice may be calculated from the expression

$$F = C_1 C_2 ptb$$

where

C_1 is a coefficient for nose inclination,
C_2 is a second coefficient, for pier geometry,
p is an ice pressure,
t is the ice thickness, and
b is the pier width at the level of the ice (both mm)

The coefficient, C_1, has the values given in Table 11.4.

Table 11.4

Angle of pier nose from vertical	C_1
0–15°	1.0
15–30°	0.75
30–45°	0.5

Coefficient C_2 is a function of the ratio b/t (see Table 11.5).

Table 11.5

b/t	C_2
0.5	1.8
1.0	1.3
1.5	1.1
2.0	1.0 etc.

Values of p vary with the condition of the ice:

(a) For ice in small cakes where break-up occurs at melting temperatures, $p = 690$ kPa (100 psi).
(b) For ice in large pieces that are internally sound, but break-up is at melting temperatures, $p = 1380$ kPa (200 psi).
(c) For initial movement of the ice sheet as a whole, or large sheets of sound ice, $p = 2070$ kPa (300 psi).
(d) For break-up or major ice movement at temperatures significantly below the melting point, $p = 2760$ kPa (400 psi).

As an example, for a pier angle less than 15°, $t = 0.75$ m and $b = 1.5$ m, both C_1 and C_2 are unity, and

$$F = ptb = 1.125p$$

It lies, therefore, in the range 776 to 3105 kN, depending on the condition of the ice. Although large, this is still much smaller than the figures for ship impact quoted in the previous section. As mentioned in Section 4.4, the Canadian Standards Association's CSA-S6 (1978) is currently under review. Neill (1976) also quotes values from a 1967 code of the USSR.

Chapter 12

Wind, earthquake and temperature effects

12.1 Wind velocities

The British Standard, BS 6399: Part 2: 1995, specifies the wind speed at a particular site as a function of five parameters: the basic wind speed, with factors to provide for altitude, direction, season of the year, and a probability factor. The basic wind velocity varies with geographic location, and the code includes a map of the United Kingdom and Ireland with contours showing basic wind speeds. These contours have minima (or low points) centred about Oxford (20 m/sec), Perth (23 m/sec) and Dublin (23 m/sec), with higher values between and outside these points. There is, for example, a basic wind velocity of about 24 m/sec in a region between the centres and extending down to Land's End, with about 26 m/sec along the north-west coast of Scotland, rising to as high as 30 m/sec around the Shetland Islands. Similar maps of wind velocities may be found for other countries and regions and in other sources, but it is of interest here to note that the map for the United Kingdom and Ireland included in British Department of Transport highway bridge code, BD37/88 (Section 4.2), issued in 1988, differs from that in BS 6399. The reason for this is clearly that BS 6399 is later. It reflects more recent work, chiefly by Cook (1982, 1983, 1985) and Cook and Prior (1987), which will be used largely in the following section. The two-volume book, Cook (1985), is a valuable reference. Some other recent works are those by Simiu and Scanlan (1996), Dyrbye and Hansen (1997) and Liu (1991); all deal extensively with the use of probability density distributions in the analysis of wind data. The American Society of Civil Engineers published a state-of-the-art report on wind loading and structural response in 1987.

The Normal (or Gaussian) Distribution was defined in Section 3.3, with a probability density, p, of a function, x, given by

$$p = \frac{1}{\sqrt{2\pi}\sigma} e^{-(x-\bar{x})^2/(2\sigma^2)}$$

For the case with the mean, $\bar{x}$, equal to zero, and the standard deviation, σ, equal to unity, this became

$$p = \frac{1}{\sqrt{2\pi}} e^{-x^2/2}$$

The Weibull distribution is commonly used for the statistical distributions of wind speed. It has a probability density function, p, given by

$$p = ckx^{k-1}e^{-cx^k}$$

where

c is a spread parameter, and
k is a shape parameter

The cumulative distribution function, P, is defined as the integral of p between two boundary values of x, commonly from $-\infty$ to a particular value. For the Weibull distribution, this becomes

$$P = 1 - e^{-cx^k}$$

The cumulative probability of exceedance is $Q = 1 - P$. Then, for the Weibull distribution,

$$Q = e^{-cx^k}$$

Some typical Weibull probability density distributions are shown in Fig. 12.1(a).

Instead of considering all values of x, one may choose to record only extreme values, where $\hat{x}(n)$ is defined as the largest of n values. Then these values also have statistical distributions. The present problem is concerned with the prediction of upper bound values of the wind velocity, V. The theory of extreme values was developed initially by Fisher and Tippett, and then by Gumbel, and their names have been applied to distributions used in the analyses of wind records. A Gumbel distribution is one whose cumulative probability distribution, for large x, approaches

$$P = 1 - e^{-g(x)}$$

where

$g(x)$ is some function

The Weibull distribution satisfies this definition, not only for large x, but also over the whole range. The Fisher-Tippett Type I (FT1) distribution has

$$P = e^{-e^{-y}}$$

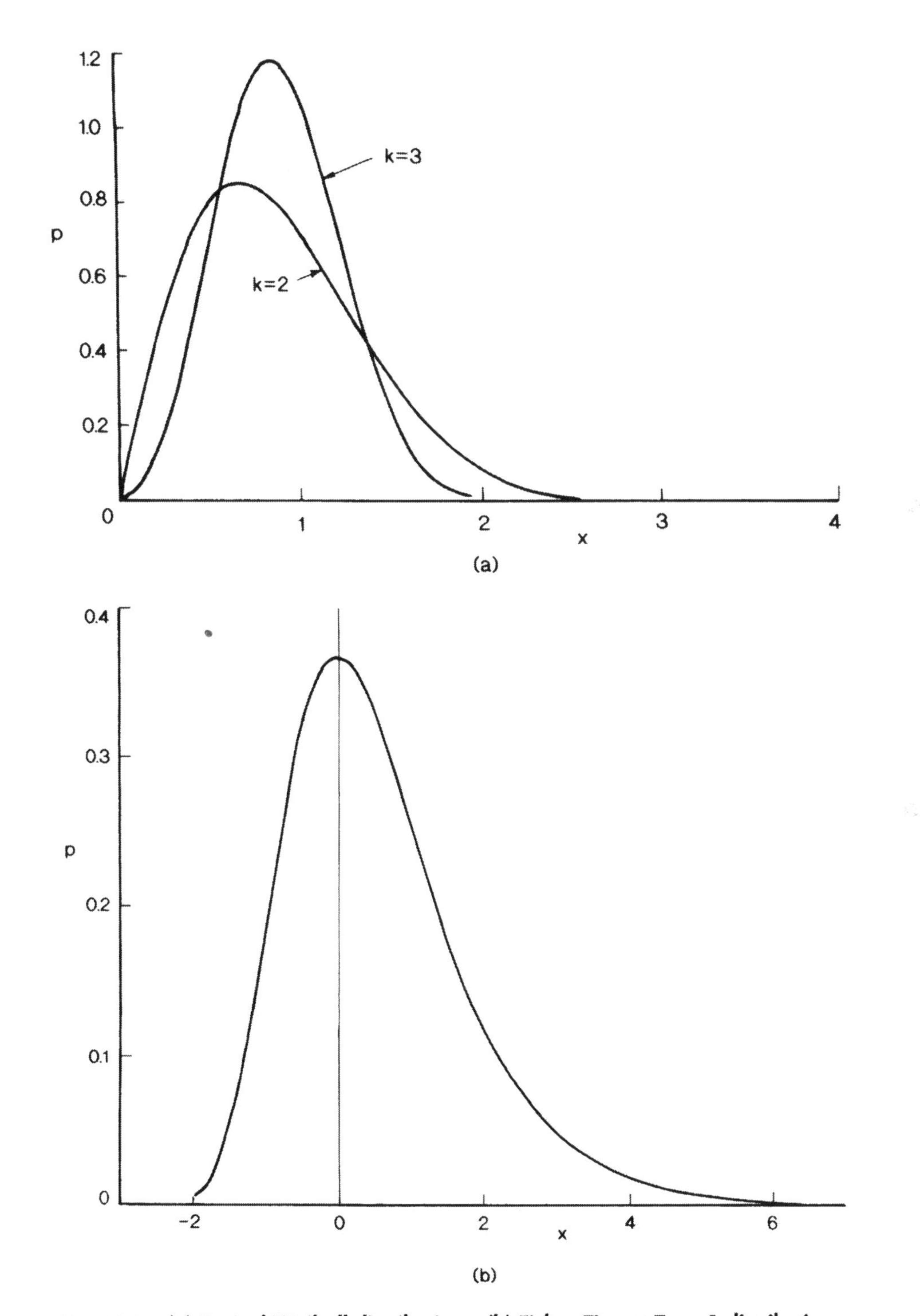

Fig. 12.1 (a) Typical Weibull distributions; (b) Fisher-Tippett Type 1 distribution

where

$y = a(\hat{x} - U)$,
U is the mode, and
$1/a$ is called the dispersion

The plot of the corresponding probability density function against y is shown in Fig. 12.1(b). This distribution may be inverted to give

$y = -\ell n(-\ell n\ P)$, with
$\hat{x} = y/a + U$

Consider a set of data values, $\hat{x}$, with cumulative probabilities, P. Then P may be plotted to a scale equivalent to $y = -\ell n(-\ell n\ P)$. The plot of these values of $P(=y)$ against $\hat{x}$ will be linear if the values satisfy the FT1 distribution. Then U and $1/a$ may be obtained and the graph extended to give y and hence x for larger values of P, or smaller values of the exceedance probability, $Q = 1 - P$.

Cook and Prior (1987) used a method proposed by Cook (1982) and based on the FT1 model to estimate wind speeds at various locations in the United Kingdom, for a 50-year Average Recurrence Interval. In doing this they used two procedures: (a) the direct extrapolation of the velocity V from observed values, and (b) the conversion of observed values of V to pressure, q, using the relation, $q = \rho V^2/2$, where ρ is the air density, followed by the extrapolation of q and then reconversion to V. They found their model to be better when based on q. The resulting contours of wind speeds (Cook and Prior 1987: 385) are those included in BS 6399: Part 2: 1995: Figure 6, Annex B. They represent estimated values of the hourly mean speed.

Cook and Prior repeated these analyses for 12 directions at each site, separated by a plan angle of 30°. They found that the underlying directional characteristics were the same for all sites in the United Kingdom. Accordingly, BS 6399: Part 2: 1995 multiplies the basic wind velocity, called V_b, by a direction factor, S_d, where this equals unity for winds from a bearing of about 240–260° (clockwise from north), reducing to about 0.73 for bearings from 30–120°.

BS 6399: Part 2 also specifies a seasonal factor S_s, that may be relevant for construction over short periods. For example, for construction periods of four months, from January to April, May to August and September to December, this factor is 0.98, 0.73 and 0.96. The worst period for construction in the United Kingdom is from December to January.

There is also a specified probability factor, S_p, that can be used to change the required exceedence probability. The basic Average Recurrence Interval in BS 6399 is 50 years, corresponding to an annual risk of exceedance, $Q = 0.02$. For other values of Q, the basic wind speed should be multiplied by S_p, where

$$S_p = \sqrt{\frac{5 - \ell n[-\ell n(1 - Q)]}{5 - \ell n[-\ell n\ 0.98]}}$$

The British highway bridge code, BD37/88, specifies a 120-year return period (Cl.5.3.1). To achieve this result, with $Q = 0.0083$,

$$S_p = 1.048$$

It is also necessary to allow for forces from wind gusts. BS 6399: Part 2: 1995 quotes formulae for gust wind speeds based on Cook (1985). The gust velocity, $V_{max,}$ active over a gust duration, t (secs), is given approximately by multiplying the mean wind velocity, V_o, by a gust factor, g_t, where $g_t = 0.42\ell n(3600/t)$.

However, the gust velocity is local in character and it is necessary to estimate the gust speed which 'will envelope the structure or component to produce the maximum loading thereon' (BS 6399: Part 2: 1995: Annex F). This standard, with particular reference to buildings, identifies the empirical relationship,

$$t = 4.5a/V$$

where a is a characteristic dimension of the structure. The older BD37/88 tabulated the gust factor against the horizontal loaded length and height above ground. The question of bridge height will be discussed shortly, but consider the case with a length of 100 m. Suppose, also, for this example, a mean wind velocity of 24 m/sec. Then the above expressions give

$$t = 18.8 \text{ sec}$$
$$g_t = 2.21$$

For a loaded length of 100 m, BD37/88 listed factors that varied over the range 1.35 to 1.98 for heights from 5 to 200 m above ground level, or from 1.60 to 1.76 for heights from 20 to 50 m. Although the size of this factor must be reviewed in its application to a particular bridge, these orders of magnitude suggest that its effects may be considerable, particularly when it is remembered that wind force is proportional to V^2.

This discussion has so far not treated bridge height, as the situation with a bridge may differ markedly from that for a building. The code BD37/88 gave contours of wind velocity for Great Britain and Ireland that were 'appropriate to a height above ground level of 10 m in open level country and a 120-year return period'. The velocities shown in its contour map (its Fig. 2) differed somewhat from those in BS 6399: Part 2: 1995, both in magnitude and distribution. The specified velocities were about 26 m/sec at Oxford, 33 m/sec north of Edinburgh, and 30 m/sec at Dublin. For comparison, figures from the later code were quoted earlier as 20, 23 and 23 m/sec, although these needed to be multiplied by 1.048 to allow for the change in specified recurrence interval. They then become 21, 24 and 24 m/sec. They also are based on wind velocities measured at a height of 10 m above open, level terrain.

Wind velocity varies with height, between zero at a plane some distance, d, above the ground to the so-called gradient height, $z_{g,}$ defined as 'the lowest height at which the wind is unaffected by the rough surface terrain' (Cook 1985:

Part 1: 77ff, 138ff). Height, d, defines the 'interfacial layer' in which the overall net flow is zero. In open country, this layer is shallow, with d approaching zero. In built-up areas, d approaches the average building height and the 'aerodynamic ground plane' is elevated. For this reason, d is called the 'zero-plane displacement'. The total height to z_g defines the 'atmospheric boundary layer'. Deaves and Harris (1978; Deaves 1981; Harris and Deaves 1980) have proposed a semi-empirical relationship connecting the velocity V_M (the equilibrium mean velocity) with height, z. For z less than 300 m, this becomes

$$V_M = 2.5u[\ell n\left(\frac{z-d}{z_o}\right) + 5.75(z-d)/z_g]$$

where

u, z_g and z_o are as given below

The term, u, is called the basic friction velocity and may be expressed in terms of V, the velocity at height, $z - d = 10$ m, corresponding to standard meteorological observations of velocity. Cook (1985: Part 1: 221) gives

$$u = 0.06886V$$

This value is used to estimate the gradient height, z_g:

$$z_g = u/6f$$

where

f varies with latitude and is called the Coriolis parameter (rad/sec)

Its general form is $1.454 \times 10^{-4} \sin\theta$, where θ is the latitude, but for the United Kingdom, where the contours of wind velocity indicate that V also varies with latitude, an approximate expression is

$$f = 4.5 \times 10^6 V(\text{rad/m})$$

Then

$$z_g = 2550\text{ m}$$

The term z_o (m), or roughness length, represents the aerodynamic roughness of the terrain. Five roughness categories are listed in Table 12.1, with the corresponding values of z_o.

The standard meteorological terrain is Category 3. For this case,

$$V_M = 2.5 \times 0.06886V\,[\ell n(z/0.03) + 5.75z/2550]$$

Table 12.1

Category	*Roughness of terrain*	z_o
0	large expanses of water or flat tarmac	0.003 m
1	flat grassland, without hedges	0.01 m
2	fields with crops, fences, low boundary hedges and few trees	0.03 m
3	farmland with frequent high hedges and occasional buildings and trees	0.1 m
4	dense woodland, domestic housing	0.3 m
5	typical town centres, with chiefly four-storey buildings	0.8 m

For

$$z = 10 \text{ m},$$
$$V_M = 0.172V\,(5.809 + 0.023) = 1.004V$$

If the second term is omitted, then $V_M = 1.00V$.

Consider a bridge over a large expanse of water (category 0). Then $z_o = 0.003$; $d = 0$.

$$V_M = 0.172V\,[\ell n(z/0.003) + 5.75z/2550]$$

For $z = 10$ m, $V_M = 1.4V$;
for $z = 30$ m, $V_M = 1.609V$;
for $z = 100$ m, $V_M = 1.83V$.

In all cases, the contribution of the second term is small: 0.4% at 10 m, 0.7% at 30 m and 2.2% at 100 m. It may be omitted in most practical cases. It follows also that a variation in z_g does not significantly affect velocities.

This UK example has been chosen because it reflects recent work in the area. Dyrbye and Hansen (1997: 35f) quote from Eurocode I (see Section 4.6) which defines a reference wind velocity as the 10-minute mean wind velocity at 10 m above a terrain with a roughness length of 0.05 m, and an average recurrence interval of 50 years. This reference wind differs from that used by BS 6399; although the recurrence interval is the same, the British Standard refers to the hourly mean wind speed (see for example, Cook and Prior 1987: 385). Some of the Eurocode values are: Southern Sweden, 24–26 m/sec; Denmark 27 m/sec; coastal areas of Germany, 32 m/sec; and coastal areas of Greece and the Greek islands, 36 m/sec. Dyrbye and Hansen query the range of the first three of these figures, for, 'in the Baltic Region, the wind climate is generally homogeneous'. The closest contour in the British data is around Norwich, with a basic wind speed of 24 m/sec.

The AASHTO specification (1994: Cl.3.8) specifies a base design wind velocity of 160 km/hr (44.4 m/sec), adjusted for heights in excess of 10 m by the expression

$$V_{DZ} = 2.5V_o\,(V_{10}/V_B)\ell n(z/z_o)$$

where

> V_{DZ} is the design wind velocity (equivalent to V_M above),
> V_o is the friction velocity (= u),
> V_{10} is the wind velocity 10 m above the ground (= V),
> V_B is the base wind velocity, equal to 160 km/hr (44.4 m/sec),
> z is the height, and
> z_o is a friction length (as above)

The terms V_o and z_o are tabulated for a few cases. For example, for 'open country', V_o is 13.2 km/hr (3.67 m/sec) and z_o is 0.07 m. The latter figure lies between the roughness categories 2 and 3 given above. Rearranging the AASHTO expression,

$$V_{DZ} = 2.5 V_{10}(V_o/V_B)\ \ell n(z/z_o)$$

or, for

$$V_o = 3.67 \text{ m/sec},\ V_B = 44.4 \text{ m/sec}$$
$$V_{DZ} = 0.206 V_{10}\ \ell n(z/z_o)$$

This has the same form, but is a little larger (0.206/0.172 = 1.2×) than the expression for V_M given above, with the term in z_g omitted. The difference results at least partly from the larger recurrence interval used in bridge design. There is also some difference in the treatment of gusts. The AASHTO Specification does not specify gust velocities, but includes the effect of gusts in the expression relating wind pressure to wind velocity. Whereas the stagnation pressure associated with a wind velocity of 160 km/hr (44.4 m/sec) is 1.23×10^{-3} MPa (code commentary), the specified pressures are somewhat larger:

trusses, columns and arches,	windward side	0.0024 MPa
trusses, columns and arches,	leeward side	0.0012 MPa
beams		0.0024 MPa
large flat surfaces		0.0019 MPa

but not less than 4.4 kN per metre for a beam, or for a windward chord member, and not less than 2.2 kN/m for a leeward chord. These may be adjusted in proportion to the ratio $(V_{DZ}/160)^2$, with V_{DZ} in km/hr.

Although Cook (1985: Part 1: 89) includes tornadoes in his wind types in the United Kingdom, the relative severity of these occurrences is small compared with other countries. The Australian wind code (AS1170.2 1989), for example, specifies higher design wind velocities in coastal areas north of latitude 27°S, to allow for the effects of tropical cyclones. Cyclone Tracy passed over the city of Darwin in December 1974. The maximum gust wind velocity (Walker 1975) was estimated as 70 m/sec.

Simiu and Scanlan (1996: 77f) give data on the variation of wind speed

Table 12.2

Height (m)	*Mean velocity*	*Maximum 1-min velocity (m/sec)*
11.3	14.5	22.8
22.9	18.7	29.1
45.7	24.1	35.8
108.2	29.1	42.9

Table 12.3

Height (m)	*Wind speed (m/sec)*
9.1	32.5
59.7	39.5
191.4	47

and height for six tropical cyclones, the first two in the American region and the other four in Northern Australia. The first of these was the American hurricane Carol (1954) during its decaying stages. Recorded wind speeds were as given in Table 12.2.

For the Australian cyclone Beverley (March 1975), the values of the maximum wind speeds over a period of 10 min, as given in Table 12.3, were recorded.

In general, the wind speeds displayed a logarithmic variation with height, of the kind discussed previously, but there is evidence that this is not always the case.

Cook (1985: Part 1: 127) discusses the *probability* of occurrence and concludes that, in most regions, cyclone incidence is too low to allow reasonable estimates, with the exception of Hong Kong, 'where the contribution from typhoons dominates all other mechanisms and the recorded annual maximum wind speed is nearly always from this source'. It must be remembered that tropical cyclones have very local effects, their paths are erratic, and their intensity diminishes rapidly as they pass from the sea to the land.

Holmes (1990), in a commentary on the Australian Standard for wind loads (AS1170: Part 2: 1989), quotes the maximum gust velocities at a height of 10 m for tropical cyclones of varying intensities, as listed in Table 12.4.

Table 12.4

Category	*Description*	*Central pressure (HPa)*	*Maximum gust velocity (m/sec)*
1	mild	>990	20–30
2	moderate	970–985	35–45
3	severe	950–965	50–60
4	very severe	930–945	65–75
5	catastrophic	<920	80–90

Table 12.5

Average Recurrence Interval (ARI) (years)	*V (m/sec)*
20	45
50	51
1000	70

Table 12.6

ARI Years	*V (m/sec)*
20	38
50	41
1000	50

Based on these values, they suggested the velocities given in Table 12.5 for coastal regions in the north of Australia.

For comparison, Table 12.6 gives typical design wind speeds elsewhere in Australia.

These are noticeably higher than values quoted earlier for the United Kingdom, which were for a 50-year value of the ARI, and re-emphasise the point that design wind velocities must be viewed geographically. Simiu and Scanlan (1996: 122ff) also discuss the probability of occurrence of tornado winds in the United States.

12.2 Wind forces

The basic expression for wind force is

$$F = C_p A(\rho V^2/2)$$

where

C_p is a pressure coefficient,
A is an area,
ρ is the air density, and
V is the design wind velocity

The air density, ρ, varies with temperature and pressure. Cook (1985: Part 1: 125, 127) suggests the use of $\rho = 1.225$ kg/m^3 for temperate regions, and 1.122 (corresponding to a temperature of 25°C, and an atmosphere pressure of 960 MPa) in tropical cyclones; the corresponding stagnation pressures, $\rho V^2/2$, are $0.613V^2$ and $0.561V^2$ (Pa).

The British code for highway bridges, BD37/88, gives values of a drag coefficient, C_D, for many practical cases. Figure 12.2 shows typical examples of

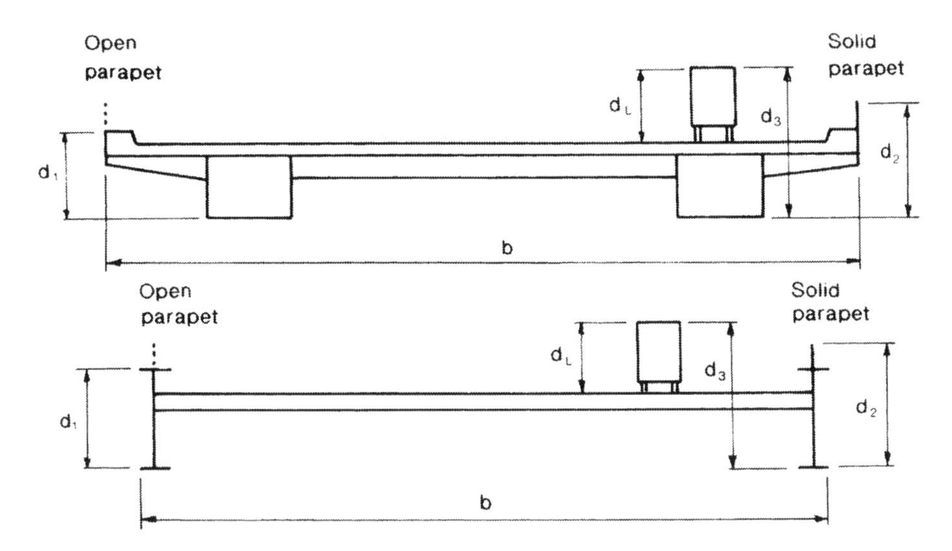

Fig. 12.2 Dimensions used to compute wind pressure in BD37/88

what is called 'a superstructure with a solid elevation'. A live load of height, d_L, may be present over part of the structure, where d_L is taken as 2.5 m for a highway, 3.7 m above rail level for a railway, and 1.25 m for a footway or cycle track. The depth of the bridge plus live load is d_3. The structural depth is regarded as d_1 (at the left) for the case with open parapets, and d_2 where the parapets are solid. Then the area A is taken as the projected area in elevation, using dimensions d_1, d_2 or d_3, whichever is appropriate. The drag coefficient, C_D, is expressed in terms of a ratio b/d, where d is defined as follows. If the structural depth, d_1 or d_2, exceeds d_L, then $d = d_1$ or d_2. If both depths d_1 and d_2 are less than d_L, then for the case with live load, $d = d_L$. Figure 5 of the code plots C_D against b/d. For most structures, C_D lies between 1.3 and 2.76, with some typical values listed in Table 12.7.

The limit of 1.3 corresponds to $b/d = 6.0$. For foot or cycle track bridges, C_D is to be not less than 2.0, where the limit corresponds to b/d a little in excess of 1.1. A value of C_D equal to 2.0 gives

$$F/A = 1.226V^2 \text{ Pa}$$

Table 12.7

b/d	C_D
0.2	1.87
0.4	2.2
0.65	2.76
1.0	2.14
1.5	1.75
2.0	1.5

Table 12.8

Solidity ratio	C_D
0.1	1.9
0.2	1.8
0.3	1.7
0.4	1.7
0.5	1.6

For a design wind velocity of 40 m/sec, this equivalent pressure is 1.96 kPa. The corresponding load, F, would be relatively small compared with the capacity of the structure to transmit it to the supports.

BD 37/88 also provides for bridges with an open superstructure, such as the older, metal truss bridges. The effective area, A, is evaluated as the sum of contributions from the windward and leeward trusses, the deck and the parapets, except that the deck area screened by the windward truss is deleted, as is the leeward truss area screened by the deck, with similar provisions for the screening of parapets. Drag coefficients for a truss depend on two factors: (a) the solidity ratio, which is equal to the ratio of the actual area seen in elevation divided by the overall space occupied by the truss; and (b) the shape of truss members. For flatsided members, C_D has the values given in Table 12.8.

There is also some shielding allowed of the leeward truss, dependent on the solidity and spacing ratios, where the latter equals the distance between the centrelines of the trusses, divided by the truss depth. A few of the code values are listed in Table 12.9.

In many actual cases, this shielding ratio would be in the range, 0.8 to 1.0. This data is also useful in the case of other structures, such as cable-stayed or suspension bridges. For a circular stranded cable, $C_D = 1.2$. The deck has a value of $C_D = 1.1$. The height of any live load on the deck is considered to be the same as for a superstructure with a solid elevation, but with a value of $C_D = 1.45$ for unshielded parts of the live load.

These provisions do allow for wind loads on traffic in the sense that the traffic adds to the effective area of the bridge in elevation. The effect of the wind on the traffic itself is another matter; a driver can compensate for the effects of a steady wind, but difficulty arises in three ways: (1) a high vehicle may be overturned by wind; (2) gusts may apply sudden forces to a vehicle, causing it to

Table 12.9 Values of C_D

Spacing ratio	*Solidity ratio*			
	0.1	*0.2*	*0.3*	*0.4*
1 or less	1.0	0.9	0.8	0.6
2	1.0	0.9	0.8	0.65
3	1.0	0.95	0.8	0.70

be difficult to control; and (3) moving vehicles may be at one moment shielded by the structure from the effects of wind and then move suddenly into open space.

Gawthorpe (1994) describes a number of incidents where trains have been blown over; for example, in 1903 on the Leven Viaduct in the north-west of England, and in 1986 on a narrow gauge railway across the 40 m-high Amarube Bridge. Narita and Katsuragi (1981) cite another example. In October 1976 a small, unloaded truck was overturned as it left a bridge in Japan, at a time when the maximum wind speed was estimated to be in excess of 30 m/sec. Field observations and wind tunnel studies showed that the direction of wind flow was affected by the local topography. Traffic leaving the bridge in the right-hand lane had, on their left, a fall of about 50 m to the sea, with a steep cutting on their right rising to about 20 m, with a length of 180 m. An access road joined the highway from the left, opposite this cutting, and it was about here that the accident occurred. The wind tunnel tests showed that for certain wind directions, the transverse component of wind changed suddenly in direction at the access road, from coming from the driver's left to a direction from his right.

Baker and Coleman have carried out analytical and wind-tunnel investigations of the possibility of vehicles being overturned or failing in other ways at high winds. An initial study (Baker 1986) indicated that a particular, unladen bus could be overturned, theoretically, by wind velocities of the order of 20 m/sec, although 'operational experience would suggest ... perhaps about 30 m/sec'. Later work (Coleman and Baker 1990) described wind tunnel tests on a 1/50th scale model of 'a typical large articulated tractor-trailer combination' (the equivalent full-scale height would be 2.85 m) travelling on a box girder bridge, with two kinds of flow: a low-turbulence wind-tunnel flow, and a turbulent flow intended to provide 'a simulation of atmospheric turbulence'. The equivalent full-scale maximum gust velocity for the turbulent wind appears to have been about 25 m/sec. Overturning in this wind had a probability of occurrence of 20 hours per year. Other work in this area can be found in Baker (1994) and Hucho (1987). Cooper (1981) and Baker (1981) have also done work on the stability of trains.

Wyatt (T.A. *Recent British Developments: Windshielding of bridges for traffic*, 159–70, in Larsen 1992) has a useful, recent discussion of the shielding of traffic from the effects of wind. He refers to experience on the Severn and Forth suspension bridges, and quotes current rules for the Severn bridge that require lane closures to be considered for gust speeds in excess of 15 m/sec, and applied if gust speeds reach 18 m/sec. Similarly, closure of the bridge to vulnerable traffic, such as holiday caravans, high-sided light commercial vehicles, and high-sided unladen heavy vehicles, is considered at gust speeds of 20 m/sec and put into effect at 22.5 m/sec. There were also earlier speed restrictions, to as low as a 32 km/h vehicle speed for gust velocities in excess of 23 m/sec, with provisions to restrict traffic to only one lane in each direction 'to increase the margin for lateral course deviations'. About 1987, these restrictions meant that lane closures were applied on this bridge about 15 times per year, totalling some 170 hours, with closure to vulnerable vehicles averaging some 60 hours per year. It can only be concluded that these restrictions constitute a failure by the bridge to perform its proper function, and it is significant to observe that Wyatt then goes on to discuss traffic protection.

The wind barriers considered by him were porous, made up of horizontal slats of various widths at a centre-to-centre spacing of 300 mm. The corresponding porosities (ratio of openings to total area) varied from 0.2 to 0.8. Barriers of various heights were considered, mounted on model bridge sections for wind tunnel tests. The aim was to reduce the overturning moment on a vehicle 4.4 m high by about 50%. One suitable configuration had a porosity of 0.4 and a height of 3 m. Figure 12.3(a), taken from Larsen (1992: 161), shows the hori-

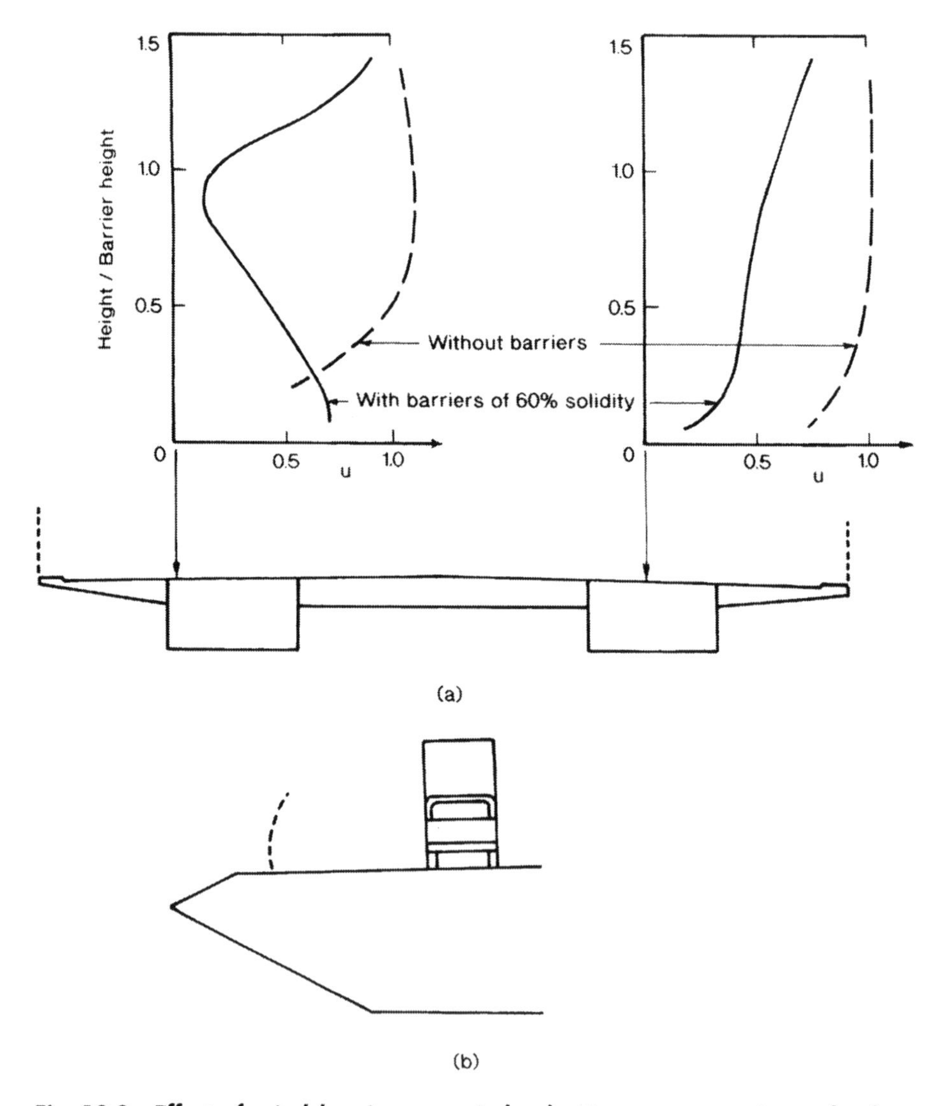

Fig. 12.3 Effect of wind barriers on wind velocities at two sections of a box girder bridge (Redrawn from Larsen 1992: 161)

zontal wind velocity plotted against height at two locations, with and without the barriers. Velocities are scaled against the approach wind velocity, V, with heights scaled against the barrier height. The deck width is about 36 m.

Referring first to the graphs shown as 'without barriers', it is evident that wind velocities to a height of about 4.5 m are close to V. The graphs 'with barriers' differ. For the upwind graph, the velocity reduces from about $0.7V$ close to the deck to a minimum of about $0.125V$ just below the top of the barrier, and then increases towards V at $1.5 \times$ the barrier height. The downwind curve is different in shape, with a velocity of about $0.45V$ at one half the barrier height.

Wind screens are also discussed in the same volume by K.H. Ostenfeld and A. Larsen ('Bridge engineering and aerodynamics', 3–22, in Larsen 1992: 9, 14), with particular reference to Denmark's Great Belt bridge; the proposed barriers are shown in Fig. 12.3(b). They also conclude that the porosity of the barriers should be in the range 0.4 to 0.5, and state further that screens 'of 0.5 porosity can be arranged on 'streamlined' box girders with little if any penalty to the aerodynamic stability. An appropriate air-gap must however be allowed for between the bottom member of the screen and the deck . . .' (see Section 12.4). They also show a cross-section proposed by Aztiz and Anderson for an 'ultra-long' span suspension bridge across the Straits of Gibraltar. This has a deck completely enclosed in an ellipse, with a horizontal major axis equal to about three times its height. The deck surface lies at about one-third of the height above the base. The entire space below the deck forms a structural tube. Much of the top of the ellipse forms an upper flange, connected to the lower tube by vertical trusses at each side. The traffic is completely enclosed between the lower tube, the upper flange and the trusses.

Gawthorpe (1994) observes that 'problems with cars and smaller goods vehicles are mainly of the Course Deviation type'. A simple numerical example may be used to give some idea of the nature of the problem. Consider the case of a moving car of mass 1200 kg, with an area in the side view of 4.5 m^2 and an assumed transverse drag coefficient of 1.2. A transverse wind velocity of 20 m/sec would give, on the basis of the simplest assumptions, a transverse force of 660 N. It may be expected that a rolling support would provide only a small resistance to transverse load. Consider a time span of 1 sec. Then an acceleration of 660/1200, or 0.55 m/sec^2, acting over this period, on a body with no initial transverse velocity would cause a transverse displacement of 0.27 m. If the car has a longitudinal velocity of 100 km/h, or 27.8 m/sec, its longitudinal displacement would be 27.8 m. For this length, the previous formulae would give a gust factor, $g_t = 2.67$, corresponding to a transverse wind velocity of 53 m/sec and a transverse force of 4700 N, an increase of 4040 N from the figure used above. Two cases arise. If the car moves from a completely shielded area into a gust of 53 m/sec, then in 1 sec, its transverse displacement is increased from 0.27 to 1.96 m. If, instead, it is the change in force that is significant, then the displacement would be 1.69 m. It is not claimed that these estimates are accurate, but they are sufficient to show that a real problem may exist. Wyatt (in Larsen 1992: 164) refers to the problem but indicates that further work needs to be done. He also (p.160) states that 'about half the wind-related accidents on the Severn bridge are believed to result on passing the

Table 12.10

Beaufort number	*Description*	*Wind speed (m/sec)*	*Effects*
5	Fresh breeze	8.0–10.7	Limit of agreeable wind on land
6	Strong breeze	10.8–13.8	Difficult to walk steadily
7	Near gale	13.9–17.1	Inconvenience felt when walking
8	Gale	17.2–20.7	Great difficulty with balance in gusts
9	Strong gale	20.8–24.4	People blown over by gusts

tower'; that is, because of intermittent shielding and the deviation of vehicles from a regular path.

Another area of interest is in the application of wind forces to pedestrians. Melbourne (in Aynsley, Melbourne and Vickery 1977: 154ff) has described a situation where wind caused people to be blown over in a building forecourt in Melbourne.

> The wind was gusting regularly up to 20 m/sec which caused all people in the area great difficulty with balance.... On separate occasions two girls were unbalanced to the extent that they came down on their hands and knees.... The maximum gust velocity ... was 23 m/sec rising from about 12 m/sec in 2–3 sec. It is concluded ... that wind gusts of 20 m/sec will unbalance and seriously inconvenience a number of people and wind gusts of 23 m/sec and over will cause people to be blown over. Obviously a person can stand in a much higher steady wind velocity but it appears to be the unexpected nature of the peak gust which catches most people off guard.

Melbourne also quotes observations by Penwarden (1973) concerning the effects of winds classified on the Beaufort scale. Some of these are listed in Table 12.10. The wind speeds are peak gust velocities. The data supports the earlier observations.

12.3 Buffeting

The term 'buffeting' is usually applied to effects that result when one structure is built in the wake of another (Cook 1985: Part 2: 362; Houghton and Carruthers 1976: 181; but see Dyrbye and Hansen 1997: 155 and Simiu and Scanlan 1996: 461). As far as bridges are concerned, one early study was by Scruton, Woodgate and Alexander (1955; see also Scruton 1981 and Scruton and Flint, 1964) in connection with the design of the third Runcorn–Widnes bridge (Anderson 1964; O'Connor 1971a: 499, 501). The first bridge here was a railway bridge with lattice trusses, still in use in 1964. In 1905 an adjacent transporter bridge was opened for highway traffic, but this needed to be replaced. The first proposal for its replacement was a suspension bridge, with a main span of 314 m; indeed, Telford had in 1814 proposed a 305 m-span suspension bridge at this site (Smiles 1862: Vol. 2: 446). Wind tunnel tests were carried out for this new proposal. To quote Anderson (1964):

> Tested alone, the suspension bridge sectional model was found to be aerodynamically stable. The surprising feature of the results came when it was tested alongside the model of the existing railway bridge. The suspension bridge oscillated severely, owing to the buffeting effect of vortices set up in the windstream by the bluff shape of the railway bridge. The oscillation was most severe at a wind speed of 60 miles/h (26.8 m/sec). The maximum amplitude was 6 in. (152 mm) when no train was on the railway bridge, but this could be increased to 12 in. (305 mm) by the presence of a train. As far as the Author is aware, this is the first case of aerodynamic buffeting arising in connection with two adjacent bridges.

The proposed suspension bridge was replaced by a cantilever arch with a 330 m central span.

Scruton and Flint (1964: 679f), in a general discussion of 'Wind-Excited Oscillations of Structures' present 'a typical diagram of the variation of buffeting amplitude with wind speed', taken from studies carried out for the Runcorn–Widnes and Tamar Suspension Bridges. It shows a maximum amplitude of about 325 mm at a wind speed of 23 m/sec (both figures differ a little from those given by Anderson). However, the significant point lies in the variation of amplitude with wind speed. For a speed of 19 m/sec, the amplitude was 61 mm, reducing to 20 mm at 15 m/sec. There was a similar but less marked reduction for wind speeds in excess of 23 m/sec: to 185 mm at 27 m/sec, and to 107 mm at 38 m/sec; it then increased slightly to about 142 mm at 61 m/sec. In other words, 23 m/sec was, in effect, a critical wind velocity. Houghton and Carruthers (1976: 182), referring to the same case, state 'the maximum buffet amplitudes ... increased linearly as the horizontal separation was increased up to 40 m'. Presumably they then reduced.

This buffeting phenomenon is complex and in particular cases may need to be explored by wind tunnel tests. There is, however, a basic component that needs to be described. The simplest cross-section of the upstream structure is a circular cylinder. This case is important in chimney design, and has been adequately explored (see, for example, Cook 1985: Part 1: 35; Dyrbye and Hansen 1997: 110ff; Scruton and Flint 1964; Simiu and Scanlan 1996: 148ff). At low velocities the flow remains attached to the cylinder (Reynold's Number, $RN \approx 1$). For a larger velocity ($RN \approx 20$) the flow remains symmetrical, with eddies on either side of the central plane immediately beyond the cylinder. At higher velocities ($RN > 30$), these eddies are shed in the form of vortices that leave the cylinder alternately on either side of the central plane. The resulting vortex trail is commonly referred to as the von Karman vortex street, but it was Strouhal who, in 1878, observed its regularity. The Strouhal Number (SN), dimensionless, is defined as

$$SN = \frac{fD}{V}$$

where

> f is the frequency of full cycles of this vortex shedding,
> D is a characteristic dimension, here the cylinder diameter, and
> V is the approach velocity

For a circular cylinder, SN is constant and close to 0.2 for RN from 30 to at least 10^7 (Scruton and Flint 1964: Figure 2; Simiu and Scanlan 1996: Figure 4.4.4).

Reynold's Number was defined previously (Section 11.3) as

$$\text{RN} = VL/\nu$$

where

> L is a characteristic dimension, and
> ν is the kinematic viscosity

For air at 20°C, ν is of the order of 15×10^6 m²/sec. Then

$$\text{RN} = 67000VL$$

A Reynold's Number of 10^7 corresponds to $VL = 1493$ m²/sec, or for $L = 30$ m, $V = 50$ m/sec. Many practical structures lie in this range. The alternate vortices cause pressures on the cylinder that have an alternating transverse force resultant. Similarly, structures in the wake are subject to transverse forces (perpendicular to V) with a frequency,

$$f = \text{SN}.V/D$$
$$= 0.2V/D \text{ for a circular cylinder}$$

Consider, for example, the case with $V = 20$ m/sec; then f has the typical values given in Table 12.11.

Simiu and Scanlan (1996: 152) list values of S for a range of structural sections; S ranges from 0.114 to 0.20. For a plate parallel to the stream flow, with a thickness equal to 0.03 × the width, $S = 0.156$. It follows that for a bridge section, one would also expect S to lie in about this range.

Table 12.11

D(m)	*f (cycles/sec)*
1	4
5	0.8
10	0.4
20	0.2
30	0.13

Now the point of significance about these forces is not so much their magnitude but their frequency. If this frequency coincides with that of the structure, then structural vibrations may initiate and then be progressively amplified, with their limiting value controlled by structural damping. Scruton and Flint (1964: 681) point out that 'oscillations occur when the energy input due to wind exceeds that dissipated by damping'. Goswami, Scanlan and Jones (1993) discuss 'Vortex-Induced Vibration of Circular Cylinders'; they refer to the Scruton number which relates fluid parameters to the level of mechanical damping, and plot experimental values of the maximum amplitude of vibration against this number. This treatment will not be repeated here, for major bridge cross-sections are not circular. It may, however, be relevant in considering component parts of a structure, such as the towers of a cable-stayed girder bridge. Indeed, one of the cases discussed by Scruton and Flint (1964) was concerned with structural vibrations of the towers of the Severn River suspension bridge (completed in 1966), for the earlier Forth Road Bridge (1964) (see O'Connor 1971a: 373, 379–83) suffered oscillations in its towers during erection. With chimney-stacks, vibrations are often controlled by the addition of aerodynamic spoilers, whose addition 'is often more feasible or more acceptable than changes of the natural frequencies or increase of the structural damping' (Scruton and Flint 1964: 688ff). Scruton was one of those (with Walshe) who developed the idea of welding projecting plates helically around a chimney to act as spoilers, as is commonly seen today.

Another case of interest to the bridge designer lies in the oscillation of cables in major bridges, but here the problem is influenced also by the formation of rain rivulets. So-called 'rain-wind induced vibration' is discussed by Matsumoto et al. (1995). A similar problem has occurred on the recent Glebe Island bridge in Sydney, which has plastic tubes encasing and protecting the cables. Unfortunately, certain combinations of wind and rain cause these tubes to vibrate, hitting the cables and creating a noise disturbance to local residents.

12.4 Aerodynamic instability

Under some conditions a bridge deck may move vertically under the action of a horizontal wind. Oscillations develop that may prove to be catastrophic, as in the failure of the first Tacoma Narrows bridge in 1940 (Section 2.4). The problem is called *aerodynamic instability*. Although it had been observed earlier (see for example Provis 1842; Russell 1841), modern research into this aspect of bridge design effectively began in 1940. It affects chiefly long-span bridges, notably suspension bridges with a relatively low vertical stiffness. A few reference works are Larsen (1992), and chapters in Cohen and Birdsall (1980); Dyrbye and Hansen (1997); Kolousek et al. (1984) and Simiu and Scanlan (1996). There have been conferences on the subject, such as by the Institution of Civil Engineers, London (1981). O'Connor (1971a) refers briefly to the problem and gives statistics for twenty suspension bridges, such as the Golden Gate bridge, San Francisco, completed in 1937 with spans of 343, 1280 and 343 m; the Verrazano Narrows bridge, New York, of 1964, with 370, 1298 and 270 m spans; and the 1966 Severn bridge, with spans of 305, 988 and 305 m and an unusual deck cross-section. Larsen (1992) includes papers on the recently completed Great Belt bridge, joining the Danish islands of Funen and Zealand, with a total length of 18 km, and suspension spans of 535, 1624 and 535 m; and the Japanese Akashi Kaikyo suspension bridge, with spans of 960, 1990 and 960 m. A useful summary of recent Japanese suspension bridge construction may be found in Burden (1991), with a map showing their locations (see also Larsen 1992: 72). Many form part of the project to link the island of Shikoku with Honshu, such as the Kurushima bridges, which form a sequential series of three suspension bridges with common intermediate anchorages, and spans of 150, 610 and 170 m; 230, 1010 and 250 m; and 250, 1030 and 370 m; totalling 4070 m. These examples show maximum spans increasing from 1280 to 1298 m between 1937 and 1964, and then, more recently, to 1624 and 1990 m. There have been also other large spans, such as across the Humber, in the United Kingdom (1980, 1410 m).

The analysis of aerodynamic instability in suspension bridges dates from about 1940, but the related problem of instability in aircraft structures was observed somewhat earlier; for example, by Fokker with his D-8 single-wing aircraft during the 1914–18 war. It is appropriate, therefore, to refer also to such books as Bisplinghoff et al. (1955, see his historical note: 3ff); Bisplinghoff and Ashley (1962); Dowell et al. (1995); Fung (1955); Scanlan and Rosenbaum (1968). In aircraft design, the term commonly used for the phenomenon is *aeroelasticity*.

This problem also extends far beyond the scope of this book, and combines both structural and wind effects; but a brief introduction may be useful. Some of the relevant factors are:

(a) the vertical and torsional stiffnesses of the bridge, and the consequent values and relative magnitudes of its natural frequencies;
(b) the aerodynamic characteristics of the deck cross-section;
(c) the wind velocity; and
(d) the level of structural damping.

The typical suspension bridge has main cables stiffened by a structural system at deck level whose purpose is three-fold: to transmit loads from the deck to hangers connected to the main cables, to reduce local deformations in the deck, and to control aerodynamic movements.

Traditionally, the most common form of stiffening system has been the truss, with two primary stiffening trusses placed in the vertical plane below the main cables, and connected by the deck and some horizontal wind bracing system. In some earlier bridges, such as the Golden Gate, there was only one horizontal bracing system. It is now normal to use two horizontal trusses, in the planes of the upper and lower chords of the main trusses, to give what may be called a closed system that is torsionally stiff. The list by O'Connor (1971a:

373, 4) of twenty bridges, dated from 1931 to 1967, had sixteen with stiffening trusses. Of the remainder, one had a prestressed concrete box girder, the Severn bridge (1966, 988 m) had what may be called a tubular girder of streamlined cross-section, and two – at Rodenkirchen (1954, 378 m) and the original Tacoma Narrows bridge (1940, 853 m) – had plate girders (with depths of 3.3 and 2.4 m respectively). The cross-section of the Tacoma Narrows bridge is shown in Fig. 12.4(a). The distance centre-to-centre of the stiffening girders is 11.9 m.

The tubular section used for the Severn bridge was a direct outcome of the Tacoma Narrows failure; see Fig. 12.4(b). It is of welded steel, with a 4.5 m footpath on either side of a roadway of 19.8 m total width. The width including the paths is 31.9 m. The depth is 3.05 m. This tubular stiffening system was torsionally stiff and of a form shown to be aerodynamically suitable. It is evident, however, that traffic on the bridge is exposed to wind, for the footpaths are set down and the crash barriers consist only of wire ropes. Many later bridges have followed the example of the Severn bridge, such as the Lillebælt bridge in Denmark, completed in 1970 with a main span of 600 m (Fig. 12.4(c)); the bridge over the Bosporus near Istanbul in Turkey, 1973, 1074 m; and the Humber bridge, 1980, 1410 m.

With this background, it is of interest to observe the structural form of two major bridges completed recently. The Great Belt bridge (1624 m span) has the tubular cross-section shown in Fig. 12.4(d), whereas the Akashi Kaikyo bridge, whose 1990 m span is now the largest in the world, has returned to the stiffening truss (Fig. 12.4(e)).

In the period following 1970, the common method used to study the interaction between aerodynamic and structural parameters was by wind tunnel tests on a sectional model, supported so as to reproduce natural vibration frequencies of the complete structure (in a vacuum). Figure 12.5 shows a spring-mounted sectional model, whose length perpendicular to the paper is finite, but sufficient to model the aerodynamic behaviour of the section. The deck and stiffening members are shown in idealised form, but would in fact be accurate models of the full-scale cross-section. It may be shown (O'Connor 1971: 425f) that this model may vibrate in a vacuum with two basic modes: (a) vertical translation, and (b) torsion, with natural frequencies (cycles per second), N_v and N_θ, given by:

$$N_v = \frac{1}{2\pi}\left(2\frac{k_s}{M}\right)^{\frac{1}{2}}$$

$$N_\theta = \frac{1}{2\pi}\left(\frac{k_s b^2}{2I_o}\right)^{\frac{1}{2}}$$

where

k_s is the stiffness of each supporting spring,
M is the total mass, and
I_o is the mass moment of inertia for rotations about 0

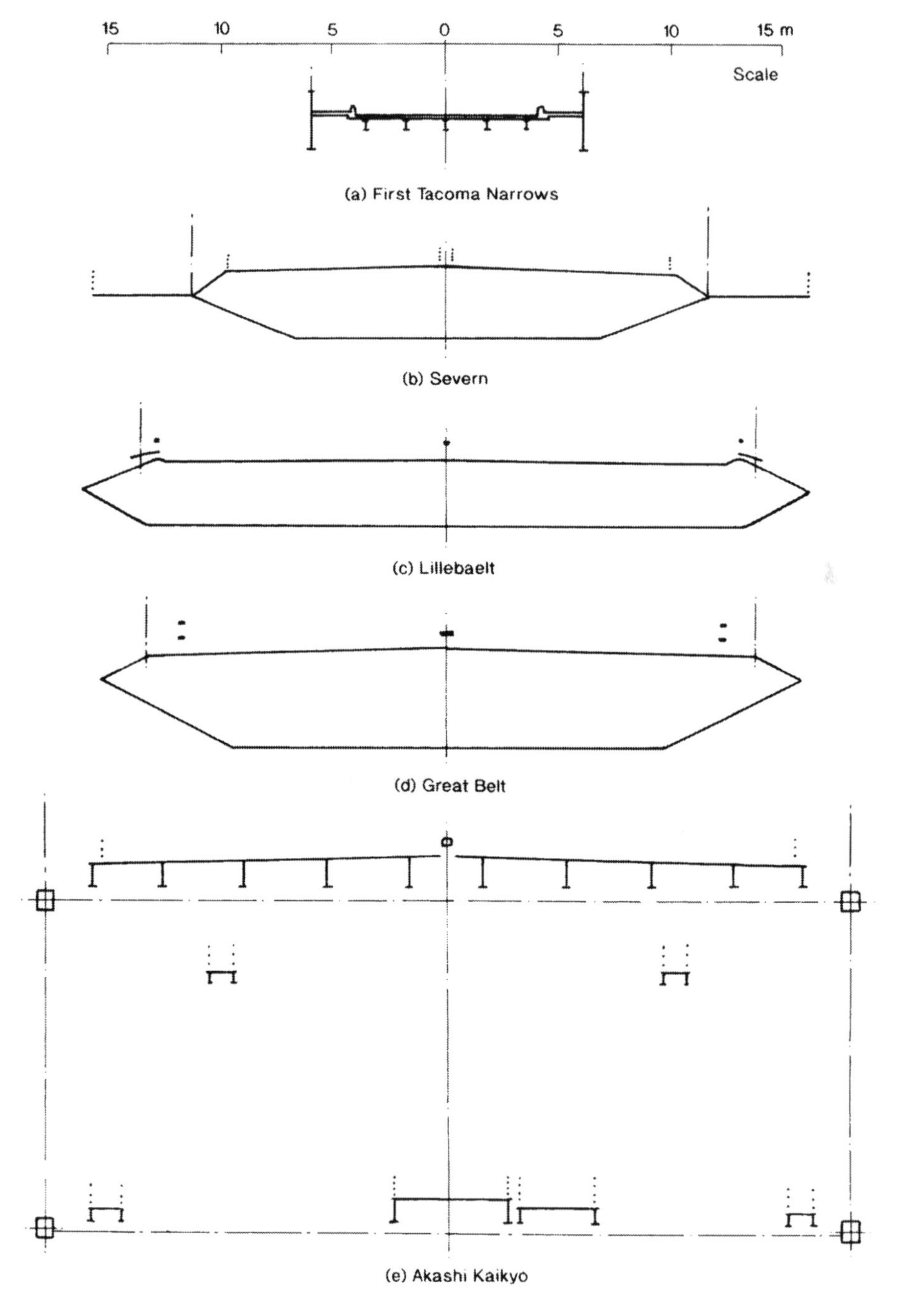

Fig. 12.4 Deck cross-sections for five notable stiffened suspension bridges (all drawn to the scale shown)

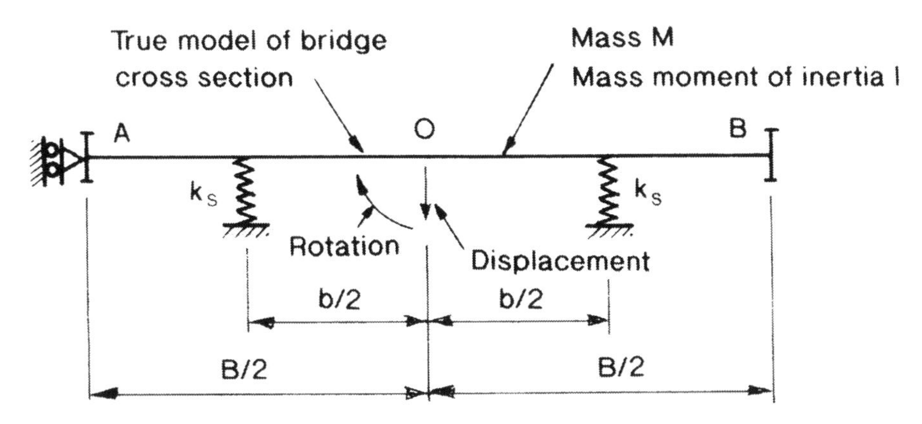

Fig. 12.5 Spring-mounted sectional model

It follows that

$$\frac{N_\theta}{N_v} = \left(\frac{b^2 M}{4I_o}\right)^{\frac{1}{2}}$$

For undamped, steady state vibration in either mode, the motion is simple harmonic, with a typical displacement given by:

$$v_A = v_{AO} \sin 2\pi\ Nt$$

where

v_{AO} is the half-amplitude of displacement v at point A,
t is the time, and
N is the frequency, N_v or N_θ

The model will be built to reproduce the deck cross-section to some scale. The parameters, M, I, k_s and b (the spring spacing) may then be chosen so as to reproduce to an appropriate scale N_v and N_θ, the vibration frequencies of the complete structure. In doing this, the following six dimensionless parameters must be kept constant, for aerodynamic similarity between the model and its prototype (Scruton 1952):

(a) $\dfrac{m}{\rho B^2}$ and $\dfrac{I_o}{\rho B^4}$

(b) $\dfrac{V}{N_v B}$ and $\dfrac{V}{N_\theta B}$

(c) δ_v and δ_θ.

The first two relate the inertias of the model and prototype, with ρ being the air density. In the second pair, the values V/N_v and V/N_θ represent the movement of a particle of air in the undisturbed airstream during the time required for one complete oscillation. The third pair require damping, as expressed by the logarithmic decrements, δ_v and δ_θ, to be the same. The model width, B, represents the model scale. For this scale, the model inertias, m and I, may be determined from the corresponding values for the prototype. The second pair of scaling parameters allow the wind velocity, V, to be determined, while the third pair define the level of damping to be provided by dashpots or other devices at the model springs.

Prototype values of N_v and N_θ, and estimates of δ_v and δ_θ, may be determined by conventional structural analyses. Provided that the parameters for similarity are satisfied, the model frequencies may differ from those in the prototype, but the ratio, N_θ/N_v must be the same. This ratio becomes, therefore, an important parameter in the assessment of aerodynamic behaviour. Values of N_v and N_θ, and their ratio, are listed in O'Connor (1971a: 439) for particular bridges. Some of these are listed in Table 12.12.

The figures for the Golden Gate bridge, which has stiffening trusses, reflect the fact that, when originally built, it had upper laterals only. After the Tacoma collapse, a lower lateral system was added, with an increase in the frequency ratio from 1.25 (as Tacoma) to 1.96. The Verrazano Narrows bridge also has stiffening trusses and upper and lower laterals; its frequency ratio is 1.92, and the range, 1.9–2.0, appears to be typical for suspension bridges of this age and type. The Forth bridge (1964, 1006 m) has the somewhat higher value of 2.78, but there is then, the major increase to 3.97 in the tubular Severn bridge.

Structural damping is also of some importance. Movement of the First Tacoma Narrows bridge was anticipated before its erection and monitored in the period from its opening on 1 July to its failure on 7 November 1940 (Farquharson 1949). Certain remedial measures were carried out: central ties were inserted between the main cables and the stiffening girders about 1 June, end buffers were inserted between the main towers and the roadway on 28 June, and tie-downs in the side spans on 4 and 7 October 1940. The second of these was essentially a damping system; Farquharson writes, 'as the span moved longitudinally, oil was forced through a pipe line, the flow being checked by a needle valve'. The wind velocity at failure was about 19 m/sec (42 mph).

Wind tunnel tests on section models of the Severn bridge before its construction indicated 'a possible minor instability caused by eddy shedding at a

Table 12.12

Bridge	N_v	N_θ	N_θ/N_v
First Tacoma Narrows	8.0	10.0	1.25
Golden Gate (upper laterals only)	5.6	7.0	1.25
Golden Gate (upper and lower laterals)	5.6	11.0	1.96
Verrazano Narrows	6.2	11.9	1.92
Forth	7.6	21.1	2.78
Severn	7.7	30.6	3.97

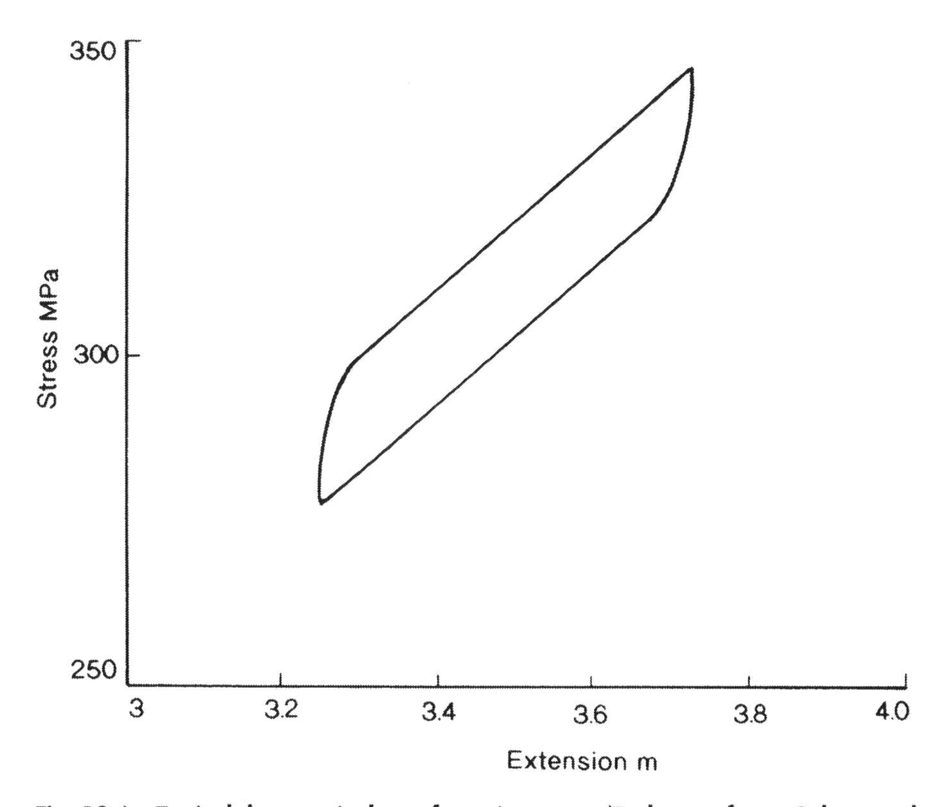

Fig. 12.6 Typical hysteresis loop for wire rope (Redrawn from Cohen and Birdsall 1980: 18)

narrow band of wind speeds around 78 mph [35 m/sec]' (Brown, W.C. *Long-span Suspension Bridges: A British Approach*, 1–26, in Cohen and Birdsall 1980: 18). The complete structure was welded, with little structural damping. It was decided to use inclined hangers (65° to the horizontal) whose axial loads would change during oscillation of the bridge. Figure 12.6 shows a typical hysteresis loop; the separation of the loading and unloading stress–strain curve for the cable implies a loss of energy, and damping.

The deck of a suspension bridge is commonly fabricated in units, lifted successively into place by equipment mounted on the main cables. For some structures it would be possible to attach these units together in such a way that they are hinged about a transverse axis, with relative rotations controlled by dashpots and suitable springs; for example, in footbridges of long span. Tuned mass dampers may also be used (Livesey and Sondergaard 1996; Vorobieff 1990; Wheeler 1982).

The non-structural aspect of behaviour is more complex, and less suited to mathematical analysis. In wind tunnel studies of sectional models, it has been common to carry out tests at successively larger velocities and note the logarithmic increment or decrement, δ, where this is defined as

$$\delta = \ell n(v_{r+1}/v_r)$$

with v_r and v_{r+1} being successive amplitudes of the same sign. Typically, δ is small for low wind velocities, but then increases markedly for velocities above a certain level, which may be called the critical velocity. The resulting movement has amplitudes that increase with time, and commonly include both vertical and torsional components. The phenomenon is called *flutter*.

The complexity of the problem may be illustrated by a few examples. The Lillebælt bridge has what is conventionally called a streamlined box-girder section, chosen as a result of wind tunnel tests by Selberg at the University of Trondheim (Larsen 1992: 66) (see Fig. 12.4(c)). What is not immediately apparent is that the deck plate joins the upper, sloping end plates by a curved transition, and curved wind flaps are mounted above this transition. Their purpose is to improve the flow over the corner and to minimise vortex shedding excited oscillations. Another example is shown in Fig. 12.3(b), which shows wind screens proposed to protect traffic on the Great Belt bridge. As mentioned earlier, wind tunnel tests showed that porous screens could be arranged on this streamlined box girder with little or no penalty to the aerodynamic stability. There is, finally, the stratagem of making part of the deck itself porous. The Mackinac bridge at the head of Lake Michigan in the USA (1958, 1158 m; O'Connor 1971a: 378ff) has a steel grating deck. In the outer lanes it is filled with concrete, but is left open in the two central lanes. The Forth bridge (1964, 1006 m; O'Connor 1971a: 379f) has a 2.9 m wide median strip roofed with an open grid. There are also open gaps, of the order of 4 m wide, between the roadway and the footpaths. Figure 12.7 shows curves for critical wind speed for two sections studied in connection with the design of the Great Belt bridge (Larsen 1992). The second of these, called the 'slotted box section', has an open slot at the median strip, causing an increase in total deck width from about 33 to 38 m. Decks with these forms were evaluated for suspension bridges with main spans from 2000 to 5000 m (side spans = main

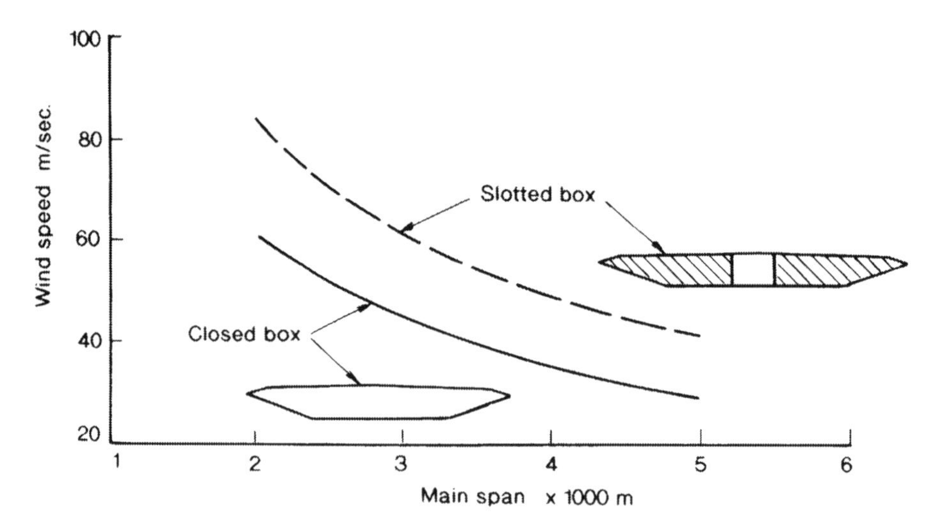

Fig. 12.7 Critical wind speeds for suspension bridges with closed and slotted streamlined box girder decks (Redrawn from Larsen 1992: Fig. 3.4)

span/3; cable sag = span/9; four roadway lanes). It can be seen that, for a main span of 2000 m, the critical wind speed is increased by the slot from about 61 to 84 m/sec, an increase of 38%. The diagram is also of interest for its prediction of lower critical velocities for greater span lengths. It is difficult to see how the behaviour of this range of sections could be predicted without model tests.

On the other hand, it is possible that wind tunnel tests of sections with controlled incidence angles could be used to accumulate data for theoretical analyses, and this work may be extended to include dynamic models, as in the work by Reinhold (Larsen 1992: 262ff). Much of this work refers to basic studies of aeroelasticity by Scanlan (such as in Dowell et al. 1995; Dyrbye and Hansen 1997: 143ff; Larsen 1992: 47ff; Scanlan 1981; Scanlan and Rosenbaum 1968; Simiu and Scanlan 1996: 246ff). Some of the earliest work in this area was by Smith and Vincent (1950).

This discussion has referred chiefly to sectional wind tunnel studies of the kind initiated by the Tacoma bridge failure. An alternative is the use of 'taut strip models' as developed chiefly by Davenport (Davenport A.G., King, J.P.C. and Larose, G.L., *Taut strip model tests*, 113–24, in Larsen 1992). The basis of the method is to replace the spring supports of the sectional model by highly tensioned wires, rods or tubes, running parallel to the deck, and extending over the full width of the wind tunnel, with the deck model also extended over this length. Consider a single wire of length L, straight, with an initial tension T, where T is large. Take, as an example, a relatively small central deflection, v. Then each half of the wire becomes inclined at an angle $(2v/L)$. In each half, the vertical component of T becomes $T(2v/L)$. The total central vertical force is, therefore,

$$P = 4Tv/L$$

with an equivalent stiffness given by

$$P/v = 4T/L$$

Davenport (Larsen 1992: 114) gives expressions for the corresponding lowest natural frequencies of vibration if a full-length model, of the kind shown in Fig. 12.5, is mounted not only on two wires, but on parallel members which add bending and torsional stiffness to the assembly. As with the sectional model, it is possible to adjust T, the wire spacing b, and the stiffness of the adjunct members, so as to model the fundamental frequencies of a bridge, and then test the whole in a wind tunnel. There are certain differences between a taut strip model and a sectional model, notably:

(a) the taut strip model can vibrate in other modes, with frequencies other than the fundamental frequencies; and
(b) the length of the taut strip model is such that it may experience variations in wind pressure along its length due to turbulent flow.

In both regards, the taut strip model would appear to have benefits over those given by the sectional model.

There is, however, a further alternative: if the wind tunnel is sufficiently large, a model of the complete bridge may be tested, and this has been done, for example, in the design of the Akashi Kaikyo bridge (Larsen 1992).

Much of this work is experimental; its consequences have a direct application and need not be further discussed here. The selection of wind speed is, of course, critical, but this has been discussed in a general manner in Section 12.1. A further discussion, of direct relevance to long span bridges, may be found in Larsen (1992).

12.5 Earthquake loads

The most common earthquakes are described as tectonic and arise typically from sudden movements or slips at geological fault zones, caused by long-term movement of the earth's crust (Bolt 1993: 98; Dowrick 1987: 38). The origin of the earthquake is local, but its effects move rapidly out in all radial directions from the epicentre by the transmission of strain waves, with axial strains in the radial direction and shear strains, perpendicular to the radius and either vertical or horizontal. The radial direct strain and vertical shear strain waves are called the P and S waves (Primary and Secondary waves) and that with horizontal shear strains, the Love wave. There is also a fourth, the Rayleigh wave, that is not unlike an ocean wave, with particle movements forming elliptical orbits about a horizontal axis at right angles to the direction of the moving wave. Theoretical expressions are available for the transmission velocities of these waves (e.g. Newmark and Rosenblueth 1971: 82ff), and also for the dependence of displacement magnitudes on distance from the epicentre.

The ground movements cause both vertical and horizontal accelerations, velocities and displacements of the foundations of a bridge. If they were all defined and known as functions of time, it would not be difficult to perform a suitable dynamic analysis of the bridge, so as to determine its response. The problem is that the nature of these movements varies, and it is difficult to prescribe a suitable design event. There are, however, other effects also that may be troublesome. In particular, saturated cohesionless soils may be subject to liquefaction, causing them to flow, and reducing their strength under load; Dowrick (1987: 203f) and Bolt (1993: 164f) discuss the problem. There is also the possible formation of fissures in the ground, causing permanent changes in distance between specific points, contributing to the risk of bridge failure by the separation of its supports.

Various scales have been used to assess the magnitude (M) of an earthquake. The Richter scale is based on records taken by a standard seismograph and is defined as 'the logarithm to base ten of the maximum seismic-wave amplitude (in thousandths of a millimetre) recorded ... at a distance of 100 kilometres from the earthquake epicentre' (Bolt 1993: 118ff). Distance from the epicentre is estimated from the difference in the first arrival times of the P and S waves. Knowing this distance, the measured amplitude may be converted to one at a distance of 100 km. Dowrick (1987: 6) refers to the Richter magnitude as the *local magnitude* and gives the largest value ever recorded on this scale as 8.9

(or possibly 9.0). Another measure of earthquake magnitude is the Modified Mercalli Intensity Scale. It is a 'felt' scale and has values I to XII (in Roman numerals), where each is verbally defined. For example, I means 'not felt except by a very few under exceptionally favourable circumstances'; II – 'felt by persons at rest, on upper floors, or favourably placed'; III – 'felt indoors, hanging objects swing, vibration similar to passing of light trucks, duration may be estimated, may not be recognised as an earthquake'; and so on, to XII – 'damage nearly total, large rock masses displaced, lines of sight and level distorted, objects thrown into the air' (Bolt 1993: App. C; Dowrick 1987: App. B). Bolt adds to these qualitative descriptions numerical values for the 'average peak velocity' and 'average peak acceleration'. Some of his values are:

IV	10–20 mm/sec	(0.015–0.02) g
VII	80–120 mm/sec	(0.10–0.15) g
X–XII	more than 600 mm/sec	more than 0.60 g

where g is the acceleration due to gravity (9.81 m/sec^2).

Bolt (1993: Apps A, B) lists major world and American earthquakes and for a few of these it is possible to compare observations on the Richter and Modified Mercalli scales. These are listed below.

1899	Alaska, Yakutat Bay	7.8 and 8.6	XI
1906	San Francisco earthquake and fire	8.25	XI
1971	San Fernando, California	6.5	VIII–XI
1989	Santa Cruz Mountains, California	7.0	X
1992	Landers, California	7.5	VIII

The list is sufficient to give some idea of the relationship, although some apparent discrepancies may be noted. The 1989 earthquake damaged part of the major San Francisco-Oakland Bay bridge (completed in 1936 with a main span of 704 m). In the earthquake as a whole, 63 lives were lost, most due to the collapse of the upper level of a two-level elevated freeway in Oakland. It is important to realise that structural collapse may lead not only to financial loss and inconvenience, but to the loss of life. The stepwise collapse of bridges with multiple spans of small to moderate size, due to one end of each span falling from its pier, was evidenced also in a bridge near Sakarya, Turkey, during the August 1999 earthquake, and in Taiwan in an earthquake in the following month. Both had intensities of 7.6 (preliminary value) on the Richter scale.

The size of an earthquake can also be gauged by its effect on the simple, one degree of freedom structure shown in Fig. 12.8(a). Mass m, which could be the elevated floor of a single storey building, is supported on elastic columns, such that the combined stiffness against horizontal displacements of the mass relative to the ground is k_s. The columns are assumed to be without mass, and in this initial analysis, the dashpot shown in the figure is not present. The structure can vibrate with displacements, u, given by

$$u = u_0 \sin \omega t$$

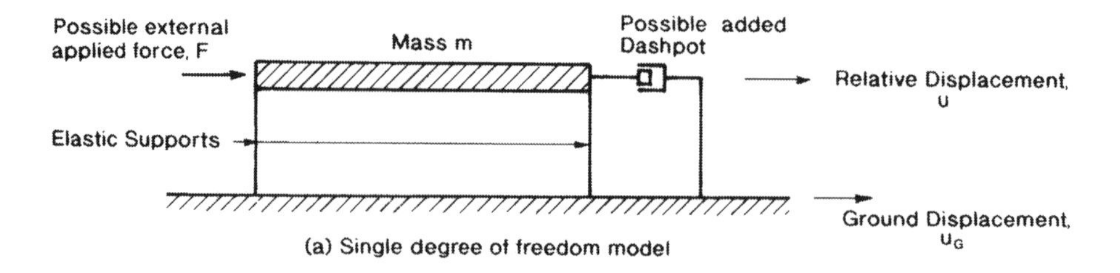

(a) Single degree of freedom model

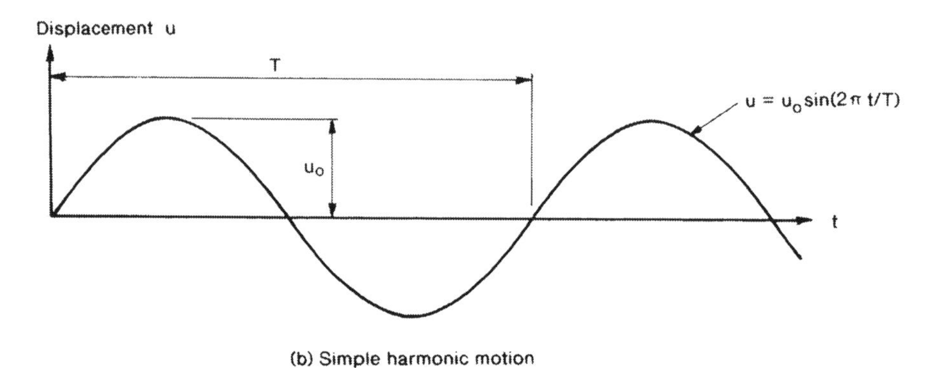

(b) Simple harmonic motion

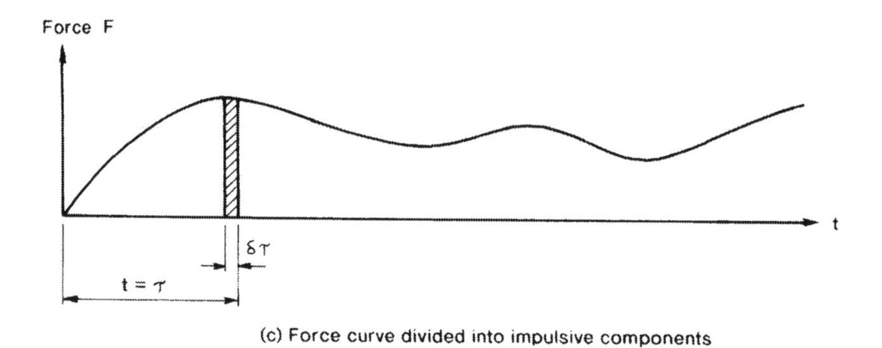

(c) Force curve divided into impulsive components

Fig. 12.8 (a) Single degree of freedom model used in earthquake studies; (b) simple harmonic motion; (c) impulsive components

The corresponding accelerations are

$$\ddot{u} = -u_o\,\omega^2 \sin \omega t$$

The equivalent horizontal force applied to mass, m, is $m\ddot{u}$. The equal and opposite forces applied to the tops of the columns will cause displacements equal to

$$\frac{mu_o\omega^2 \sin \omega t}{k_s}$$

For this to be a correct solution, these must be the same as those expressed initially as

$$u = u_o \sin \omega t$$

It follows that

$$u_o = m u_o \omega^2 / k_s$$

The initial equation of motion may be expanded to the form

$$u = u_o \sin (2\pi t/T)$$

where T is the period.

Then u is zero at $t = 0$ and T, as shown in Fig. 12.8(b). It follows that,

$$\omega = 2\pi/T$$

$$\frac{m\omega^2}{k_s} = \frac{m(2\pi/T)^2}{k_s} = 1$$

Hence,

$$T = 2\pi\sqrt{m/k_s}$$

The inverse of T, or $1/T$, is the frequency, f (cycles/sec, or Herz).

Consider the application to the mass, m, of an impulsive force, F, constant over a short time, δt. F will cause an acceleration, F/m. If the mass is initially at rest, then at the end of this time it will have a small velocity,

$$\Delta \dot{u} = F\delta t/m$$

Following δt, F is zero, and the motion will be simple harmonic. Referring to Fig. 12.8(a), the initial velocity is

$$\dot{u} = u_o \omega$$

Equating this to $\Delta \dot{u}$,

$$u_o = \Delta \dot{u}/\omega = \frac{TF\delta t}{2\ \pi m}$$

Figure 12.8(c) shows a graph of a longer term variable force F, that may be applied to the system. This can be replaced by a series of impulsive actions, such as that shown, with force $F(\tau)$ at time τ, assumed to be constant over the time interval, $\delta\tau$. The combined effect of these impulses can be written,

$$u(t) = \int_0^t \frac{F(\tau)}{m\omega} \sin \omega(t - \tau) d\tau$$

This integral is called the Duhamel integral, and may be evaluated by numerical methods.

In the case of an earthquake, the forces, F, applied to the mass, m, result from ground movements, u_g. $F(\tau)$ may be replaced by $m\ddot{u}_g(\tau)$, where $\ddot{u}_g(\tau)$ represents the ground acceleration. This may be found for particular earthquakes, and substituted in the above expression.

More normally, the elastic system of Fig. 12.8(a) is replaced by the damped system, with the dashpot added. The damping force is proportional to the velocity; it opposes the motion and may be written as $-C\dot{u}$. The damping ratio, d, is given by

$$d = \frac{C}{2m\omega} = \frac{TC}{4\pi m}$$

where T is the undamped natural frequency.

The displacements of the damped structure may still exhibit periodicity, but with a revised value, ω_D, given by

$$\omega_D = \omega\sqrt{1 - d^2}$$

The corresponding period, T_D, is

$$T_D = T(1 - d^2)$$

For small values of C, the effect of d on the period is small. For larger values, T_D is reduced. A particular value of C is said to provide critical damping. If the mass, m is displaced and then released, it will simply move back to the initial position without oscillation. The corresponding value, C_{cr} is called critical damping, and is given by

$$C_{cr} = 2\sqrt{mk_s}$$

It should be noted that

$$d = \frac{TC}{4\pi m}$$

with

$$T = 2\pi\sqrt{m/k_s}$$

that is,

$$d = \frac{C\sqrt{m}}{2m\sqrt{k_s}} = \frac{C}{2\sqrt{mk_s}} = \frac{C}{C_{cr}}$$

For a damped system, the term within the Duhamel integral should be multiplied by $e^{-d\omega(t-\tau)}$, where the use of ω in place of ω_D is an approximation

Table 12.13 Values of $\dot{u}$ (m/sec)

	d		
T	0	0.02	0.05
0.2	0.57	0.25	0.16
0.4	1.02	0.46	0.31
0.6	1.06	0.58	0.39
1.0	0.93	0.61	0.44
2.0	0.83	0.66	0.51

that is acceptable for low levels of damping. This analysis can be extended to compute maximum values of such functions as u and $\ddot{u}$ for particular earthquakes, and for particular levels of the damping ratio, d. Graphs of such functions against T, the natural period of a structure, are another form of representing the magnitude and nature of a particular earthquake. A typical example may be found in Dowrick (1987: 21) for the 1940 El Centro earthquake, rated as X on the Modified Mercalli scale. This graph shows the maximum velocity, $\dot{u}$, called the spectral velocity, plotted against T. Some of the values are shown in Table 12.13.

The trend is for $\dot{u}$ to vary linearly with T, for T greater than 0.8 sec and d equal to or greater than 0.02. The effect of damping may be seen from the following values, all at $T = 3.0$ sec: (0.0, 0.82), (0.02, 0.69), (0.05, 0.57), (0.10, 0.38), (0.20, 0.33) and (0.40, 0.24); where these are in the order, d and $\dot{u}$. This analysis has been presented in the context of assessing earthquake magnitudes. It has also a structural significance, and this may be pursued, as in the text by Newmark and Rosenblueth (1971).

Earthquake magnitudes, for a particular probability of recurrence, vary greatly from region to region. Maps showing the locations of major incidents may be found in the above sources, and codes of practice tend to define by region the magnitude of earthquake that should be considered. This again extends beyond the scope of the present work, but a recent draft code, the Eurocode 8 (DD ENV 1998: 1996) will be used to illustrate modern practice. Its general title is 'Design provisions for earthquake resistance of structures', and Part 2 deals particularly with bridges.

It states first that a reference return period of 475 years has been used. Regions have been defined essentially in terms of a single parameter, the effective peak ground acceleration, a_g, in rock and firm soil. Seismic zones with a_g less than $0.04 \times g$ (where g is the acceleration due to gravity) are described as those where the provisions of this code need not be observed, and zones with a_g less than $0.10 \times g$ are called low seismicity zones, where simplified design procedures may be followed.

The vertical components of movement are specified in terms of the structural period T; for T less than 0.15 sec, vertical movements are taken as 0.70 x: the horizontal movements; for T greater than 0.5 sec, a factor of 0.5 is used; with linear interpolation between 0.15 and 0.5 sec.

The code then defines an 'elastic response spectrum' of the same kind as

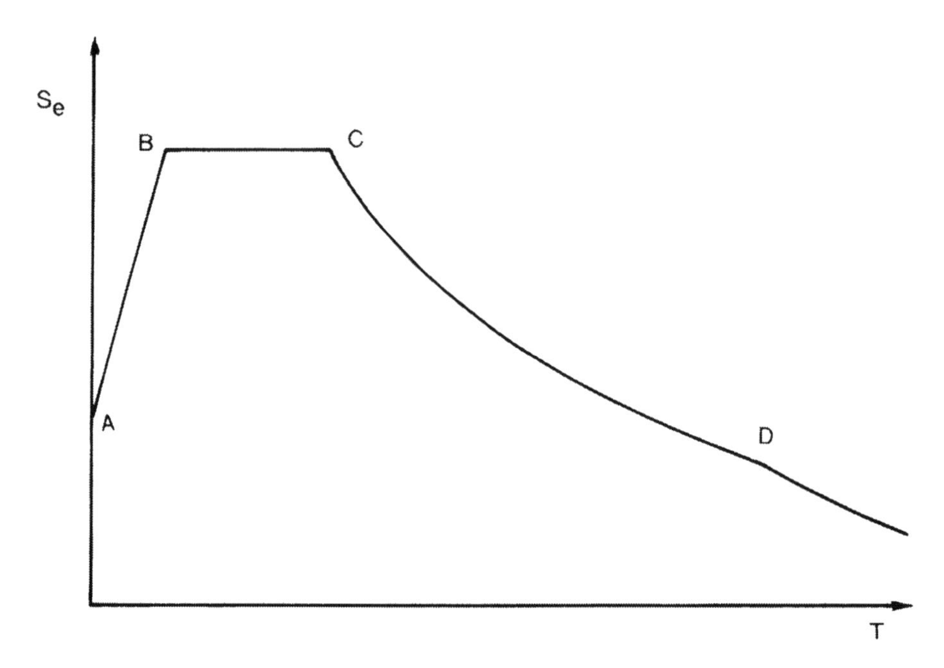

Fig. 12.9 Elastic response spectrum, as specified in Eurocode 8, DD ENV 1998-1-1: 1996: Fig. 4.1 (Redrawn)

that described above. Its basic form is shown in Fig. 12.9, where the vertical ordinate, S_e, represents acceleration, and the horizontal is T, the 'vibration period for a single degree of freedom system'. This spectrum is divided into four zones, by the points A, B, C and D; where BC is called 'the constant spectral acceleration branch'; and D, 'the beginning of the constant displacement range of the spectrum'. Values of the periods, T_B, T_C and T_D are listed in a table and expressions are given for the vertical ordinate, S_e, in terms of the following parameters:

a_g is the design ground acceleration,
S is the soil parameter,
η is a damping factor, equal to $\sqrt{7/(2 + d)}$, where d is the damping ratio in per cent,
β_o is an amplification factor, and
k_1 and k_2 are exponents.

These parameters are tabulated for three sub-soil classes, where class A is rock, or a stiff deposit of sand, characterised by a shear wave velocity, v_s, of at least 400 m/sec at a depth of 10 m; B is a deep deposit of medium dense sand, with v_s equal to at least 200 m/sec at a depth of 10 m, increasing to at least 350 m/sec at a depth of 50 m; and C is a loose cohesionless soil.

For class A, the code specifies the following values: $S = 1.0$; $\beta_o = 2.5$; $k_1 = 1.0$; $k_2 = 2.0$; $T_B = 0.1$ sec; $T_C = 0.4$ sec; and $T_D = 3.0$ sec. In the region

BC, S_e is $(a_g S\eta\beta_o)$; or, for class A subgrades, S_e is $2.5a_g$ for a structural damping ratio of 5%, and this applies for structural periods between 0.1 and 0.4 sec. Not all of the code provisions will be repeated here, but to give one further example, in the region CD, with periods between 0.4 and 3.0 sec,

$$S_e = a_g S\eta\beta_o [T_c/T]^{k_1}$$

Again, for class A subgrades, and a structural damping ratio of 5%, S_e is $2.5a_g \times 0.4/T$. That is, for T equal to 0.4, S_e is $2.5a_g$, and for T equal to 3.0 sec, S_e is $a_g/3$.

The code also suggests that the ground displacement, u_g, can be estimated from the expression,

$$u_g = 0.05 a_g S T_C T_D$$

Once more, for class A subgrades,

$$u_g = 0.06 a_g$$

The code then allows the designer either to generate an artificial accelerogram (a history of ground accelerations versus time), or to select a minimum of five recorded accelerograms, such that these should match the specification of the elastic response spectrum. Indeed, the clauses quoted previously can be seen simply as provisions which define the magnitudes of appropriate accelerograms, either taken directly from records or chosen from suitable past experience. These accelerograms then become the input data for a dynamic analysis of the structure. Part 2 of the code goes on to assist these analyses by specifying such matters as the choice of suitable element stiffnesses.

Another important reference may be found in the proceedings of a workshop on the *Earthquake resistance of highway bridges* issued by the US Applied Technology Council in November 1979. It includes useful material, such as contours of effective peak accelerations and velocities for the USA (pp 131–4), a zoning map for Japan (pp 49–51), and discussions of analysis and design that go beyond the scope of this book.

12.6 Temperature effects

Two classes of temperature effects occur in bridges:

(a) overall temperature changes must be considered in the design of moving bearings and in the choice of their location; and
(b) differential temperature effects may occur such that at a particular time, the temperature at one point in a structure will not be the same as at another, and these temperature differences may cause locked-in stresses and possible failure.

Concerning the first, in a single-span bridge, it is conventional to permit longitudinal, horizontal movement in the bearings at one end of a span, so that the bridge may expand or contract freely under the action of temperature changes or other related matters, such as concrete shrinkage or creep, and elastic strains in the structure under load. They also permit foundation movements. In the case of a multi-span bridge, it may be rendered continuous over its full length, or over a number of spans, and these longitudinal movements may in the limit, add to one location. Alternatively, if the bridge consists of a number of simply-supported spans, there may be a moving bearing at one end of each span.

Each provision for movement must extend across the full structure. As a result, there will be not only moving bearings that support the bridge, but deck expansion joints. It is a simple fact that one of the most common causes of maintenance in highway bridges lies in the performance of these deck joints, or in consequences that follow when rainwater from the deck passes through these joints. For these reasons, designers may well choose to limit the number of expansion joints. An example is the Gateway bridge, Brisbane, shown in Fig. 12.10, which was completed in 1986 with a main span of 260 m. The main girders are structurally continuous; temperature effects cause longitudinal movements at the pier tops and these are designed so as to have sufficient flexibility to tolerate these movements without distress. This decision, however, gives rise to other problems; if the girders are anchored longitudinally at the abutments, then temperature movements will accumulate over greater distances than a single span length. On the other hand, it may be decided that both end supports should provide for longitudinal movement. The design of the main piers must then strike a balance between provision for expansion and the retention of sufficient strength to transmit longitudinal traffic forces to the ground.

Concerning the second class, the development of locked-in stresses due to differential temperatures caused damage to the major Newmarket Viaduct, Auckland, in the period following its completion in 1966 (Buckle and Lanigan 1971; see also Leonhardt et al. 1965; Priestley 1972; White 1979). Temperatures may vary within a cross-section; they may, for example, vary with depth. If this variation is linear, and if the structure is determinate, then it may adopt a deflected shape without the development of stresses due to these temperature differences. If either of these conditions is not satisfied, then stresses will be developed due to temperature. These may take the form of longitudinal direct stresses. For example, to take the case of a bridge continuous over three spans, as the Gateway bridge (Fig. 12.10), in the mornings it may be expected that temperatures in the upper flange will be higher than in the lower. If the structure were freed from its central piers, these temperatures, when considered alone, would cause the girder to rise. It is, in fact, restrained from doing so and added downward reaction components will be applied to the structure at its intermediate supports. Temperature differentials of this kind will, therefore, tend to cause restraint stresses that are tensile in the lower flange. These may not be a problem at the supports themselves, but will add to other design stresses at mid-span. Not only so, but the hold-down reactions developed at the intermediate piers will cause vertical end reactions that add to the end shears in the members. The combination of these effects – the effects of non-linearity in the temperature dis-

Fig. 12.10 The Gateway Bridge, Brisbane, completed in 1986 with a main span of 260 m, is an example of a continuous girder bridge with temperature movements accommodated by pier deflections (O'Connor photograph)

tributions, and the effects of restraint forces – may well cause cracking in a concrete structure, and possible greater distress.

Again, temperatures will vary geographically, but an example of temperature specification may be found in the British code BD37/88. This gives isotherms, or contours, of maximum and minimum shade temperatures in Britain and Ireland for a 120-year return period. They vary from 37 to −24°C, a surprising variation. For a return period of 50 years, this variation is reduced, with both extremes modified by 2°C. Tables are given that relate bridge temperatures to these shade temperatures; for example, for a shade temperature of +37°C,

Table 12.14

	Positive case (°C)	*Negative case (°C)*
At top face	24	−6
0.1 m below top	14	
0.2 m below top	8	
0.3 m below top	4	
0.5 m below top		0
At bottom	0	0

Table 12.15

	Positive case (°C)	*Negative case (°C)*
At top face	+13.5	−8.0
0.2h below top		−1.5
0.3h below top	+3.0	
0.45h below top		0
0.6h below top	0	
0.45h above bottom		0
0.3h above bottom	0	
0.2h above bottom		−1.5
At bottom face	+2.5	−6.3

the maximum effective bridge temperature varies from 46°C to 36°C, where the highest figure is for a bridge with a steel deck on steel girders, and the lowest for a bridge with concrete deck and girders. The corresponding minimum temperatures range from −14°C to −28°C.

BD 37/88 code also gives data on non-linear temperature variations across the depth of a bridge: for 'positive temperature differences' with a maximum temperature at the upper face, and for 'negative temperature differences', where the situation is reversed in sign. For a bridge with a steel deck on steel girders, the specified temperatures are as given in Table 12.14.

Typical figures for full concrete construction, with a total depth, $h = 1$ m, are as given in Table 12.15.

These assume a 40 mm thickness of wearing surface for the steel structure, and 100 mm for the concrete. For other thicknesses, different temperatures are specified. Temperatures are assumed to vary linearly between the stated values, but the overall plots depart from a single straight line.

Bibliography

Agarwal, A.C. and Cheung, M.S. (1987) Development of loading – truck model and live-load factor for the Canadian Standards Association CSA-S6 code, *Canadian Journal of Civil Engineering*, 14: 58–67.

Agarwal, A.C. and Wolkowicz, M. (1976) *Ontario Commercial Vehicle Survey, 1975*, Ontario Ministry of Transportation and Communications, Interim Report, Engineering Research and Development Branch.

Allen, D.E. (1992) Canadian Highway bridge evaluation: reliability index, *Canadian Journal of Civil Engineering*, 19: 987–91.

Allen, D.N. de G. and Southwell, Sir R. (1949) Relaxation methods applied to engineering problems XIV, plastic straining in two dimensional stress-systems, *Philosophical Transactions of the Royal Society Series A* (242): 379–414.

AASHO [American Association of State Highway Officials] (1931, 1973 etc) *Standard Specifications for Highway Bridges*, Washington D.C.

—— (1961,2) Road Test *National Research Council, Highway Research Board, Special Reports* 61A-G, including 61D: Bridge Research; Report 7: Summary Report.

AASHTO [American Association of State Highway and Transportation Officials] (1989) *Guide Specification for Bridge Railings*, Washington D.C.

—— (1990) *A Policy of Geometric Design of Highways and Streets*, Washington DC.

—— (1991) *Guide Specification and Commentary on Vessel Collision Design of Highway Bridges*, Washington D.C.

—— (1994) *AASHTO LRFD Bridge Design Specifications*, Washington D.C.

AREA [American Railway Engineering Association] (1990) *Manual for Railway Engineering* (2 vols), Washington D.C.

Anderson, J.K. (1964) *Runcorn-Widnes Bridge*, Institution of Civil Engineers, London, Proceedings, 29: 535–70.

Anderson, T.L. (1995) *Fracture Mechanics: Fundamentals and Applications*, CRC Press, Boca Raton, Ann Arbor.

Andreski, A.T. (1987) *The Effect of Scour and Debris Impact on Bridge Piers*, MSc/DIC thesis, Imperial College, London.

Apelt, C.J. (1960) Flow loads on bridge piers, *Institution of Engineers, Australia, Brisbane Division, Technical Papers*, 1 (2): 1–13.

—— (1986) Flow loads on submerged bridges, *Institution of Engineers, Australia, Queensland Division, Technical Papers,* 27 (19): 17–33.

Apelt, C.J. and Isaacs, L.T. (1968) Bridge piers — Hydrodynamic force coefficients. *American Society of Civil Engineers, Journal of the Hydraulics Division*, 94 (HY1): 17–30.

Apelt, C.J. and Piorewicz, J. (1986) *Breaking Wave Forces on Vertical Cylinders,* The University of Queensland. Department of Civil Engineering, Research Report CE71.

Asplund, S.O. (1955) Probabilities of traffic loads on bridges, *American Society of Civil Engineers, Proceedings*, 81 (285): 585-1–12.

Australian and New Zealand Railway Conferences (1974) *Railway Bridge Design Manual.*

Austroads (1992 etc.) *Australian Bridge Design Code*, Sydney.

—— (1994) *Waterway Design: A Guide to the Hydraulic Design of Bridges, Culverts and Floodways*, Sydney.

—— (1998) *Economics of Higher Bridge Design Loadings*, Project H96, Technical Report, Sydney.

Austroads, Australasian Railway Association and Standards Australia (1996) *Australian Bridge Design Code: Railway Supplement*, Austroads, Sydney.

Aynsley, R.M., Melbourne, W. and Vickery, B.J. (1977) *Architectural Aerodynamics*, Applied Science Publishers, London.

Baker, C.J. (1981) Ground vehicles in high cross-winds on trains. *American Society of Mechanical Engineers, Journal of Fluid Mechanics*, 103: 170–8.

—— (1986) A simplified analysis of various types of wind-induced road vehicle accidents, *Journal of Wind Engineering and Industrial Aerodynamics*, 22: 69–85.

—— (1994) The quantification of accident risk for road vehicles in cross winds, *Journal of Wind Engineering and Industrial Aerodynamics*, 52: 93–107.

Baker, Sir J. and Heyman, J. (1969–71) *Plastic Design of Frames* (2 vols), Cambridge University Press, Cambridge.

Baker, Sir J., Horne, M.R. and Heyman, J. (1956) *The Steel Skeleton* (2 vols), Cambridge University Press, Cambridge.

Baker, Sir J. and Roderick, J.W. (1938) An experimental investigation of the strength of seven portal frames, *Institute of Welding, Journal*, 5: 206ff.

Bakhmeteff, B.A. (1932) *Hydraulics of Open Channels*, McGraw-Hill, New York.

Bakht, B. and Jaeger, L.G. (1990) Bridge evaluation for multi-presence of vehicles, *American Society of Civil Engineers, Journal of Structural Engineering*, 116 (3): 603–18.

Bakht, B. and Moses, F. (1988) Lateral distribution factor for highway bridges, *American Society of Civil Engineers, Journal of the Structural Division*, 114 (8): 1785–803.

Barber, E.H.E., Bull, F.B. and Shirley-Smith, H. (1971) *Report of Royal Commission into the Failure of West Gate Bridge*, Government Printer, Melbourne.

BD37/88 (1988) *Departmental Standard BD37/88: Loads for Highway Bridges*, British Department of Transport, London.

Beedle, L.S. (1958) *Plastic Design of Steel Frames*, Wiley, New York.

Behan, J.E. and O'Connor, C. (1982) Creep buckling of reinforced concrete columns, *American Society of Civil Engineers, Journal of the Structural Division*, 108 (ST12): 2799–818.

Benjamin, J.R. and Cornell, C.A. (1970) *Probability, Statistics and Decision for Civil Engineers*, McGraw-Hill, New York.

Bez, R., Cantieni, R. and Jacquemoud, J. (1987) Modeling of highway traffic loads in Switzerland, *International Association for Bridge and Structural Engineering, Periodica*, (3): 153–68.

Billing, J.R. (1984) Dynamic loading and testing of bridges in Ontario, *Canadian Journal of Civil Engineering*, 11 (4): 833–43

Billing, J.R. and Green, R. (1984) Design provisions for dynamic loading of highway bridges, *Transportation Research Record*, (950): 94–103.

Bisplinghoff, R.L. and Ashley, H. (1962) *Principles of Aeroelasticity*, Wiley, New York.

Bisplinghoff, R.L., Ashley, H. and Halfman (1955) *Aeroelasticity*, Addison-Wesley, Reading, Massachusetts.

Blockley, D. (ed.) (1992) *Engineering Safety*, McGraw-Hill, New York.

Bolt, B.A. (1993) *Earthquakes*, Freeman, New York.

Boyce, W.H. (1987) Cyclone Namu and the Ngalimbiu Bridge: Did it fall or was it pushed? *First National Structural Engineering Conference, Melbourne*, 26–28 August 1987, Proceedings: 36–41.

Boyce, W.H. (1989) Cyclone Namu and the Ngalimbiu Bridge: Design implications, *International Conference on Case Histories of Structural Failures, CHSF89, Singapore*, 20–22 March 1989, Proceedings: C-11–23.

Boyd, G.M. (ed.) (1970) *Brittle Fracture in Steel Structures*, Butterworth, London.

Bradley, J.N. (1960) *Hydraulics of Bridge Waterways*, US Department of Commerce, Bureau of Public Roads, Washington, D.C.

Bradley, J.N. (1978) *Hydraulics of Bridge Waterways*, US Department of Transportation, Federal Highway Administration, Washington, D.C.

Brady, A.B. (1896) Low-level bridges in Queensland, *Institution of Civil Engineers, London, Minutes of Proceedings*, 124: 323–34.

Brameld, G.H. (1995) *Austroads/Railways of Australia: Bridge Design Code; Section 2 — Load Factors*, Queensland University of Technology, Brisbane.

Breusers, H.N.C. and Raudkivi, A.J. (1991) *Scouring*, Hydraulic Structures Design Manual, Balkema, Rotterdam.

Bridge Aerodynamics, Proceedings Institution of Civil Engineers Conference, 25–6 March 1981, Thomas Telford, London.

BS 131: Part 1: 1961 (1996) *The Izod Impact Test of Metals*, British Standards Institution, London.

BS 153: 1923, 1958 etc *Specification for Steel Girder Bridges*, British Standards Institution, London.

BS 3680: 1992 *Measurement of Liquid Flow in Open Channels*, British Standards Institution, London.

BS 5400: 1978 etc *Steel, Concrete and Composite Bridges*, British Standards Institution, London.

BS 6399: Part 2: 1995 *Loading for Buildings, Part 2: Code of Practice for Wind Loads*, British Standards Institution, London.

BS 6779 *Highway Parapets for Bridges and Other Structures, Part 1 1992: Specification for vehicle containment parapets of metal construction, Part 2 1991 Specifications for vehicle containment parapets of concrete construction, Part 3 1994 Specification for vehicle containment parapets of combined metal and concrete construction*, British Standards Institution, London.

BS EN 10020: 1991 *Definition and Classification of Grades of Steel*, British Standards Institution, London.

BS EN 10045.1: 1990 *Charpy Impact Test on Metallic Materials: Test Method (V- and U-notches)*, British Standards Institution, London.

Buckland, P.G. (1981) Recommended design loads for bridges (Committee on Loads and Forces on Bridges of the Committee on Bridges of the Structural Division). *American Society of Civil Engineers, Journal of the Structural Division*, 107 (ST7): 1161–213.

Buckland, P.G. (1982) Bridge loading: research needed (Committee on Loads and Forces on Bridges of the Committee on Bridges of the Structural Division). *American Society of Civil Engineers, Journal of the Structural Division*, 108 (ST5): 1012–20.

Buckland, P.G. and Bartlett, F.M. (1992) Canadian highway bridge evaluation: A general overview of Clause 12 of CSA Standard CAN/CSA-S6–88, *Canadian Journal of Civil Engineering*, 19: 981–6.

Buckland, P.G., Navin, F.P.D., Zidek, J.V. and McBryde, J.P. (1980) Proposed vehicle loadings for long-span bridges, *American Society of Civil Engineers, Journal of the Structural Division*, 106 (ST4): 915–32.

Buckland, P.G. and Sexsmith, R.G. (1981) A comparison of design loads for highway bridges, *Canadian Journal of Civil Engineering*, 8: 16–21.

Buckle, I.G. and Lanigan, A.G. (1971) Transient thermal response of box girder bridge decks, *Third Australasian Conference on the Mechanics of Structures and Materials*, Auckland.

Burden, A.R. (1991) Modern Japanese suspension bridge design, *Institution of Civil Engineers, London, Proceedings*, 90 (1): 157–77.

Calow, P. and Petts, G.E. (eds) (1992–...) *The Rivers Handbook: Hydrological and ecological principles* (2 vols), Blackwell, Oxford.

Cantieni, R. (1984a) Dynamic load testing of highway bridges, *International Association for Bridge and Structural Engineering, Periodica*, (3): 57–72.

—— (1984b) Dynamic load testing of highway bridges, *Transportation Research Record*, (950): 141–8.

—— (1987) Dynamic load testing of a two-lane highway bridge, *International Conference, 'Traffic Effects on Structures and Environment'*, Strbske Pleso, Czechoslovakia, 1–3 December *1987*.

—— (1992) *Dynamic Behaviour of Highway Bridges under the Passage of Heavy Vehicles*, Swiss Federal Laboratories for Materials Testing and Research (EMPA), Dübendorf, Switzerland.

Casado, C.F. (no date) *Historia del Puente en España: Puentes Romanos*, Madrid.

Cement and Concrete Association of Australia (no date) *The Safety Shape (G39)*, Sydney.

Chan, T.H.T. (1988) *Highway Bridge Impact*, PhD thesis, The University of Queensland.

Chan, T.H.T. and O'Connor, C. (1986) The Use of Bridges to Measure Dynamic Wheel Loads, *10th Australian Conference on the Mechanics of Structures and Materials*, Adelaide, 599–604.

—— (1990a) Wheel loads from highway bridge strains: Field studies, *American Society of Civil Engineers, Journal of Structural Engineering*, 116 (7): 1751–71.

—— (1990b) Vehicle model for highway bridge impact, *American Society of Civil Engineers, Journal of Structural Engineering*, 116 (7): 1772–93.

Chanson, H. (1999) *The Hydraulics of Open Channel Flow: An Introduction*, Arnold, London.

Chatterjee, S. (1991) *The Design of Modern Steel Bridges*, BSP Professional Books, Oxford.

Cheremisinoff, P.N., Cheremisinoff, N.P. and Cheng, S.L. (eds) (1987–8) *Civil Engineering Practice* (5 vols), Technomic, Lancaster, Pa.

Chettoe, C.S. and Adams, H.C. (1933) *Reinforced Concrete Bridge Design*, Chapman and Hall, London.

Chitty, G.B., Grundy, P. and McTier, H. (1990) Dynamic stresses in short span railway bridges, *Institution of Engineers, Australia, Structural Engineering Conference*, Adelaide, 3–5 October, 1990: 213–8.

Chow, V.T. (1959) *Open-channel Hydraulics*, McGraw-Hill, New York.

Cohen, E. and Birdsall, B. (eds) (1980) Long-span bridges (O.H. Amman Centenial Conference), *Annals of New York Academy of Sciences*: 352.

Cohen, H. and Hoel, L.A. (1990) *Truck weight limits: Issues and options* Committee for the Truck Weight Study, Transportation Research Board, Special Report 225.

Cohn, M.Z. and Bartlett, M. (1982) Computer-simulated flexural tests of partially prestressed concrete sections, *American Society of Civil Engineers, Proceedings*, 108 (ST12): 2747–65.

Coleman, S.A. and Baker, C.J. (1990) High sided road vehicles in cross winds, *Journal of Wind Engineering and Industrial Aerodynamics*, 36: 1383–92.

Colosimo, V. (1996) Design proposals for bridge barriers, *Asia-Pacific Symposium on Bridge Loading and Fatigue*, 21–38.

Cook, N.J. (1982) Towards better estimation of extreme winds, *Journal of Wind Engineering and Industrial Aerodynamics*, 9: 295–323.

—— (1983) Note on directional and seasonal assessment of extreme winds for design, *Journal of Wind Engineering and Industrial Aerodynamics*, 12: 365–72.

—— (1985) *The Designer's Guide to Wind Loading of Building Structures* (2 vols), Butterworth, London.

Cook, N.J. and Prior, M.J. (1987) Extreme wind climate of the United Kingdom, *Journal of Wind Engineering and Industrial Aerodynamics*, 26: 371–89.

Cooper, R.K. (1981) The effect of cross-winds on trains, *American Society of Mechanical Engineers, Journal of Fluid Mechanics*, 103: 170–8.

CSA-S6 (1974, 1978 etc) *Design of Highway Bridges*, Canadian Standards Association, Rexdale, Ontario.

CSA-S408 (1981) *Guidelines for the Development of Limit States Design*, Canadian Standards Association, Rexdale, Ontario.

Csagoly, P.F., Campbell, T.I. and Agarwal, A.C. (1972) *Bridge Vibration Study*, Ontario Ministry of Transportation Report, RR181.

Csagoly, P.F. and Dorton, R.A. (1973) *Proposed Ontario Bridge Design Load: RR186*, Ministry of Transportation and Communication, Research and Development Division, Ontario.

—— (1977) *The Development of the Ontario Bridge Code*, Ministry of Transportation and Communication, Ontario.

—— (1978a) *Truck Weights and Bridge Design Loads in Canada*, American Association of State Highway and Transportation Officials, Annual Meeting, Louisville, Kentucky.

—— (1978b) The development of the Ontario Highway Bridge Design Code, *Transportation Research Record*, (665): 1–12.

Darvall, P. Le P. and Allen, F.H. (1987) *Reinforced and Prestressed Concrete*, MacMillan, Melbourne.

Das, P.C. (ed.) (1997) *Safety of Bridges*, Thomas Telford, London.

Deaves, D.M. (1981) Terrain-dependence of longitudinal rms velocities in the neutral atmosphere, *Journal of Wind Engineering and Industrial Aerodynamics*, 8: 259–74.

Deaves, D.M. and Harris, R.J. (1978) *A Mathematical Model of the Structure of Strong Winds: CIRIA Report 76*, Construction Industry Research and Information Association, London.

Devore, J.L. (1995) *Probability and Statistics for Engineering and the Sciences*, Wadsworth, Belmont, Cal.

DIN1072: (1967) *Road and Foot-Bridges: Design Loads*, Deutsche Institut für Normung, Berlin.

Dolan, C.W. (1986) Analysis and design of reinforced concrete guideway structures:

Report by ACI Committee 358, *American Concrete Institute Journal, Proceedings*, 83 (5): 838–68.
Dorton, R.A. and Bakht, B. (1984) The Ontario Bridge Code: Second Edition, *Transportation Research Record*, (950): 88–93.
Dorton, R.A. and Csagoly, P.F. (1977) *The Development of the Ontario Bridge Design Code, 1977 National Lecture Tour for the Canadian Society for Civil Engineering, Structural Division*, Ministry of Transportation and Communication, Ontario.
Dowell, E.H., Crawley, E.F., Curtiss Jr., H.C., Peters, D.A., Scanlan, R.H., and Sisto, F. (1995) *A Modern Course in Aeroelasticity*, Kluwer Academic, Dordrecht, The Netherlands.
Dowling, N.E. (1972) Fatigue failure predictions for complicated stress-strain histories, *Journal of Materials*, 7: 71–87.
Dowrick, D.J. (1987) *Earthquake Resistant Design for Engineers and Architects*, Wiley, New York.
Duczmal, Z.R. (1988) A procedure for obtaining the dynamic response of a composite road bridge, *Eleventh Australasian Conference on the Mechanics of Structures and Materials,* Auckland.
—— (1989) *Dynamic Non-Linear Vehicle-Bridge Interaction*, PhD thesis, The University of Queensland.
Duczmal, Z.R. and Swannell, P. (1985) A computer model for non-linear bridge–vehicle systems. *Conference on Computational Techniques and Applications,* University of Melbourne.
Dyrbye, C. and Hansen, S.O. (1997) *Wind Loads on Structures,* Wiley, New York.
Earthquake Resistance of Highway Bridges, Proceedings of Workshop 29–31 Jan. 1979, US Applied Technology Council, California.
Effects of Studded Tires: National Co-operative Highway Research Program, Synthesis of Highway Practice 32 (1975), Transportation Research Board, Washington D.C.
Eurocode 1 (1994 etc) *ENV1991 — Basis of Design and Action on Structures, Part 1: Basis of Design* (Sept. 1994); *Part 3: Traffic Loads on Bridges* (Mar. 1995), CEN [European Committee for Standardisation].
Eurocode 8 (1998) *Design Provisions for Earthquake Resistance of Structures, Part 1.1 General Rules — Seismic Actions and General Requirements for Structures; Part 2 Bridges*, British Standards Institution, London.
ECMT [European Conference of Ministers of Transport] (1994) *Light Rail Transit Systems*, OECD Publication Service, Paris.
Farquharson, F.B. (1949) Aerodynamic Stability of Suspension Bridges: with Special Reference to the Tacoma Narrows Bridge, Part 1: Investigations Prior to October, 1941, *University of Washington, Bulletin 116.*
Farraday, R.V. and Charlton, F.G. (1983) *Hydraulic Factors in Bridge Design*, Hydraulics Research Station, Wallingford, Thomas Telford, London.
Fenwick, J.M. (1985) Definition of design loads in the NAASRA Bridge Design Code. *Seminar on Practices and Developments in Bridge Design,* Brisbane, 18–19 September 1985, Australian Road Research Board and NAASRA: 1–14.
Fisher, J.W. (1977) *Bridge Fatigue Guide: Design and Details*, American Institute for Steel Construction, Chicago.
—— (1981) *Inspecting Steel Bridges for Fatigue Damage*, Lehigh University, Fritz Engineering Laboratory, Report Number 386–15.
Fisher, J.W., Pense, A.W., Hausammann, H. and Irwin, G.R. (1980) Quinnipiac River Bridge Cracking, *American Society of Civil Engineers, Journal of the Structural Division*, 106 (ST4): 773–89.
Frandsen, A.G. and Langsø, H. (1980) Ship collision problems, *International Association for Bridge and Structural Engineering, Proceedings*, P-31 (2): 81–101.

Fraser, D.A.S. (1976) *Probability and Statistics: Theory and Applications*, Duxbury Press, North Scituate, Mass.

French, R.H. (1985) *Open-channel Hydraulics*, McGraw-Hill, New York.

Freudenthal, A.M. (1960) *Methods of Safety Analysis of Highway Bridges*, International Association for Bridge and Structural Engineering, 6th Congress, Stockholm, Preliminary Publication: 655–64.

—— (1961) Safety, reliability and structural design, *American Society of Civil Engineers, Proceedings*, 87 (ST3, part 1): 1–16.

—— (1968a) *Combination of the Theories of Elasticity, Plasticity and Viscosity in Studying the Safety of Structures*, International Association for Bridge and Structural Engineering, 8th Congress, New York: 45–55.

—— (1968b) *Critical Appraisal of Safety Criteria and their Basic Concepts*, International Association for Bridge and Structural Engineering, 8th Congress, New York: 13–24.

Freudenthal, A.M. (ed) (1972) *International Conference on Structural Safety and Reliability, Washington, DC, April 9–11 1969*, Pergamon Press, Oxford.

Freudenthal, A.M., Garrelts, J.M. and Shinozuka, M. (1966) The analysis of structural safety, *American Society of Civil Engineers, Proceedings*, 92 (ST1): 267–325.

Frýba, L. (1996) *Dynamics of Railway Bridges*, Thomas Telford, London.

Fung, Y.C. (1955) *An Introduction to the Theory of Aeroelasticity*, Wiley, New York.

Gabel, B.J. (1982) *Minimum Cost Design of Slab on Prestressed Concrete Girder Bridge Superstructures*, MEngSc thesis, The University of Queensland.

Gawthorpe, R.G. (1994) Wind effects on ground transportation, *Journal of Wind Engineering and Industrial Aerodynamics*, 52: 73–92.

Gill, P.E. and Murray, W. (eds) (1974) *Numerical Methods for Constrained Optimization*, Academic Press, London and New York.

Gillespie, T.D. (1992) *Fundamentals of Vehicle Dynamics*, Society of Automotive Engineers, Warrendale, Pa.

Gillespie, T.D., Sayers, M.W. and Segel, L. (1980) *Calibration of Response-type Road Roughness Measuring Systems*, Transportation Research Board, National Co-operative Highway Research Program Report, 228.

Gim, G. (1988) *Vehicle Dynamic Simulation with a Comprehensive Model for Pneumatic Tires*, PhD thesis, University of Arizona. UMI Dissertation Services, Ann Arbor, Michigan, 1999.

Goh, C.C. and Grundy, P. (1989) *Estimation of Plastic Curvature after Shakedown of Continuous Beam*, Monash University, Melbourne, Civil Engineering Research Report 9.

Goodger, C.J. and Kohoutek, R. (1990) Dynamic Considerations of Pedestrian Bridges, *Institution of Engineers, Australia, Vibration and Noise Conference, Melbourne, 18–20 September 1990*: 349–53.

Gordon, R. and Boully, G. (1998) Bridges: Planning for the Future, *5th International Symposium on Heavy Vehicle Weights and Dimensions, Australian Road Research Board, Melbourne, 29 March–2 April 1998*: 40–50.

Goswami, I., Scanlan, R.H. and Jones, N.P. (1993) Vortex-Induced Vibration of Circular Cylinders, 1: Experimental Data, *American Society of Civil Engineers, Journal of Engineering Mechanics*, 119 (11): 2270–87.

Grassie, S.L. and Kalousek, J. (1993) Rail corrugation: characteristics, causes and treatments, *Institution of Mechanical Engineers, Proceedings, Part F, Journal of Rail and Rapid Transit*, 207 (1): 57–68.

Grundy, P. (1980) *Criteria for Evaluating Fatigue of Railway Bridges*, Monash University, Melbourne, Civil Engineering Research Report 2.

—— (1982) Fatigue Life of Australian Railway Bridges, *Institution of Engineers, Australia, Civil Engineering Transactions*, CE24 (3): 267–82.

—— (1996) Impact Factors in Railway Bridge Loading Codes. *Asia-Pacific Symposium on Bridge Loading and Fatigue, 17–18 December 1996*: 165–70.

Grundy, P. and Teh, S.H. (1986) Fatigue Strength of Beams at Cover Plate Terminations, *Institution of Engineers, Australia, Civil Engineering Transactions*, CE28 (2): 183–9.

Gupta, R.K. and Traill-Nash, R.W. (1980) Vehicle Braking on Highway Bridges, *American Society of Civil Engineers, Journal of Engineering Mechanics*, 106 (EM4): 641–58.

Hamill, L. (1995) *Understanding Hydraulics*, MacMillan, London.

—— (1999) *Bridge Hydraulics*, E & FN Spon, London.

Harman, D.J. and Davenport, A.G (1979) A statistical approach to traffic loading on highway bridges, *Canadian Journal of Civil Engineering*, 6: 494–513.

Harman, D.J., Davenport, A.G. and Wong, W.S.S. (1984) Traffic loads on medium and long span bridges, *Canadian Journal of Civil Engineering*, 11 (3): 556–73.

Harris, R.I. and Deaves, D.M. (1980) *The Structure of Strong Winds. Wind Engineering in the Eighties – Proceedings of CIRIA Conference, London, 12–13 November 1980, Paper 4*, Construction Industry Research and Information Service, London.

Harter, H.L. (1969) A new table of percentage points of the Pearson type III distribution, *Technometrics*, 11 (1): 177–87.

Henderson, F.M. (1966) *Open Channel Flow*, MacMillan, New York.

Henderson, W. (1954) British highway bridge loading, *Institution of Civil Engineers, Proceedings*, 3 (2): 325–50.

Henrickson, J.A., Wood, D.S. and Clark, D.S. (1958) The Initiation of Brittle Fracture in Mild Steel, *American Society of Metals, Transactions*, 50: 656–81.

Heywood, R.J. (1990) Multiple Presence Effects in Bridges, *12th Australasian Conference on the Mechanics of Structures and Materials, Brisbane, September 1990*: 45–51.

—— (1992a) *Bridge Live Load Models from Weigh-in-Motion Data*, PhD thesis, The University of Queensland.

—— (1992b) *A Multiple Presence Model for Bridges*, American Society of Civil Engineers Conference on Probabilistic Mechanics and Structural and Geotechnical Reliability, Denver, Colorado.

—— (1994) *Influence of truck suspensions on the dynamic response of a short span bridge over Cameron Creek*, Engineering Foundation Conference, 'Vehicle-Road and Vehicle-Bridge Interaction', The Netherlands, 5–10 June 1994.

Heywood, R.J. and Ellis, T. (1998) Australia's New Bridge Design Load: Improving Transport Productivity, *5th International Symposium on Heavy Vehicle Weights and Dimensions, Australian Road Research Board, Melbourne, Mar.29–Apr.2 1998*: 47–66.

Heywood, R.J. and O'Connor, C. (1992) A bridge design and evaluation method derived from weigh-in-motion data, *Canadian Journal of Civil Engineering*, 19: 423–31.

Hirsch, T. and Fairbanks, W.L. (1985) Bridge Rail to Contain and Redirect 80,000 lb Tank Trucks, *Transportation Research Record*, (1024): 27–34

Hitchings, W. (1979) *West Gate*, Outback Press, Melbourne.

Hobbs, P.V. (1974) *Ice Physics*, Clarendon, Oxford.

Holmes, J.D. (1990) *A Commentary on the Australian Standard for Wind Loads, AS1170, Part 2, 1989*, Australian Wind Engineering Society, Melbourne.

Hopkins, G.R., Vance, R.W. and Kasraie, B. (1975) *Scour Around Bridge Piers: Interim Report*, Federal Highway Administration, Washington, D.C.

Houghton, E.L. and Carruthers, N.B (1976) *Wind Forces on Buildings and Structures*, Arnold, London.

Hucho, W.-H. (ed.) (1987) *Aerodynamics of Road Vehicles*, Butterworth, London.
Humphrey, N. et al. (1989) *Providing Access for Large Trucks, Committee for Truck Access Study, Special Report 223*, Transportation Research Board, National Research Council, Washington, D.C.
Hwang, E.-S. and Nowak, A.S. (1991) Simulation of dynamic load for bridges, *American Society of Civil Engineers, Journal of Structural Engineering*, 117 (5): 1413–34.
Impact in Highway Bridges (1931), Final Report of the Special Committee, American Society of Civil Engineers, Transactions, (paper number 1786).
Inglis, C.E. (1934) *A Mathematical Treatise on Vibrations in Railway Bridges*, Cambridge University Press, Cambridge.
IEAust [Institution of Engineers Australia] (1981) *Track and Vehicle Dynamics*, Railway Engineering Conference, Sydney, 7–10 September 1981.
—— (1987) *Australian Rainfall and Runoff* (2 vols), Canberra.
Irwin, A.W. (1978) Human response to dynamic motion of structures, *Structural Engineer*, 56A (9): 237–44.
ISO 2631 — see Standards Association of Australia (1990).
Ivy, R.J., Lin, T.Y., Mitchell, S., Raab, N.C., Richey, N.C. and Scheffey, C.F. (1954) *Live Loading for Long-Span Highway Bridges*, American Society of Civil Engineers, Transactions, 119 (paper number 2708): 981–1004.
Jackson, C.F.V. (1900) Design and construction of steel bridge work, with particulars of a recent example in Queensland, *Institution of Civil Engineers, London, Minutes of Proceedings*, 142: 253–71 (and Plate 8).
Jempson, M.A. and Apelt, C.J. (1995) Flood loads on bridge superstructures, *Conference, Bridges into the 21st Century, 2–5 Oct. 1995, Hong Kong*: 1025–32.
Jempson, M.A., Apelt, C.J., Fenske, T.E. and Parola, A.C. (1997a) Flood loads on submerged and semi-submerged bridge superstructures. *Austroads 1997 Bridge Conference, Bridging the Millennia*, Sydney, 3–5 Dec., Vol. 2: 19–33.
—— (1997b) Debris loadings on bridge superstructures and piers. *Austroads 1997 Bridge Conference, Bridging the Millennia*, Sydney, 3–5 Dec., Vol. 2: 4–17.
Kalay, S. and Reinschmidt, A. (1989) Overview of wheel/rail load environment caused by freight car suspension dynamics, *Transportation Research Record*, (1241): 34–52.
Kennedy, D.J.L. and Baker, K.A. (1984) Resistance factors for steel highway bridges, *Canadian Journal of Civil Engineering*, 11: 324–34.
Kennedy, D.J.L., Gagnon, D.P., Allen, D.E. and MacGregor, J.G. (1992) Canadian highway bridge evaluation: load and resistance factors, *Canadian Journal of Civil Engineering*, 19: 992–1006.
Klesnil, M. and Lukas, P. (1992) *Fatigue of Metallic Materials*, Elsevier, Amsterdam.
Knott, J.F. (1973) *Fundamentals of Fracture Mechanics*, Butterworth, London.
Kobori, T. and Kajikawa, Y. (1974) Psychological effects of highway bridge vibrations on pedestrians, *Japan Society of Civil Engineering, Proceedings*, 36 (222): 15–23.
Koerte, A. (1992) *Two Railway Bridges of an Era: Firth of Forth and Firth of Tay*, Birkhäuser Verlag, Basel.
Kolousek, V. et al. (1984) *Wind Effects on Civil Engineering Structures*, Elsevier, Amsterdam.
Krebs, W. and Cantieni, R. (1997) Dynamic Vehicle/Bridge Effects, *OECD DIVINE Asia-Pacific Concluding Conference*, 5–7 November, 1997, Melbourne.
Kulicki, J.M. and Mertz, D.R. (1991) A new live load model for bridge design, *8th International Bridge Conference, June 1991, Proceedings*, 238–46.
Kunjamboo, K.K. (1982) *Vehicle-Highway Bridge Interaction*, PhD thesis, The University of Queensland.

Kunjamboo, K.K. and O'Connor, C. (1983) Truck suspension models, *American Society of Civil Engineers, Journal of Transportation Engineering*, 109 (5): 706–20.

Langer, B.F. (1937) Fatigue failure from stress cycles of varying amplitude, *American Society of Mechanical Engineers, Transactions*, 59: A-160.

Larsen, A. (ed.) (1992) *Aerodynamics of Large Bridges*, A.A. Balkema, Rotterdam.

Laursen, E.L. (1960) Scour at bridge crossings. *American Society of Civil Engineers, Journal of the Hydraulics Division*, 86 (HY2): 39–54.

Laursen, E.L. (1963) An analysis of relief bridge scour. *American Society of Civil Engineers, Journal of the Hydraulics Division*, 89 (HY3): 93–118.

Lavery, J.H. and O'Connor, C. (1957) *Collapse of Qantas Hangar, Sydney, Report to Contractors*, The University of Queensland, Brisbane.

Lay, M.G. (1984) *History of Australian Roads*, Australian Road Research Board, Melbourne.

—— (1985) *Sourcebook for Australian Roads*, Australian Road Research Board, Melbourne.

Leonard, D.R., Grainger, J.W. and Eyre, R. (1974) *Loads and Vibrations caused by Eight Commercial Vehicles with Gross Weights Exceeding 32 Tons (32.5Mg)*, Transport and Road Research Laboratory, Crowthorne, Report LR582.

Leonhardt, F., Kolbe, G. and Peter, J. (1965) Temperature differences endanger prestressed concrete bridges, *Beton und Stahlbetonbau*, 60 (7): 231–44.

Linsley, R.K., Kohler, M.A., Paulhus, J.L.H. and Wallace, J.S. (1988) *Hydrology for Engineers*, McGraw-Hill, New York.

Liu, H. (1991) *Wind Engineering: A Handbook for Structural Engineers*, Prentice Hall, Englewood Cliffs, New Jersey.

Livesey, F. and Sondergaard, T. (1996) Verification of the effectiveness of TMD's using wind tunnel section model tests, *Journal of Wind Engineering and Industrial Aerodynamics*, 64: 161–70.

Mallett, G.P. (1994) *Repair of Concrete Bridges*, Thomas Telford, London.

Marshall, W.T. and Nelson, H.M. (1990) *Marshall & Nelson's Structures*, Third edition, revised by Bhatt, P. and Nelson, H.M., Longman, Harlow.

Massonnet, Ch., Olszak, W. and Philips, A. (1979) *Plasticity in Structural Engineering: Fundamentals and Applications*, Springer-Verlag, Wien and New York.

Matsumoto, M., Saitoh, T., Kitazawa, M., Shirato, H. and Nishizaki, T. (1995) Response characteristics of rain-induced vibrations of stay cables of cable-stayed bridges, *Journal of Wind Engineering and Industrial Aerodynamics*, 57: 323–33.

Meguid, S.A. (1989) *Engineering Fracture Mechanics*, Elsevier, London.

Merrison, A.W., Flint, A.R., Harper, W.J., Horne, M.R. and Scruby, G.F.B. (1971) *Inquiry into the Basis of Design and Method of Erection of Steel Box-Girder Bridges: Interim Report*, Her Majesty's Stationery Office, London.

—— (1973) *Inquiry into the Basis of Design and Method of Erection of Steel Box-Girder Bridges: Final Report (abridged version) and Appendix 1*, Her Majesty's Stationery Office, London.

Miller, C.W. (1984) *On the Dynamic Behaviour of a Bridge-Vehicle System*, PhD thesis, The University of Queensland.

Miller, C.W. and Swannell, P. (1983) Bridge-Vehicle Interaction with Particular Reference to Six Mile Creek Bridge, *Institution of Engineers, Australia, Metal Structures Conference, Brisbane, 18–20 May*: 197–202.

Miner, M.A. (1945) Cumulative Damage in Fatigue, *American Society of Mechanical Engineers, Journal of Applied Mechanics*, 12: A159–64.

Ministère des Transports, Paris (1973) *Cahier des prescriptions communes applicables aux marchés de travaux publis relevant des servis de l'équipement.*

Minorsky, V.U. (1959) An analysis of ship collisions with reference to protection of nuclear power plants, *Journal of Ship Research*, 3 (2): 1–4.

Modjeski, R., Borden, H.P. and Monsarrat, C.N. (1919) *The Quebec Bridge over the St Laurence River: Report of the Government Board of Engineers* (2 vols).

Monahan, C.C. (1995) *Early Fatigue Crack Growth at Welds*, Computational Mechanics, Southampton.

Montes, S. (1998) *Hydraulics of Open Channel Flow*, American Society of Civil Engineers Press, Reston, Virginia.

Montgomery, C.J. and Lipsett, A.W. (1980) Dynamic tests and analysis of a massive pier subjected to ice forces, *Canadian Journal of Civil Engineering*, 7 (3): 432–41.

Moore, D.F. (1975) *The Friction of Pneumatic Tyres*, Elsevier, Amsterdam.

Moses, F. (1979) Weigh in motion system using instrumented bridges, *American Society of Civil Engineers, Journal of Transportation Engineering*, 105 (3): 233–49.

Moses, F., Lebet, J. and Bez, R. (1994) Application of field testing to bridge evaluation, *American Society of Civil Engineers, Journal of Structural Engineering*, 120 (6): 1745–62.

Mulcahy, N.L., Pulmano, V.A. and Traill-Nash, R.W. (1983) Vehicle properties for bridge loading studies, *International Association for Bridge and Structural Engineering, Periodica*, (3): 153–67.

Muller, J.F. and Dux, P.F. (1988) *Effects of Rail Defects on Prestressed Concrete Bridge Live Load Stresses*, The University of Queensland, Department of Civil Engineering, Research Report CE91.

—— (1992) Prestressed concrete railway bridge live load strains, *American Society of Civil Engineers, Journal of Structural Engineering*, 118 (2): 359–76.

Murakami, E., Kunihiro, T., Ohta, M. and Asakura, H. (1972) Actual traffic loadings on highway bridges and stress levels in bridge members, *International Association for Bridge and Structural Engineering, 9th Congress, Preliminary Report*: 675–84.

Nadai, A. (1950, 1963) *Theory of Flow and Fracture of Solids*, McGraw-Hill, New York.

Narita, N. and Katsuragi, M. (1981) Gusty wind effects on driving safety of road vehicles, *Journal of Wind Engineering and Industrial Aerodynamics*, 9: 181–91.

NAASRA [National Association of Australian State Road Authorities] (1953, 1967, 1970, 1972 etc) *Highway Bridge Design Specification*, Sydney.

—— (1968) *Vehicle Limits for Road Safety and Road Protection*, Sydney.

Neal, B.G. (1977) *The Plastic Methods of Structural Analysis*, Chapman and Hall, London.

Neill, C.R. (1964) *River-bed Scour: A Review for Bridge Engineers*, Canadian Good Roads Association, Ottawa.

—— (1973) *Guide to Bridge Hydraulics*, Roads and Transportation Association of Canada: Project Committee on Bridge Hydraulics, University of Toronto Press.

—— (1976) Dynamic ice forces on piers and piles: An assessment of design guidelines in the light of recent research, *Canadian Journal of Civil Engineering*, 3 (2): 305–41.

Neill, C.R. (ed) (1981) *Ice Effects on Bridges*, Roads and Transportation Association of Canada, Ottawa.

Newmark, N.M. and Rosenblueth, E. (1971) *Fundamentals of Earthquake Engineering*, Prentice Hall, Englewood Cliffs, New Jersey.

Norris, C.H. and Wilbur, J.B. (1960) *Elementary Structural Analysis*, McGraw-Hill, New York.

Nowak, A.S. (1991) Bridge load models, *Austroads Bridges Conference, Brisbane, 13–15 November*: 503–24.

Nowak, A.S. (1993) *Calibration of LRFD Bridge Design Code*, Department of Civil and Environmental Engineering, University of Michigan, Ann Arbor.
—— (1994) Load model for bridge design code, *Canadian Journal of Civil Engineering*, 21: 36–49.
—— (1995) Calibration of LRFD Bridge Code, *American Society of Civil Engineers, Journal of the Structural Division*, 121 (8): 1245–51.
—— (1999) *Calibration of LRFD Bridge Design Code*: Transportation Research Board, National Cooperative Highway Research Program, Report 368.
Nowak, A.S. and Grouni, H.N. (1983) Development of design criteria for transit guideways, *American Concrete Institute Journal, Proceedings*, 80 (5): 387–95.
—— (1994) Calibration of the Ontario Highway Bridge Design Code 1991 edition, *Canadian Journal of Civil Engineering*, 21: 25–35.
Nowak, A.S. and Hong, Y-K. (1991) Bridge live load models, *American Society of Civil Engineers, Journal of Structural Engineering*, 117 (9): 2757–67.
Nowak, A.S. and Lind, N.C. (1979) Practical bridge code calibration, *American Society of Civil Engineers, Journal of the Structural Division*, 105 (12): 2497–510.
Nowak, A.S., Nassif, H. and de Frain, L. (1993) Effect of truck loads on bridges, *American Society of Civil Engineers, Journal of Transportation Engineering*, 119 (9): 853–67.
O'Connor, C. (1964) *Brittle Fracture of Steel: Performance of NDIB and SAA A1 Structural Steels*, Department of Civil Engineering, The University of Queensland, Bulletin no. 4.
—— (1966) The initiation of brittle fracture in notched tension specimens, *Australian Welding Institute, Welding Fabrication and Design*, 10 (1): 49–52.
—— (1968) Control against brittle fracture in welded steel structures, *Institution of Engineers, Australia, Queensland Division Technical Papers*, 9 (17): 1–23.
—— (1969) Structural implications of fracture tests on A1 steel, *Institution of Engineers, Australia, Civil Engineering Transactions*, CE11: 40–6.
—— (1971a) *Design of Bridge Superstructures*, Wiley, New York.
—— (1971b) Cost Trends in Computer Designed Prestressed Concrete Girder Bridges, *3rd Australasian Conference on the Mechanics of Structures and Materials, Auckland, August 1971*, 1 (A4).
—— (1980) *An Appraisal of the Ontario Equivalent Base Length*, The University of Queensland, Department of Civil Engineering, Research Report CE8.
—— (1981) Ontario Equivalent Base Length: An appraisal, *American Society of Civil Engineers, Journal of the Structural Division*, 107 (ST1): 105–27.
—— (1993) *Roman Bridges*, Cambridge University Press, Cambridge.
O'Connor, C. and Chan, T.H.T. (1988a) Dynamic wheel loads from bridge strains, *American Society of Civil Engineers, Journal of Structural Engineering*, 114 (8): 1703–23.
—— (1988b) Wheel loads from bridge strains: laboratory studies, *American Society of Civil Engineers, Journal of Structural Engineering*, 114: 1724–40.
O'Connor, C., Kunjamboo, K.K. and Nilsson, R.D. (1980) Dynamic simulation of single-axle truck suspension unit, *Australian Road Research Board, Proceedings*, 10 (3): 176–85.
O'Connor, C. and Pritchard, R.W. (1982) *Service Performance of Six Mile Creek Bridge*, Australian Road Research Board Report A1R 321-1, Melbourne.
—— (1983) Dynamic behaviour of Six Mile Creek Bridge, *Metal Structures Conference, Brisbane, 18–20 May 1983*: 192–6.
—— (1984) Dynamic behaviour of Six Mile Creek Bridge, *Institution of Engineers, Australia, Civil Engineering Transactions*, CE26 (2): 89–93.
—— (1985) Impact studies on a small composite girder bridge, *American Society of Civil Engineers, Journal of the Structural Division*, 111 (3): 641–53.

OECD [Organisation for Economic Cooperation and Development] (1980) *Evaluation of Load Carrying Capacity of Bridges*, Paris.

—— (1995) *Repairing Bridge Superstructures*, Paris.

OHBDC (1979, 1983, 1991) *Ontario Highway Bridge Design Code*, Ontario Ministry of Transportation, Downsview, Ontario.

Ostenfeld, C. (1965) *Ship Collisions against Bridge Piers*, International Association for Bridge and Structural Engineering (or IVBH) Publication number 25.

Page, J. (1973a) *Dynamic Behaviour of a Single Axle Vehicle Suspension System: a Theoretical Study*, Transport and Road Research Laboratory, Crowthorne, Report LR580.

—— (1973b) *Dynamic Behaviour of Two Linked-Twin-Axle Lorry Suspension Systems: a Theoretical Study*, Transport and Road Research Laboratory, Crowthorne, Report LR581.

—— (1976) *Dynamic Wheel Load Measurements on Motorway Bridges*, Transport and Road Research Laboratory, Crowthorne, Report LR722.

Pearson, K. (1930) *Tables for Statisticians and Biometricians*, Cambridge University Press, Cambridge.

Pearson, E.S. and Hartley, H.O. (1966) *Biometrika Tables for Statisticians, Volume 1*, Cambridge University Press, Cambridge.

Penwarden, A.D. (1973) Acceptable wind speeds in towns, *Building Science*, 8: 259–67.

Perronet, J-R. (1782, 1788, 1987) *Construire des Ponts au XVIIIe Siècle*, First edition F.A. Didot, Paris, 1782; Second edition, 1788; Re-issued l'École Nationale des Ponts et Chaussées, Paris, 1987.

Peters, R.J. (1986) CULWAY – An unmanned and undetectable highway speed vehicle weighing system, *13th Australian Road Research Board and Road Engineering Association of Asia and Australasia Conference, Proceedings*, 13 (6): 70–83.

Petroski, H. (1995) *Engineers of Dreams*, Vintage Books, Random House, New York.

Priestley, M.J.N. (1972) Temperature gradients in bridges: Some design considerations, *New Zealand Engineering*, 27 (7): 228–33.

Pritchard, R.W. (1982) *Service Traffic Loads on Six Mile Creek Bridge, Queensland*, MEngSc thesis, The University of Queensland.

Pritchard, R.W. and O'Connor, C. (1984) Measurement and prediction of traffic loads on Six Mile Creek Bridge, *12th Australian Road Research Board Conference, Hobart, 27–31 August 1984, Proceedings*, 12 (2): 57–62.

Provis, W.A. (1842) Observations on the Effects Produced by Wind on the Suspension Bridge over the Menai Strait, *Institution of Civil Engineers, London, Transactions*, 3.

Queensland Water Resources Commission (1980–...) *Queensland Stream Flow Records for 1979* (3 vols), Brisbane.

Rackwitz, R. and Fiessler, B. (1978) Structural reliability under combined random load sequences, *Computers and Structures*, 9: 489–94.

Rainer, J.H., Pernica, G. and Allen, D.E. (1988) Dynamic loading and response of footbridges, *Canadian Journal of Civil Engineering*, 15 (1): 66–71.

Ransom, A.L. (2000) *Assessment of Bridges by Proof Load Testing*, ME thesis, Queensland University of Technology, Brisbane.

Report of Royal Commission into the Failure of Kings Bridge (1963) Govt Printer, Melbourne.

Robinson, J.R. (1964) *Piers, Abutments and Formwork for Bridges*, Crosby Lockwood, London.

Rockey, K.C. and Evans, H.R. (1981) *The Design of Steel Bridges*, Granada, London.

Russell, J.S. (1841) On the Vibration of Suspension Bridges and other Structures, and the Means of Preventing Injury from this Cause. *Royal Scottish Society of Arts, Transactions*, 1.

Saul, R. and Svensson, H. (1982) On the theory of ship collisions against bridge piers, *International Association for Bridge and Structural Engineering, Proceedings*, (P-51/82): 29–40.

Sayers, M.W. (1989) Two quarter-car models for defining road roughness: IRI and HRI, *Transportation Research Record*, (1215): 165–72.

—— (1995) On the calculation of international roughness index from longitudinal road profile, *Transportation Research Record*, (1501): 1–12.

Scanlan, R.H. (1981) *State-of-the-art Methods of Calculating Flutter, Vortex-induced and Buffeting Response of Bridge Structures, Report FHWA/RD-80/050*, US Federal Highway Administration, Washington, DC.

Scanlan, R.H. and Rosenbaum, R. (1968) *Aircraft Vibration and Flutter*, Dover, New York.

Scott, W.L. (1931) *Reinforced Concrete Bridges*, Crosby Lockwood, London.

Scruton, C. (1952) An experimental investigation of the aerodynamic stability of suspension bridges, with special reference to the proposed Severn Bridge, *Institution of Civil Engineers, London, Proceedings*, 1: 189–222.

—— (1981) *An Introduction to Wind Effects on Structures*, Oxford University Press, Oxford.

Scruton, C. and Flint, A.R. (1964) Wind-excited oscillations of structures, *Institution of Civil Engineers, London, Proceedings*, 27: 673–702.

Scruton, C., Woodgate, L. and Alexander, A.J. (1955) *The Aerodynamic Investigation for the Proposed Runcorn–Widnes Suspension Bridge*, National Physics Laboratory, Aerodynamics Report 29.

Selby Smith, P. (1996) New Australian bridge code provision for bridge loading and fatigue. *Asia-Pacific Symposium on Bridge Loading and Fatigue, 17–18 December 1996*: 171–7.

Selby Smith, P. and Bowmaker, G. (1995) *Stage 4: Final Report to the Railways of Australia*, Maunsell, Melbourne.

Shaw, P.A. (1993) *Static Analysis and Dynamic Simulation of Linked Multi-axle Truck Suspension Units*, PhD thesis, The University of Queensland.

Shaw, P.A. and O'Connor, C. (1986) On the static analysis and dynamic simulation of linked multi-axle truck suspension units. *10th Australasian Conference on the Mechanics of Structures and Materials, Adelaide, August 1986*: 635–40.

Shepherd, R. and Frost, J.D. (1995) *Failures in Civil Engineering: Structural, Foundation and Geoenvironmental Case Studies*, American Society of Civil Engineers, Technical Council on Forensic Engineering.

SIA160 (1989) *Action on Structures*, Swiss Society of Engineers and Architects, Zurich.

Simiu, E. and Scanlan, R.H. (1996) *Wind Effects on Structures: Fundamentals and Applications to Design*, Wiley, New York.

Simons, D.B. and Lewis, G.L. (1971) *Flood Protection at Bridge Crossings*, Colorado State University.

Skinner, F.W. (1908) *Types and Details of Bridge Construction, Part III Specifications and Standards for Short Railroad Spans*, McGraw Publishing Company, New York.

Smiles, S. (1862) *Lives of the Engineers* (3 vols), Murray, London; reprinted David and Charles, Newton Abbot, 1968.

Smith, D.W. (1976) Bridge failures, *Institution of Civil Engineers, London, Proceedings*, 60 (1): 367–82.

Smith, F.C. and Vincent, G.S. (1950) *Aerodynamic Stability of Suspension Bridges: with Particular Reference to the Tacoma Narrows Bridge, Part 2: Mathematical Analyses*, University of Washington, Bulletin 116.

Smith, J.O. and Sidebottom, O.M. (1965) *Inelastic Behaviour of Load Carrying Members*, Wiley, New York.

Sommer, H. (ed.) (1995) Durability of high performance concrete, *Proceedings of Rilem International Workshop, Vienna, 14–15 February 1994.*

Soroushian, P. and Choi, K. (1987) Steel mechanical properties at different strain rates, *American Society of Civil Engineers, Journal of Structural Engineering*, 113 (ST4): 663–72.

Soroushian, P., Choi, K. and Alhamad, A. (1986) Dynamic constitutive behaviour of concrete, *American Concrete Institute, Journal*, 83: 251–9.

South African Transport Services (1983) *Bridge Code*, Johannesburg.

Spangler, E.B. and Kelley, W.J. (1965) GMR Road Profilometer: A method for measuring road profiles, *Highway Research Record*, (121): 27–54.

Springenschmid, R. (ed.) (1995) *Thermal Cracking in Concrete at Early Ages. Proceedings of Rilem International Symposium, Munich, 10–12 October 1994*, E and FN Spon, London.

Standards Association of Australia, or Standards Australia (1963) *CA35–1963: SAA Code for Prestressed Concrete*, Sydney.

—— (1989) *AS1170.2–1989: SAA Loading Code, Part 2: Wind Loads*, Sydney.

—— (1990) *AS2670.1 (ISO 2631/1–1985): Evaluation of Human Exposure to Whole-body Vibration, Part 1: General Requirements*, Sydney.

Stanley, H.C. (1898) Re-erection of the Albert Bridge, Brisbane, *Institution of Civil Engineers, London, Minutes of Proceedings*, 132: 288–301.

Steel Box Girder Bridges, Proceedings of an International Conference 13–14 Feb. 1973, Institution of Civil Engineers, London.

Stevens, N.J. (1998) *Victoria Bridge Light Rail Investigation*, Nick Stevens Consulting (unpublished).

Still, P.B. and Jordan, P.G. (1980) *Evaluation of the TRRL High-Speed Profilometer*, Transportation and Road Research Laboratory, Crowthorne, Report LR922.

Still, P.B. and Winnett, M.A. (1975–6) *Development of a Contactless Displacement Transducer*, Transport and Road Research Laboratory, Crowthorne, Report LR690.

Swannell, P. and Miller, C.W. (1987) Theoretical and experimental studies of a bridge-vehicle system, *Institution of Civil Engineers, London, Proceedings*, 83 (2): 613–35.

Swannell, P., Miller, C.W. and Duczmal, Z.R. (1985) A Computer Model for Bridge-Vehicle Systems. *Institution of Civil Engineers, Second International Conference on Civil and Structural Engineering, London.*

Sweatman, P.F. (1983) *A Study of Dynamic Wheel Forces in Axle Group Suspensions of Heavy Vehicles*, Australian Road Research Board, Special Report 27.

—— (1987) Suspension research and its implementation, *Australian Road Research Board, Symposium on Heavy Vehicle Suspension Characteristics, Vermont South*: 19–42.

Tabsh, S.W. and Nowak, A.S. (1991) Reliability analysis of highway girder bridges, *American Society of Civil Engineers, Journal of Structural Engineering*, 117 (8): 2373–88.

The Quebec Bridge: Opening Booklet (1918), St Lawrence Bridge Co. Ltd.

Thomas, J. (1972) *The Tay Bridge Disaster: New Light on the 1879 Tragedy*, David and Charles, Newton Abbot.

Thurston, H. (1998) *Building the Bridge to Prince Edward Island*, Nimbus, Halifax.

Tilly, G.P., Cullington, D.W. and Eyre, R. (1984) Dynamic behaviour of footbridges, *International Association for Bridge and Structural Engineering, Periodica*, 2/1984.

Timoshenko, S.P. and Goodier, J.N. (1970) *Theory of Elasticity*, McGraw-Hill, New York.

Timoshenko, S.P. and Young, D.H. (1965) *Theory of Structures*, McGraw-Hill, New York.

Tipper, C.F. (1962) *The Brittle Fracture Story*, Cambridge University Press, Cambridge.

Todd, J.D. (1981) *Structural Theory and Analysis*, Macmillan, London.

Transit New Zealand (1994) *Bridge Manual*, Wellington, New Zealand.

Turkstra, C.J. and Madsden, H.O. (1980) Load combinations in codified structural design, *American Society of Civil Engineers, Journal of the Structural Division*, 106 (ST12): 2527–43.

US Federal Highway Administration (1993) *Evaluating Scour at Bridges, Hydraulic Engineering Circular Number 18*, Washington, D.C.

US Interagency Advisory Committee on Water Data (1982) *Guidelines for Determining Flood Flow Frequency, Bulletin 17B*, US Department of Interior, Hydrology Subcommittee, Office of Water Data Coordination.

Van Den Broek, J.A. (1948) *Theory of Limit Design*, Wiley, New York.

Varney, R.F. and Galambos, C.F. (1965) Field dynamic loading studies of highway bridges in the US, 1948–1965, *Highway Research Record*, (76): 285–305.

Vorobieff, G. (1990) Evaluation of auxiliary mass damper connected to pedestrian bridge, *Institution of Engineers, Australia, 2nd National Structural Engineering Conference, Adelaide, 3–5 Oct. 1990*: 33–7.

Waddell, J.A.L. (1908) *De Pontibus: a Pocket Book for Bridge Engineers*, Wiley, New York.

Walker, G.R. (1975) *Cyclone Tracy: Effect on Buildings*, James Cook University of North Queensland, Townsville.

Wang, T.L., Shahawy, M. and Huang, D.Z. (1993) Dynamic response of highway trucks due to road surface roughness, *Computers and Structures*, 49 (6): 1055–67.

Wheeler, J.E. (1980) Pedestrian-induced vibration in footbridges, *Australian Road Research Board, Proceedings*, 10 (3): 21–35.

—— (1982) Prediction and control of pedestrian-induced vibrations in footbridges, *American Society of Civil Engineers, Journal of the Structural Division*, 108 (ST9): 2045–65.

Whelan, J., Seaton, E. and Dax, E.C. (1976) *Aftermath: the Tasman Bridge Collapse: Criminological and Sociological Observations*, Australian Institute of Criminology, Canberra.

White, I.G. (1979) *Non-linear Differential Temperature Distributions in Concrete Bridge Structures: A Review of Current Literature*, Cement and Concrete Association, Wexham Springs.

Wilson, E.M. (1983) *Engineering Hydrology*, MacMillan, London.

Woodward, R.J. and Williams, F.W. (1988) Collapse of Ynys-y-Gwas Bridge, West Glamorgan, *Institution of Civil Engineers, London, Proceedings*, 84: 635–69.

Yoe, L.F. (1986) *Longitudinal Profile Measurement Using Profilometer*, MEngSc thesis, The University of Queensland.

Index